Oil Mist Lubrication:

Practical Applications

Oil Mist Lubrication:
Practical Applications

Heinz P. Bloch, P.E.
and
Abdus Shamim, Ph.D.

Published by
THE FAIRMONT PRESS, INC.
700 Indian Trail
Lilburn, GA 30047

Library of Congress Cataloging-in-Publication Data

Bloch, Heinz P., 1933
Oil mist lubrication: practical applications / Heinz P. Bloch and Abdus Shamim.
p. cm.
Includes bibliographical references and index.
ISBN 0-88173-256-7
1. Oil mist lubrication. I. Shamim, Abdus, 1964-. II. Title.
TJ1073.6.B556 1998 621.8'9--dc21 97-53166
CIP

Oil mist lubrication: practical applications by Heinz P. Bloch and Abdus Shamim.

Published by The Fairmont Press, Inc.
700 Indian Trail
Lilburn, GA 30247

Printed in the United States of America

10 9 8 7 6 5 4 3 2 1

ISBN 0-88173-256-7 FP

ISBN 0-13-975210-2 PH

Distributed by Prentice Hall PTR
Prentice-Hall, Inc.
A Simon & Schuster Company
Upper Saddle River, NJ 07458

Prentice-Hall International (UK) Limited, London
Prentice-Hall of Australia Pty. Limited, Sydney
Prentice-Hall Canada Inc., Toronto
Prentice-Hall Hispanoamericana, S.A., Mexico
Prentice-Hall of India Private Limited, New Delhi
Prentice-Hall of Japan, Inc., Tokyo
Simon & Schuster Asia Pte. Ltd., Singapore
Editora Prentice-Hall do Brasil, Ltda., Rio de Janeiro

To June
and
Rima

Contents

Preface

In a way, not much has changed since the First Edition of this text was released in 1987. It's surely as true as it was in 1987 that there is very little *of substance* on the face of the earth that someone hasn't already done, experienced, or even written about. Oh yes, there has been an "information explosion," but even an optimist will have to agree that it hasn't been a knowledge explosion, let alone a measurable increase in wisdom.

That said, it would be presumptuous to claim an exception for the technical topic of oil-mist lubrication. Oil mist lubrication has been applied since the 1930's and what many an old-timer has known for years we had condensed into the 1987 text. Still, in the late 1980's, a significant step forward was initiated when Texas A&M University, in College Station, received a state grant to study the theory of oil mist lubrication. As the younger member of this co-author team (A.S.) started to work on the project in fulfillment of requirements towards M.S., and later Ph.D. degrees in mechanical engineering, it became evident to the recent Exxon retiree (HPB) that the First Edition of this text no longer represented a comprehensive write-up of current knowledge. So, the two of us teamed up and decided that the impressive findings that were documented at Texas A&M are certain to be of interest and value to a very wide maintenance-technical, and also academic audience.

Here, then, is the result: A revised and thoroughly expanded Second Edition, which includes the key findings of these efforts and extends the pool of knowledge from the primarily practical to previously unknown theoretical aspects of oil-mist technology.

We owe thanks to the same companies and individuals who were instrumental in making the original, 1987 edition possible: Alemite Division of Stewart-Warner (Charlotte, North Carolina), Lubrication Systems, Inc. (Houston, Texas), and C.A. Norgren Company (Littleton, Colorado). Their kind permission to make liberal use of, and to synthesize into book format, major portions of their copyrighted material are gratefully acknowledged, as are the diligent efforts of Alemite retiree Donald M. Bornarth in reviewing the initial manuscript. Very special thanks are reserved for Dr. Fred Kettleborough, whose guidance and direction as faculty advisor at Texas A&M was instrumental in channeling everyone's energies in the right direction.

May the reader benefit from our efforts, and also from the generous input we received from others.

Heinz P. Bloch, P.E.
Montgomery, Texas

Abdus Shamim, Ph.D.
Cincinnati, Ohio
January, 1998

Part 1: Practical Applications

Chapter 1

The Impact of Oil-Mist Lubrication

The application of centralized liquid lubrication to machines is not new. Conventional circulating, pump-driven liquid lubrication systems have been in successful use for decades. They were designed and implemented because they offered many advantages and did away with much of the "human" element of machinery lubrication. The time-consuming task of manual point-by-point lubrication was thus eliminated, as was the requirement for making all lubrication points accessible for maintenance.

These liquid-oil circulating systems, while providing advantages, were nevertheless found to impose additional handicaps upon machine designs. They can be expensive, since pumps, filters, reservoirs, and piping are required. Machine assembly is more difficult, and the pump itself becomes an extra maintenance item. Moreover, the development of extremely high-speed grinders and spindles demonstrated that a liquid-oil environment can cause certain bearings to run hot. The added heat input often results in accelerated oxidation and a sharp decrease in the service life of the lubricant.

It was then realized that the distribution of lubricant by conveying oil particles in compressed air overcomes many of the drawbacks of a circulating liquid system. No pump and no return piping is required. Using the energy of this compressed air stream allows the generation of a fine, dry, smoke-like fog of oil particles. This dry-oil fog can be conveyed over distances up to 150 m, or close to 500 ft, through piping or tubing. Keeping the flow velocity below 7 meters per second (roughly 22 feet per second) results in very little of the oil condensing in the piping or tubing. At the point of lubrication, an application fitting or reclassifier nozzle will either meter this dry fog or change it to a wet fog, which is applied directly to the machine element to be lubricated.

For many years now, centralized oil-mist application has permitted the continuous lubrication of numerous machine elements, requiring only one common lubricator to be maintained per system. Oil mist has been used to lubricate bearings of all types, gears, chains, slides, ways, and other devices requiring a thin film of oil for lubrication. Soon after World War II, machine tool builders began to design oil-mist lubrication into their finest and costliest machines (Figure 1-1). Textile mills, steel mills, mine operators, paper and rubber factories (Figures 1-2 through 1-5) have applied it to web and processing equipment, conveyors, mobile equipment, shaker screens, vibrators, crushers, centrifuges, kilns, pulverizers, dryers, and a host of other equipment. The results have been excellent. In the highly conservative petrochemical industry, entire billion-dollar complexes have adapted oil mist to thousands of new and existing rolling element bearings in pumps and electric motors (Figures 1-6 and 1-7). Oil mist lubrication has also been used to lubricate railway rails and wheel flanges, resulting in up to 400% reduction in wheel flange wear [66].

Figure 1-1. Since the late 1940s, sophisticated machine tools have been equipped with oil-mist lubrication *(Source: C.A. Norgren Company.)*

These industrial users have established that oil mist has significant advantages over most other methods of lubrication. Continuous application of oil can be matched closely to actual bearing requirements. The low rate of lubricant application made possible with oil mist provides continuous lubrication without the necessity of designing a circulating system. This reduces the manufacturing and installation cost of many designs and improves housekeeping by reducing oil consumption, particularly where oil seal maintenance is a problem. Heat generation due to lubricant friction is reduced to a minimum in applications where a liquid oil sump is no longer maintained. We call this a pure mist, or dry sump oil mist application mode, to distinguish from wet sump applications where oil mist is used as a bearing housing purge. The pros and cons of the two application modes are dealt with later.

Figure 1-2. Dry can stacks in textile mills make extensive use of oil-mist lubrication. *(Source: C.A. Norgren Company.)*

Figure 1-3. Backup roll bearings in major steel mills are oil-mist lubricated. *(Source: Alemite Division of Stewart-Warner Corporation.)*

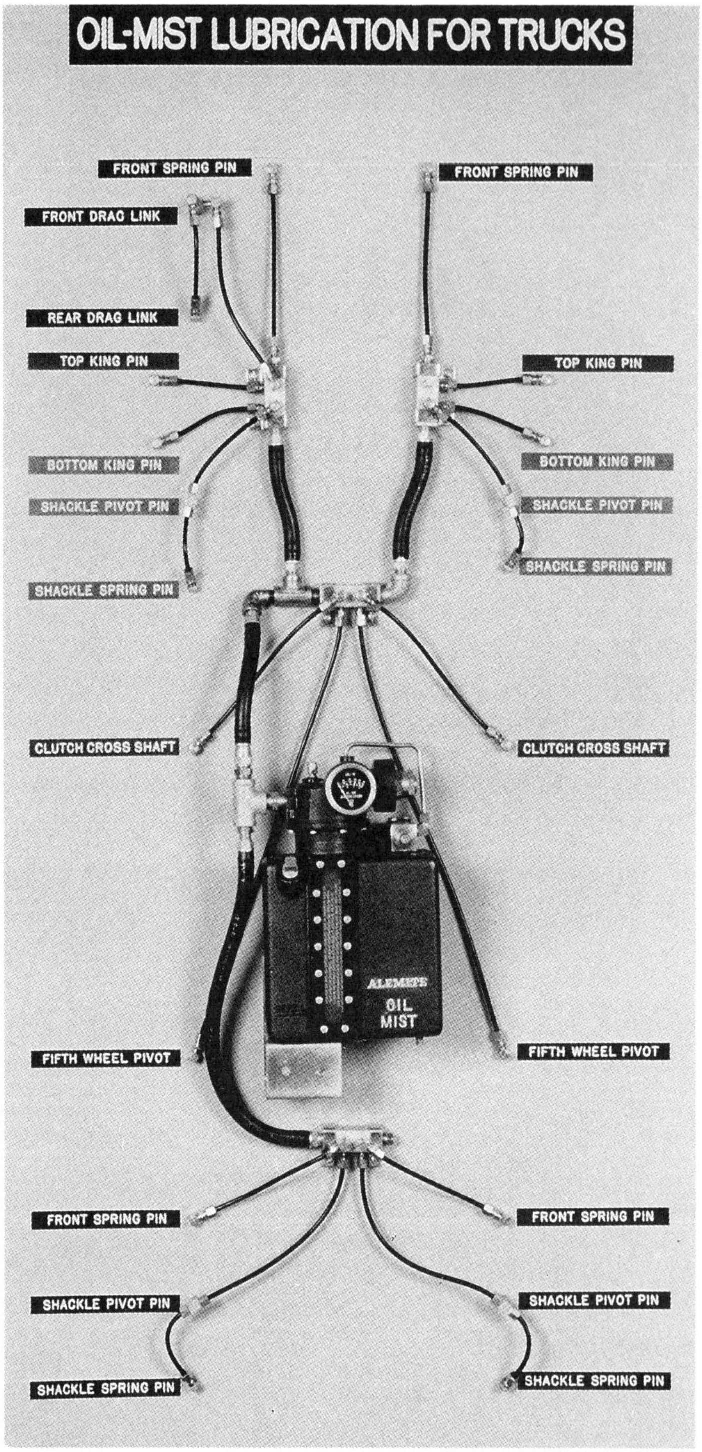

Figure 1-4. Super-sized mining trucks often come equipped with oil-mist lubrication. *(Source: Alemite Division of Stewart-Warner Corporation.)*

Figure 1-5. Paper machinery makes extensive use of oil-mist lubrication. *(Source: C.A. Norgren Company.)*

The carrier air used to distribute the oil can provide additional benefits, particularly in designs where grease lubrication was previously considered appropriate. The air maintains the bearing housing or bearing enclosure under slight positive pressure, and the outward airflow prevents the entrance of contaminants that could seriously limit bearing life. This property makes oil mist lubrication ideally suited to provide lubrication in chemically corrosive environment or in environment where hard particles may enter the housing and cause wear to the load bearing surfaces [67,68]. By far the most common of these contaminants is airborne water vapor.

Oil Mist Prevents Moisture Intrusion

In a paper presented in 1976 at the 31st Annual Meeting of the American Society of Lubrication Engineers in Philadelphia, PA, four researchers reported on water-accelerated bearing fatigue in oil-lubricated ball bearings [1]. Their contribution cited specifically how L. Grunberg and D. Scott [2] had earlier investigated the acceleration of pitting failure by water in the lubricant. This earlier research had established that:

Figure 1-6. Oil-mist consoles supply lubrication to pumps and electric motors in a modern petrochemical plant. *(Source: Lubrication Systems Company.)*

Figure 1-7. Oil-mist lubricated multistage centrifugal pump train in a U.S. petrochemical plant. *(Source: Lubrication Systems Company.)*

"The presence of water in the lubricant greatly accelerated the pitting failures of ball bearing steel and gross contamination with water could easily halve the mean life of a bearing. In recent years, other workers, including Schatzberg and Felsen [3], Ciruna and Szieleit [4], and Fein [5], have studied the deleterious influences of water on rolling contact lubrication. They all report that small amounts of water can significantly reduce the fatigue life of rolling contact elements."

The harmful effect of water on rolling contact fatigue life in lubricating oils is staggering. Armstrong et al. [1] continue:

"Using a base mineral oil dried over sodium, Grunberg and Scott found the fatigue life at a water content of 0.002% was reduced 48% and, at 6.0% water, it was reduced 83%. Schatzberg and Felsen showed a reduction of 32-43% for squalene containing 0.01% water. Ciruna and Szieleit report about an 80% drop with a moist air environment contacting dried mineral oil. The other investigators similarly report reductions from 29 up to 73%, depending on the type of lubricant and the amount of water contamination in the oil."

Finally, they provide a good theoretical explanation for the deleterious effects of water in lubricating oil [1]:

"The detailed mechanism(s) for reduction of fatigue life by water in a lubricant is not completely understood but is concerned with aqueous corrosion. There is much evidence that the water breaks down and liberates atomic hydrogen. This results in hydrogen embrittlement and markedly increases the rate of cracking of the bearing material."

Because there are no moving parts in the basic oil-mist components, and because the system pressure is very low, oil mist is a reliable lubrication method. Proper lubrication system operation can be interlocked with machine operation or an alarm system, assuring adequate lubrication.

The savings due to lower preventive maintenance labor requirements, equipment repair cost avoidance, and reductions in unscheduled production outage events have been significant and cannot possibly be overlooked by a responsible manager or cost-conscious manufacturing facility. Oil-mist systems have become incredibly reliable and can be used not only to lubricate operating equipment, but to preserve standby, or totally deactivated ("mothballed") equipment as well.

These facts make a compelling case for oil-mist lubrication. However, as briefly stated earlier, there are two different ways of applying oil mist, the wet sump method and the dry sump method.

Wet Sump Vs. Dry Sump Oil-Mist Applications

Oil mist was initially applied in the machine tool industry. By the mid-1950s, petrochemical plants began to apply oil-mist lubrication to general purpose machinery. They proceeded cautiously, using "wet sump" or oil-mist purge techniques. With this application method, a dispersion of fine particles of oil fog in air is conveyed into the vapor space above the oil level in a typical bearing housing. The primary result of this oil-mist purge was to establish a positive pressure that reduced the entry of solid atmospheric contaminants and water vapor. This extended bearing life and oil replacement frequency.

In the 1960s serious experimentation showed the feasibility of applying "dry sump" oil mist to antifriction bearings in virtually all categories of rotating machinery in petrochemical plants. The dry sump or pure mist method refers to the conveying of the same dispersion of extremely small particles of oil fog in air into a bearing housing from which oil has been drained completely. This method was found to be indispensable in high-speed grinding spindles as early as 1937. Wet-sump lubrication is rare in the machine tool industry, although extremely large roll-neck bearings in steel mills still rely on it.

Nevertheless, dry sump oil-mist methods were not readily accepted at first. Even a mechanically inclined person may have difficulty visualizing how "a puff of oily air" could provide superior lubrication for high-speed antifriction bearings. However, the merits of dry sump oil-mist lubrication have since been thoroughly proven and are well documented in the literature. Dry sump (or pure) oil-mist lubrication excels over wet sump (or purge) oil-mist lubrication by allowing higher bearing operating temperatures. Usually, dry-sump bearings operate cooler then wet-sump, because the generation of frictional heat within sump oil is eliminated. Furthermore, oil rings, which are prone to cause lube oil deterioration, and their frictional heating, are also eliminated. With dry sump lubrication, only fresh, unoxidized lube oil reaches the points to be lubricated. This decreases the potential for bearing damage due to the continual reuse of moisture- or debris-containing lube oil.

It should be emphasized that wet sump lubrication does not offer these same advantages. This is evident from pump bearing failure statistics assembled by Charles Towne of Shell Oil Company [6]. Mr. Towne reviewed accurate records that had been assembled for 191 oil-mist-lubricated refinery pumps over periods ranging from 1-1/2 to 3 years. Some of the pumps were dry sump, others wet sump lubricated. Calculating weighted averages, 5.3% of the dry-sump-lubricated bearings failed each year. However, 16.8% of the wet-sump-lubricated bearings failed in the same time period.

Many U.S. petrochemical plants are now using dry sump oil-mist as the standard lubrication method. They screen the applicability of oil mist by using a rule of thumb from a formula quoted in literature issued by the MRC Bearing Division of SFK Industries: $K < 10^9$. K is defined as equal to DNL, where D = bearing bore in mm, N = inner ring rpm, and L = load in pounds. If K does not exceed 10^9, oil mist is considered feasible.

Of the more than ten thousand motors and pumps so lubricated in U.S. Gulf Coast plants alone, many hundreds have now been in trouble-free operation since the late 1960s. In fact, sufficient experience has accrued to single out dry sump oil-mist methods as best suited for plant-wide lubrication of entire grass-roots petrochemical complexes [7]. Economic justification is based on comparison of failure statistics for conventional versus dry sump oil-mist lubricated machinery. For an assessment of these economics, a given facility may prefer to use its own failure statistics. However, some generalized approaches are available and are offered later in this text.

Chapter 2

Operating Principles and Systems Overview

As stated in the introduction, oil mist is a centralized lubrication system in which the energy of compressed gas, usually thoroughly dry air taken from the plant instrument air supply, is used to atomize oil, which is then conveyed by the air in a low-pressure distribution system to multiple points of lubricant application. In the hydrocarbon processing industry it is standard practice to dry the air in order to prevent corrosion at the points to be lubricated. However, users report effective moisture removal using means other then dryers.

The dry compressed air is passed through a venturi or vortex, as shown in Figures 2-1 and 2-2. Oil, siphoned from a reservoir by the air flow, is atomized into a fine spray. Baffles downstream from the venturi or vortex (Figure 2-3) cause the larger oil particles to coalesce and return to the reservoir. The remaining air-oil mixture is oil mist. Oil mist contains oil particles averaging about 1-1/2 μ (1.5 μm or 0.00006 in.) in diameter. These particles can be conveyed through distribution piping at velocities up to 7 m/s (approximately 22 fps). Excessive velocity causes the mist to condense in the distribution piping (Figure 2-4).

Near points of lubricant application, such as shown in Figure 2-5, the oil-mist distribution piping usually terminates in an application fitting that acts like a restriction orifice. The airborne oil particles coalesce, i.e., they are "reclassified" or combined into larger droplets, and are "wetted-out" by impinging on a surface at sufficient speed to cause adherence. We often use the terms "application fitting" and "reclassifier" inter-

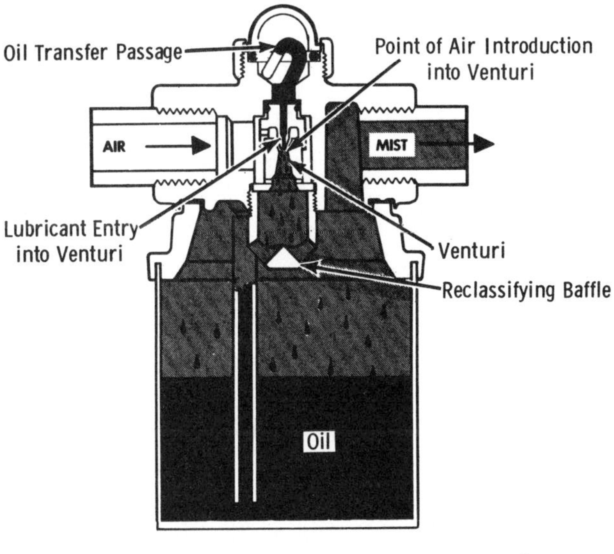

Figure 2-1. Cross-section of venturi-type oil-mist generator. *(Source Lubriquip-Houdaille.)*

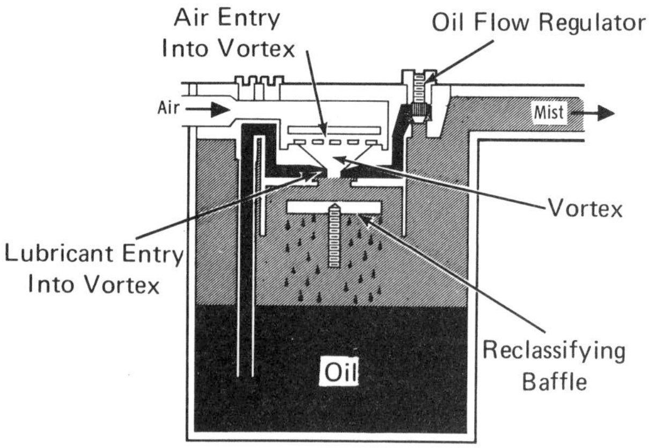

Figure 2-2. Cross-section of vortex-type oil-mist generator. *(Source: Lubrication Systems Company.)*

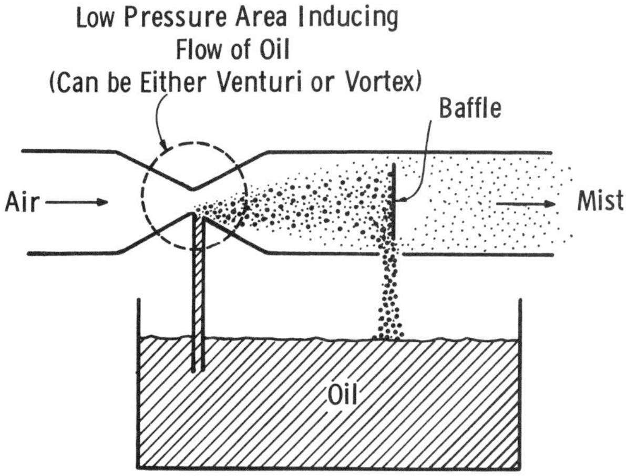

Figure 2-3. Baffles downstream from the venturi or vortex cause the larger oil particles to coalesce and return to the reservoir. *(Source: Reference 8, Chevron Research Company.)*

changeably, although, strictly considered, the former should function only as a metering orifice, while the latter should aim to change the droplet size.

One company (Alemite) uses "application fitting" as the collective term including mist, spray, condensing, condensed spray, and pressure jet fittings. Note that the fitting-type nomenclature relates to the output. Mist fittings only meter flow, and their output is essentially of the same quality as the input—a "dry" mist. The others meter flow too, but they also reclassify the mist to another form—wet spray, large droplets, etc.

Oil mist can be introduced into enclosed housings without prior coalescence or recombination of droplets, if the lubricated surfaces are running within a certain speed range. The speed of the gears, chains, and rolling element bearings may cause sufficient "wet-out" on these parts to provide good lubrication. However, for most lubrication points, the application fitting must perform the "wetting-out" function. Sufficient pressure drop through an application fitting creates enough mist velocity (typically, 27 m/s or approximately 90 fps) to cause turbulent flow. If the length of the passage in the fitting is much longer than the passage diameter (from 6 diameters minimum length for 8 in. H_2O [2 kPa] pressure drop, to 20 diameters minimum length for 20 in. H_2O [5 kPa] and higher pressure drop), there is a sufficiently turbulent region to cause good "wetting-out" at the

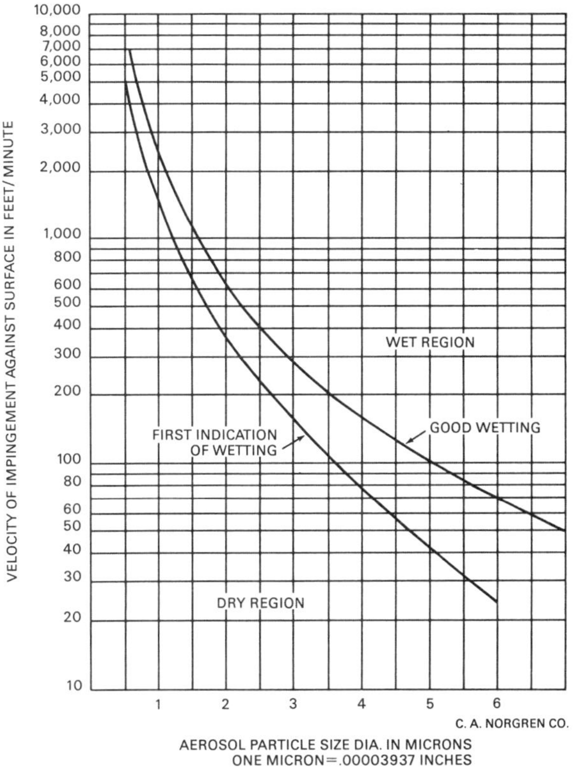

Figure 2-4. High flow velocities and/or large aerosol particle size favor wetting (condensation), i.e., conversion of oil mist to oil liquid. *(Source: C.A. Norgren Company.)*

Figure 2-5. In this continuous annealing furnace, the rolling bearings are lubricated with oil mist. Stainless steel mist-supply tubing terminates at the point of application. *(Source: DeLimon-Fluhme.)*

exit edge of the fitting. The use of baffles in an application fitting increases the "wetting-out" action even further and almost all of the oil particles are separated from the air and combined into drops of oil.

The total output of oil mist is a function of the oil-mist generator rating. This output is controlled by the size of the venturi nozzle or vortex generator and the applied air pressure. The application fittings meter, or proportion, the oil-mist generator output. The manifold pressure, or pressure in the mist distribution system, is the pressure drop across all the application fittings that is required for these fittings to pass the entire output of the generator.

Oil-mist systems are designed to maintain manifold pressures from 5 in. H_2O to 40 in. H_2O (1.25 kPa to 10 kPa) depending on the particular application. The design manifold pressure is selected for each application. It is a function of the type of application fitting used, the oil viscosity, and the speed of the surfaces being lubricated.

Basic Oil-Mist System Components

1. A source of compressed air, followed by a suitable air dryer.
2. An air-line filter to assure a clean air supply to the oil-mist generator (sometimes called mist head).
3. An air pressure regulator to control the oil mist generator atomizing air pressure.
4. An oil-mist generator, which includes a venturi nozzle or vortex generator, oil lift tube, reservoir, and oil-flow adjustment screw.
5. Mist distribution manifolds to convey the oil mist to the application fittings.
6. Mist, spray, or condensing application fittings to meter and convert the oil mist at each lubrication point.
7. A mist manifold pressure gauge (manometer) for visual indication of manifold pressure.

Accessory Oil-Mist System Components

1. A solenoid or similar on-off valve to start and stop the air supply to the oil-mist generator.
2. An oil heater to maintain the oil in the generator reservoir at the proper viscosity for good mist generation.
3. An air heater to stabilize the oil/air ratio at varying ambient temperatures or to mist heavy oils that will not atomize at the prevailing ambient temperature.

4. An oil-level switch to signal low oil level in the reservoir or to control automatic reservoir refill.

5. A mist manifold pressure switch to signal low or high manifold pressure.

6. An oil-mist detection unit to signal high or low density of oil particles in the mist.

The respective locations of the basic and accessory systems components are shown schematically in Figure 2-6. They are the same, regardless of whether the machinery is using dry sump or wet sump oil-mist lubrication.

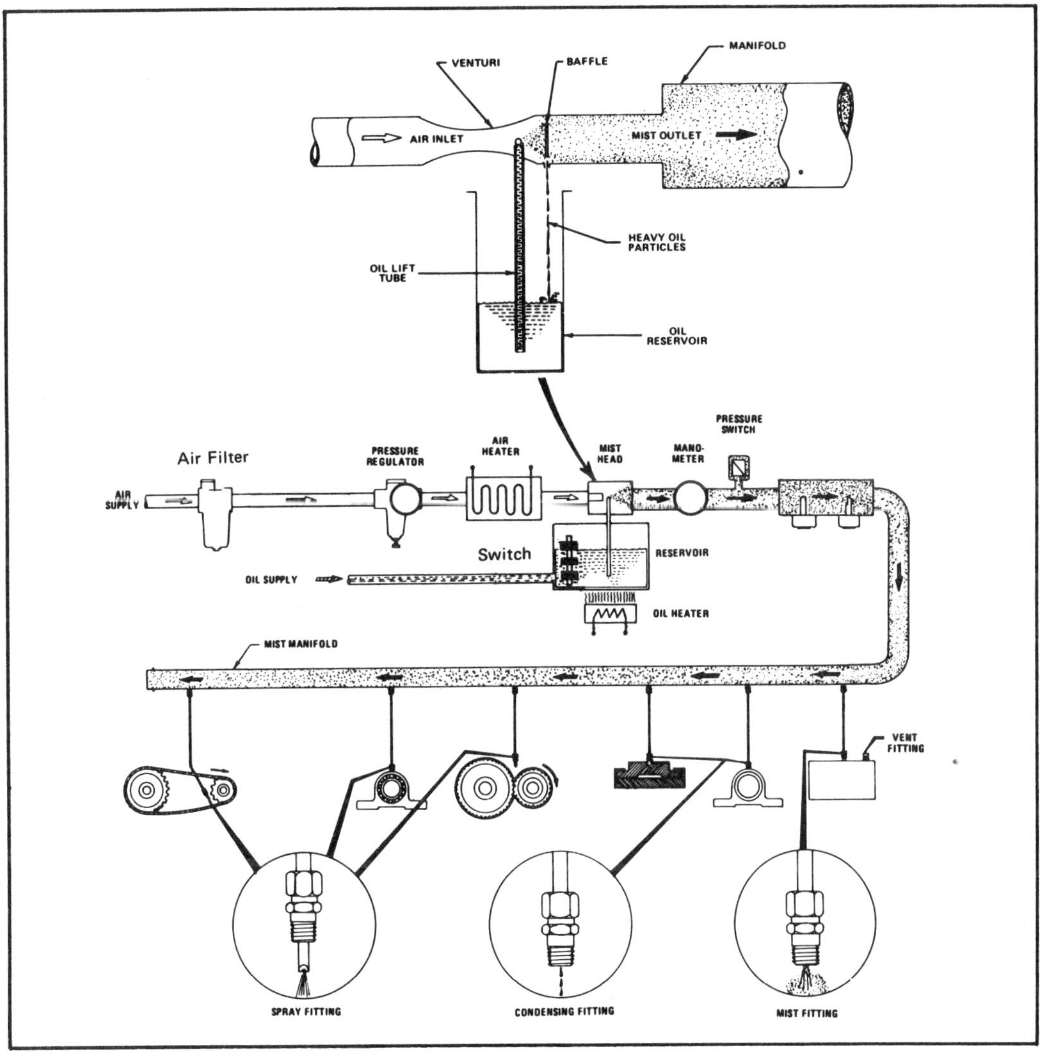

Figure 2-6. Schematic view of oil-mist system. *(Source: Alemite Division of Stewart-Warner Corporation.)*

Defining the Size of an Oil-Mist System

It will be intuitively evident to the reader that small bearings require less lubricant than large bearings and multiple-row bearings more lubricant than identical diameter single-row bearings. Similarly, a heavily loaded thrust bearing will require more oil mist lubricant than an identically dimensioned but lightly loaded bearing.

Lubrication requirements of the various bearing types and bearing sizes have been determined and catalogued over several decades. Three calculation methods are in use for defining the lubrication requirements of machine elements. These are called the bearing-inch, lubrication unit, and standard cubic feet per minute (scfm) rating systems.

Bearing-inch (B.I.)

The term "bearing-inch" has long been in use as an arbitrary means of computing lubrication requirements for machine elements. The bearing-inch basically reduces all machine elements to a common denominator. It could be stated that the term bearing-inch describes the amount of oil mist needed to lubricate a simple rolling element bearing with a bore diameter of 1 in. Or, an oil-mist system with a flow capacity of 100 B.I. could serve 100 bearings with a 1-in. bore diameter, or 50 bearings with a 2-in. bore diameter, etc.

As will be shown later, oil-mist system size can be determined after each machine element has been analyzed as to its bearing-inch requirement. The figures can be totaled to compute the actual bearing-inch requirements of the machine or machines to be lubricated. This rating is then used to select the proper oil-mist venturi, vortex generator, or mixing assembly. When selecting the venturi or vortex generator assembly, the user should be sure that the bearing-inch number falls within its specified range. Selecting a greatly oversized assembly may cause difficulty in achieving the proper air-to-oil mixture.

The bearing-inch system is intended for rolling element bearings, but formulas are available for other types of bearings. For design purposes, it assumes a rate of mist containing 0.018 $in.^3$ (0.01 fl. oz. or 0.3 ml) of oil per hour per bearing-inch.

Lubrication units

These are the metric equivalent of the bearing inch. One lubrication unit (L.U.) equals the amount of oil mist needed to lubricate a simple rolling element bearing with a bore diameter of 25 mm. Lubrication units equal shaft diameter, in millimeters, divided by 25. This expression is then multiplied by the number of rows of bearing elements (e.g., balls, rollers, needles, etc.) Two rows of bearings, each having a shaft diameter of 50 mm, would equal four lubrication units. Seven rows of bearings, each having a shaft diameter of 100 mm, would equal 28 lubrication units.

Scfm

The scfm system is based on the number of cubic feet of air passing through the mist generator per minute. For design purposes an oil/air ratio of .65 $in.^3$ (.36 fl. oz. or ~10 cm^3) of oil per hour per scfm of air is standard for at least one major manufacturer of oil mist systems. Another uses 0.4 $in.^3$ (0.22 fl. oz. or ~6.56 cm^3) of oil per hour per scfm of air. The oil is assumed to have a viscosity of 500 Saybolt Universal Seconds (SUS) at 100°F, or ~105 cSt at 40°C and is assumed to be applied at an ambient temperature of 68°F (20°C). The user decides which rating method he prefers for determining the lubrication requirements of his machinery.

Chapter 3

Lubricants for Oil-Mist Systems

By far the most important function of any lubricant is to put a reliable oil film between moving parts. This requirements is paramount, regardless of the application method selected. In other words, the application method should not unduly restrict lubricant choice. Conversely, an associated requirement would be for the lubricant to suit a specific application method. If oil-mist lubricants are chosen with the right degree of forethought, they will readily fulfill both of these primary needs.

Premium quality oil-mist lubricants are only marginally more expensive than other premium grade lubricants. A typical multiplier might place bulk quantities of mineral-base oil mist lubricants at 1.3 times a mineral-base turbine lube oil. Dibasic ester synthetic lubricants, which, as of this writing, represent the latest in sophistication for oil-mist lubes, would probably sell for two to three times the cost of premium grade mineral-base turbine oils.

In selecting oil-mist lubricants, the user must consider five major properties:

1. Performance properties, such as film strength and oxidation resistance.
2. Viscosity and ability to resist wax formation at low temperature.
3. Stability at high temperature.
4. Misting and reclassification characteristics.
5. Low toxicity.

Performance Properties

Oil-mist lubricants must be compounded or formulated to fully serve the needs of the machine component to be lubricated. This poses few problems where only rolling element bearings of a similar velocity rating, or gears or similar size, material composition, and speed must be lubricated.

However, matters get a bit more tricky when an oil is to be selected for simultaneous high performance on a variety of components. A pure mineral oil may no longer serve well in an application where gears demand anti-spall properties, or where certain alloys are susceptible to corrosion attack.

This is where experienced lube oil suppliers resort to additives in order to achieve consistently high performance over a wide spectrum of applications. Oil-mist lubricants can be given additives to inhibit oxidation, reduce the susceptibility to rust formation, control foaming, reduce component wear, allow high-pressure contact by adding extreme pressure (EP) or anti-weld agents, promote wetability of metal surfaces, and increase detergent action. It does not make economic sense to save a small percentage of the cost of lubricants by deleting desirable additives when the end result will be increased failure risk or equipment repair costs. However, the user should not assume that a super premium motor oil is suitable for oil-mist systems. While it may have many good properties, it may not mist well.

Viscosity and Ability to Resist Wax Formation

The viscosity requirements of rolling element bearings depend on the size, speed, and operating temperature of the bearing. These factors are all related, and the ideal viscosity can be estimated from the Figures 3-1A and 3-1B. These figures depict suitable viscosities as a function of dN-value (bearing diameter in mm, multiplied by rpm) and bearing operating temperatures. Figure 3-1B employs the U.S. system of measurement and tells us that a bearing with a dN-factor of 200,000 and an operating temperature of 120°F would be optimally lubricated with an oil having a viscosity slightly higher than 100 SUS at 100°F. Alternatively, Figure 3-1B could be used to tell us that if the same 200,000 dN-bearing were to reach an operating temperature of 165°F, optimal lubrication would require an oil with a viscosity of approximately 300 SUS at 100°F.

It can be stated that using a lubricant with a higher viscosity than indicated in Figures 3-1A and 3-1B will increase the frictional torque of the bearing, but will not adversely influence the life expectancy of the bearing. If we were to use a lubricant with too low a viscosity rating relative to the operating temperature experienced by the bearing, it could be said that we would risk a life expectancy impairment. This is due to breakage of the lubricating film, allowing metal-to-metal contact and severe wear.

The question as to what single lube oil viscosity should be chosen to satisfy a reasonably wide range of requirements should be answered by the machinery manufacturer in cooperation with the bearing and lube oil suppliers. In the early 1970s SKF's chief engineer Henry Keire addressed this issue for the petrochemical industry. Mr. Keire concluded that the overwhelming majority of rolling element bearings used by this industry in literally thousands of pumps, electric motors, and other machines would be well served by lubricating oils whose viscosity did not drop below 13.2 cSt at the operating temperature of the bearing. We can use this information in conjunction with the well-known ASTM temperature-viscosity chart, shown in Figure 3-2.

Using a typical light turbine oil with ISO viscosity grade 32 (32 cSt @ 40°C or 150 SUS @ 100°F) would allow the bearing to operate in the acceptable viscosity range, i.e., above 13.2 cSt, until it reached a maximum operating temperature of 64°C (147°F). Applying, instead, a premium grade oil-mist lubricant with ISO viscosity grade 100 (100 cSt @ 40°C or 550 SUS @ 100°F) would result in bearing operation above the 13.2 cSt minimum viscosity until it reached a new maximum allowable temperature of 93°C (200°F). Therefore, it is good practice to select ISO grade 100 oil-mist lubricants for lubrication of rolling element bearings in general purpose machinery, such as pumps and motors in a petrochemical plant.

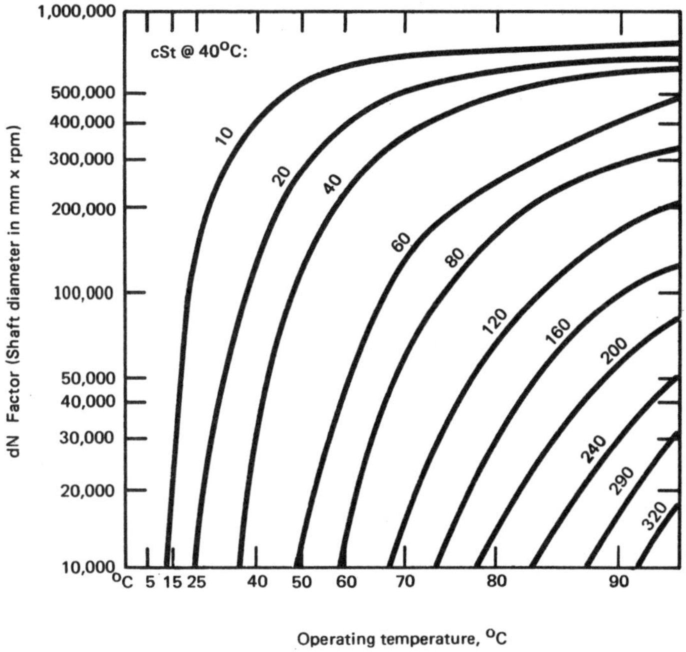

Figure 3-1. For oil-lubricated ball and roller bearings the proper oil viscosity depends on the size, speed, and operating temperature of the bearing. These factors are all related, and these graphs can be used as a general guide for determining the best viscosity recommendation.

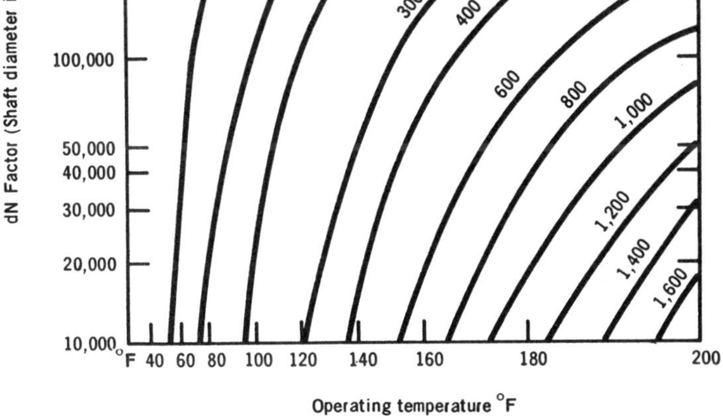

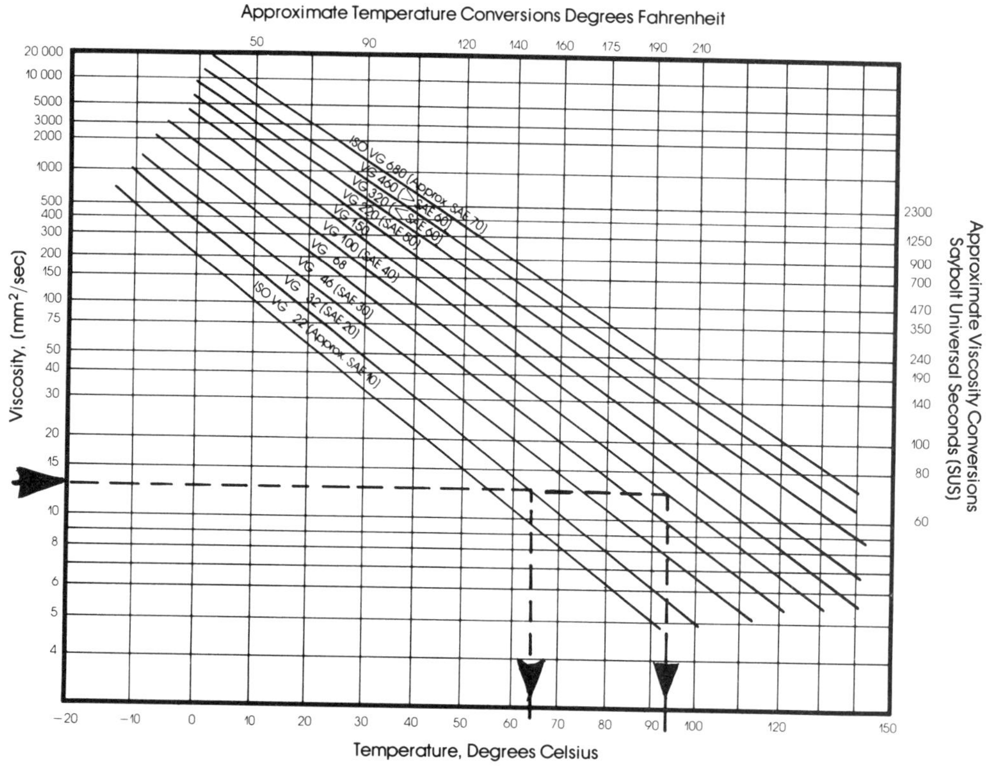

Figure 3-2. ASTM viscosity-temperature chart indicates that higher viscosity grade lube oil allows operation at higher bearing temperature than would be permitted if lower viscosity oil has been chosen.

The ability to resist wax formation is greatly influenced by the choice of lubricant base stock. Paraffinic oils are generally more susceptible to this undesirable development and plugging of reclassifiers may result due to wax crystal buildup in colder ambients. Naphthenic oils are considerably less likely to contain wax crystals and are thus preferred over paraffinic base mineral oils in oil-mist systems.

Wax formation is not possible in dibasic ester synthetics; and indeed, these lubricants are now preferred by oil-mist users who value the minimum downtime risk and lowest possible maintenance requirements that can be attributed to these superior oil-mist lubricants. The properties of a few dibasic ester synthetic lubricants often recommended to be used with oil mist lubrication are given in Table 3-1a. However, the compatibility of dibasic esters with certain engineering materials must be verified. Tables 3-1 and 3-2 give guidance in this regard.

Whether a given lube oil must be heated to allow proper misting is generally a function of its viscosity and the minimum operating temperature. Specific guidelines are given later in this text.

Table 3-3 lists generally accepted ranges of oil viscosities for lubricating broad categories of machine elements. It should be noted that unheated oil-mist systems operating

Table 3-1a. Properties of a Few Dibasic Ester Oil-Mist Lubricants *(Source: Lubrication Systems Company, Houston, Texas)*

LubOil Grade		**32**	**68**	**100**	**150**
Specific gravity at 15.6°C (60°F)		0.9176	0.9559	0.9617	0.959
Viscosity, cSt at 40°C	ASTM D 445	30.1	65.0	05.0	132.0
cST at 100°C		5.6	7.5	10.1	12.7
SSU at 100°F		153	339	510	612
SSU at 210°F		44.7	51.2	61.0	69.0
Viscosity Index		120	70	85	86
Pour Point, °C	ASTM D 97	–57	–34	–48	–30
°F		–70	–30	–55	–34
Flash point,°C ASTM D 92	246	266	246	262	
°F		475	510	475	505
Autogenous ignition temperature, °C	ASTM d 2155	404	371	407	414
°F		760	700	765	778
Coefficient of thermal expansion, $cm^3/cm^3/°C$		0.00044	0.00042	0.00042	—
Foam tendency and stability, ml	ASTM D 892				
at 23.9°C (75°F)		10/0	<5/0	<5/0	5/0
at 93.3°C (200°F)		<5/0	<5/0	<5/0	10/0
at 23.9°C after 93.3°C		<5/0	<5/0	<5/0	5/0
Demulsibility at 54.5°C (130°F), ml oil/water/emulsion	ASTM D 1401	40/38/2	44/35/1	41/37/2	43/37/0
Evaporative loss, 22 hr at 98.9°C (200°F), mass %	ASTM D 972	0.25	0.40	0.03	—
Oxidation corrosion stability, 72 hr at 175°C (347°F), Δ metal mass, mg					
copper		–0.03	+0.01	–0.01	±0.03
magnesium		–0.01	–0.01	–0.01	±0.03
iron		–0.02	+0.01	–0.01	±0.03
aluminum		0.00	+0.01	–0.01	±0.03
silver		–0.02	–0.03	–0.03	±0.03
evaporative loss, mass %		0.2	0.6	0.5	—
sludge level		0.0	0.0	0.0	0.0
total acid number increase, mg KOH/g		0.19	0.35	0.02	1.57
Copper strip corrosion 3 hr at 100°C (212°F)	ASTM D 130	1A	1A	1A	1A
Rust Prevention, 24 hr dist water	ASTM D 665	no rust	no rust	no rust	no rust
Hydrolytic stability	ASTM D 2619				
mass loss of copper, mg/cm^2		0.08	0.02	0.03	—
condition of copper strip		purple streaks	copper color	gold	—

(Continued)

Table 3-1a. (Continued)

LubOil Grade		32	68	100	150
neutralization number of water layer		0.01	0.01	0.04	—
Panel coking, 6 hr at 316°C (600°F), mg deposit		9.8	<10	5.5	—
Conradson carbon residue, mass %	ASTM D 189	0.00	0.00	0.00	0.00
Wear Testing					
4-ball, 10 kg. 1800 rpm, 65.6°C, 30 min, scar diameter, mm		0.20	0.23	0.30	0.399
Falex, 30 min at 249 kg (550 lb)		pass	pass	pass	—
Vickers Vane pump test, 65.6°C (150°F)	ASTM D 2882				
ring mass loss, mg		9.7	5.9	—	—
vane mass loss, mg		15.1	5.1	—	—
total mass loss, mg		24.8	11.0	—	—

Table 3-1. Compatibility of Dibasic Ester Oil-Mist Lubricants with Various Engineering Materials

Acceptable	Marginally Acceptable*	Unacceptable
Metals		
Steel and alloys	Cadmium	None
Aluminum and alloys	Zinc	
Copper and alloys	Lead	
Tin		
Nickel		
Inconel monel		
Silver		
Titanium		
Hard chrome		
Paints		
Epoxy	Alkyls (based finish preferred)	Acrylic
Baked phenolic	Phenolic	Latex (household)
2 component urethane	Single component urethane	Vinyl (PVC)
Moisture cure urethane	Industrial latex	Varnish
		Lacquer
Plastics		
Nylon (including filled)	Polyurethane	Polyethylene
Fluorocarbon (Teflon)	Polypropylene	Polycarbonate (Lexan)
Polyacetal (Delrin Celcon)	Polysulfone	Acrylic (Lucite, Plexols)
		Polyvinyl chloride
		ABS (Acrylonitrile/butadiene/styrene)

*Dibasic esters can form objectionable soaps with zinc and cadmium. Lead gives variable results and can show very low lead corrosion Despite these possible problems, no field problems have been reported to date.

Table 3-2. Compatibility of Dibasic Ester Synthetic Lubricants with Commonly Used Elastomeric Seals

Recommended	Not Recommended
• Fluorocarbon (Viton, Teflon)	• Polysulfide (Thiokol)
• High nitrile rubber (Buna N, NBR)*	• Ethylene propylene copolymer (EPR)
• Medium nitrile rubber (Buna N, NBR)†	• Ethylene-propylene terpolymer (EPDM)
• Fluorosilicone rubber	• Natural rubber
• Polyurethane	• Styrene-butadiene rubber (Buna S, SBR)
• Epichlorohydrin	• Butyl rubber
• Polyacrylate rubber	• Low nitrile rubber (Buna N, NBR)††
• Silicone rubber	• Polychloroprene (Neoprene)
• Chlorosulfonated polyethylene	

* Greater than 36% Acrylonitrile
† 30% to 36•% Acrylonitrile
†† Less than 30% Acrylonitrile

with a high-viscosity oil would benefit from oils with a high viscosity index (V.I.). The higher the V.I. of an oil, the less its viscosity changes with temperature. Consult the bibliography for additional information.

Figure 3-3 allows us to rapidly determine the viscosity index. Using viscosities either in SUS or cSt, place a straight edge from viscosity at 100°F through viscosity at 210°F. We can now read the V.I. on the scale. The chart may also be used to determine viscosity at either temperature, knowing the V.I. and the viscosity at one temperature.

Stability

In oil-mist systems with heaters for either the air, oil, or both, stability of the lube oil must be assured. Since heat input tends to oxidize a lubricant, the oil must be compounded with suitable oxidation inhibitors. Moreover, it is prudent to apply no more heat than is necessary to promote proper misting and, at the same time, avoid accelerated decomposition of the oil in the heated sump. Decomposition products tend to collect and clog oil sump pickup screens and similar close-clearance components. Just as wax crystals could clog reclassifiers, so could decomposition products or sludge traveling towards these small orifices. Again, dibasic ester synthetic oils proved superbly resistant to thermal decomposition in oil-mist systems.

Misting and Reclassification Characteristics

Oil-mist systems must be able to put a find dispersion of tiny oil droplets into carrier air and transport this aerosol over a practical distance of typically 500 ft. (~150 m) to points of application. At these points, the oil must be effectively "reclassified" into a

Table 3-3. Component Categories to which Oil Mist Has Been Applied

	Viscosity, Saybolt Universal Seconds			
	Max for Starting	**Min for Load Carrying**	**Desirable for Operation**	
Chains		200-2,000	800-1,500	
Cams		10,000-30,000		
Slides	100,000	80-1,000	200-500	
Slow-speed journal bearings:				
Light load			200-300	
Medium load			200-300	
Heavy load			800-500	
Rolling element bearings:				
Small and/or medium, high speed	3,000	50	80-150	
Medium size and speed	4,000	80	150-250	
Large and/or slow speed	6,000	100	250-500	
Open gearing		10,000-30,000		
Rack and pinion gears	100,000		1,000-5,000	
Enclosed reduction gears and speed increasers:			Speed, rpm	
			Above 1,000	Under 1,000
Spur gears	100,000	80-1,000	150-800	500-1,000
Bevel and spiral bevel	100,000	80-1,000	150-800	500-1,200
Helical and herringbone	100,000	80-1,000	150-1,000	800-1,500
Worm	100,000	500-5,000	1,000-5,000	
Hypoid	100,000	200-2,000	1,000-5,000	

form that will wet out on components to be lubricated.

The oil should therefore exhibit both good misting properties, i.e., allow itself to be dispersed into an air stream, and exhibit good reclassifying properties, i.e., allow itself to be effectively separated from the carrier air, without escaping to the atmosphere in mist form. A good oil-mist lubricant is compounded to meet both objectives.

Oil-mist viscosity and molecular composition are known to affect the degree of stray mist, or mist that would defy reclassification. Small amounts of high-molecular-weight polymers have been found effective in increasing the amount of oil being reclassified, and decreasing the amount of oil escaping as a mist [9]. However, the compounding or admixing of these polymers must be done judiciously. Excessive amounts may reduce the mistability of the oil and properly balancing all additives is of great importance. Again, dibasic ester synthetic lubes were found to strike this balance quite well.

Toxicity Considerations For Oil-Mist Lubricants

The publication *Threshold Limit Values for Chemical Substances and Physical Agents in the Work Environment and Biological Exposure Indices with Intended Changes for 1984-85* issued by the American Conference of Governmental Industrial Hygienists [10] lists numerous threshold limit values for long-term as well as short-term exposure. For mineral oil mist, it gives a threshold limit value—time-weighted average (TLV—TWA) of 5 mg/m^3. This is the time-weighted average concentration for a normal 8-hour workday and a 40-hr workweek, to which nearly all workers may be repeatedly exposed, day after day, without adverse affect.

The same publication gives a threshold limit value—short-term exposure limit (TLV—STEL) of 10 mg/m^3. This is the concentration to which workers can be exposed

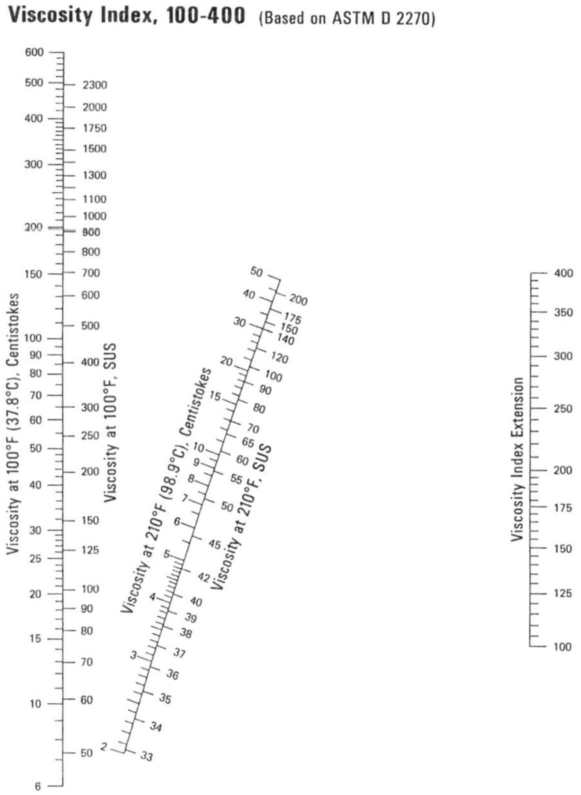

Figure 3-3. The viscosity index (V.I.) can be determined if the viscosities at 100°F and 210°F are known. *(Source: ASTM D-567.)*

continuously for a short period of time without suffering 1. Irritation. 2. Chronic or irreversible tissue damage. 3. Narcosis of sufficient degree to increase the likelihood of accidental injury, impair self-rescue or materially reduce work efficiency, provided that the daily TLV-TWA is not exceeded.

The 5 mg/m^3 is not commonly exceeded in the breathing space surrounding oil-mist lubricated machinery in outdoor locations. This is reassuring, as are the findings of a study published in the American Industrial Hygiene Association Journal [11], which reported on the respiratory function of guinea pigs exposed for one hour to submicrometer oil mists. Five oils were used: medicinal grade mineral oil, laboratory grade paraffin oil, light lubricating oil, unused motor oil, and used motor oil. The last three oils were also studied after having been reacted with sulfur dioxide. This reference states:

> "No alterations were produced by exposure to any of the oils at concentrations of 10 to 40 mg/m^3. At concentrations above 200 mg/m^3 both the reacted and unreacted light lubricating oil caused a decrease in compliance which remained throughout the post-exposure period. Although the average flow-resistance of the group was increased by the unreacted unused motor oil and by the reacted used motor oil, the variation in response was too great to make this response statistically significant. The fact that reaction of the oils with sulfur dioxide did not increase irritant action indicates that the reaction products with sulfur dioxide are not acutely irritant."

Translated into layman's language it simply means that oil mist in concentrations found around oil-mist-lubricated machinery is not a health hazard.

Stray mist need not be that great a problem. With careful system design, using mostly reclassifying fittings with mist oils, capable vendors seldom have much trouble keeping stray mist below 20% of the OSHA limit of 5 mg oil/m^3 air. At least one division of a major auto manufacturer measures stray mist at housing vents and is in the process of applying oil mist to all spindles and gear heads in all of its plants. Many, if not most, major spindle manufacturers are now designing for oil-mist lubrication, probably as a result of pressure from the automobile industry.

Nevertheless, objectionably high concentrations of oil mist could accumulate in indoor locations unless special provisions were made to prevent stray oil mist from escaping from points being lubricated. This is accomplished by one of the draw-off methods described later in Chapter 6. Alternatively, the area would have to be vented by a suitable fan, blower, or other air exchange means.

Lubricant Consumption

The number of points that can be lubricated by a given system is governed by the manufacturer's rating of the oil-mist generating assembly, i.e., the venturi or vortex dimensions, and by the requirements of the machine elements. A mist generator with sufficient output to lubricate four bearings in a large rolling mill might have too high a minimum rating to supply a hundred or more small bearings in another machine. As explained earlier, these ratings are expressed in bearing inches, lubrication units, or scfm capacity. These rating systems vary somewhat from one manufacturer to another, so there are no universal factors on convert from, say, bearing inches to scfm. (Bearing inches and lubrication units are merely different names for the same thing since L.U. formulas include factors to convert millimeters to inches.) One manufacturer, that uses the scfm system, publishes ratings on its mist generators and application fittings in both scfm and bearing inches. The relationship used is 33 B.I. = 1 scfm. Another manufacturer, that uses the bearing-inch system, published flow data for its application fittings ranging between 6 and 33 B.I./scfm.

In this text, where, for purposes of illustration, it is desired to convert from one system to the other, 33 B.I./scfm is used.

Actual oil consumption rate is a function of the particular model of mist generator and how it is adjusted, number and ratings of application fittings in the system, oil used, and air supply and oil temperatures. For general design purposes, to estimate oil consumption or delivery rates, multiply the system or application fitting rating by the manufacturer's standard oil/air ratio.

Examples:

Total system B.I. × fluid ounces oil/hour/B.I.
= fluid ounces oil consumed from generator each hour

B.I. of one application fitting × fluid ounces/hour/B.I.
= fluid ounces of oil delivered to one point each hour

System scfm × $in.^3$ oil/hour/scfm
= $in.^3$ oil/hour consumed from generator

Scfm rating of one application fitting × $in.^3$ oil/hour/scfm
= $in.^3$ oil/hour delivered to one point

Using the conversion criteria just given, an additional useful expression can be derived by converting scfm to liters per minute:

1 scfm = 28.3 liters per minute
1 liter per minute = 1.2 bearing-inches

Also, for purposes of general design and sizing of typical units operating at a header pressure of 20 in. H_2O (5 kPa), it can be assumed that 0.3 cm^3 or 0.018 $in.^3$ (0.01 fl. oz.) of lubricant are contained in the air stream furnishing lubrication to a 25 mm bore (~1-in. bore) diameter bearing in a one-hour period.

All of these consumption figures and ratings are purposely on the conservative side. In other words, we tend to deliberately err on the safe side, risking over-lubrication rather than inadequate lubrication.

The various relationships just given will readily allow us to calculate oil consumed and, hence, yearly cost of lube oil. For a graphical representation of air and oil consumption, refer to Figure 3-4.

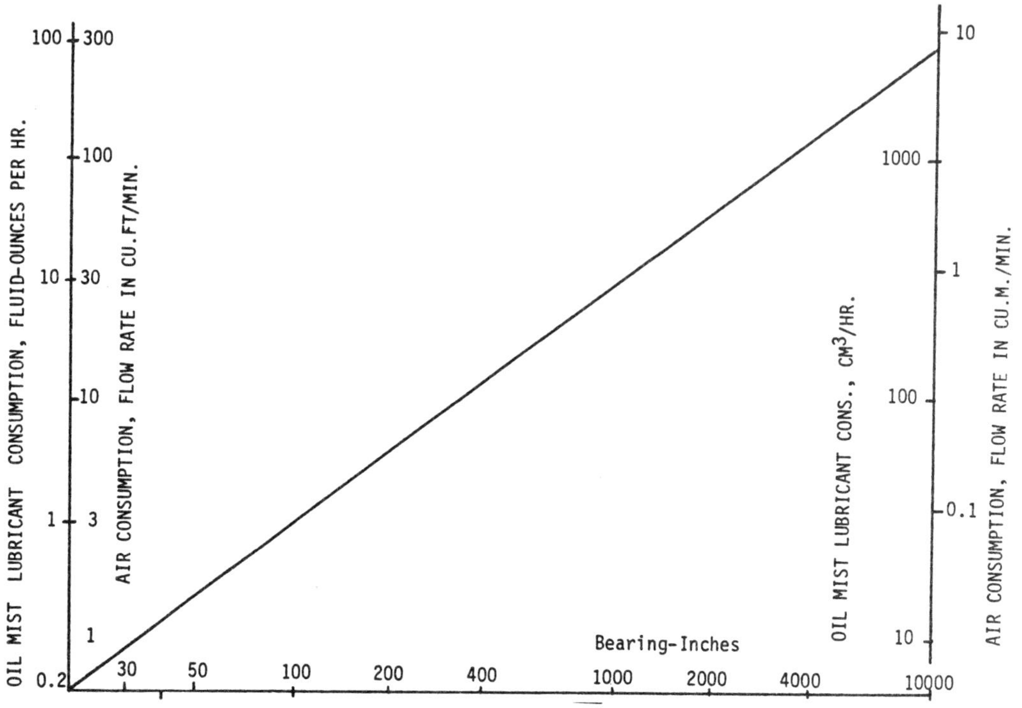

Figure 3-4. Lubricant and air consumption rates as a function of system size. *(Source: Reference 7.)*

Lubricant Cost Comparison

In 1982, a large North-American petrochemical plant encountered occasional wax formation during cold weather operation of their 17 plant-wide oil-mist lubrication systems. The plant was confronted with two options: To designate a two-man crew to perform daily, comprehensive preventive maintenance or to switch to a superior, non-waxing lube oil.

The cost impact of the two alternatives was analyzed as follows and the decision to use synthetic lube reached based on a remarkably simple comparison.

Continued Use of Wax-Forming Mineral Oil—

Combined consumption of all systems: = 127 scfm
Converting to bearing-inches (B.I.): = 127 scfm × 33 B.I./scfm
= 4,191 B.I.

Calculating lube consumption:
0.01 fl oz./B.I./hr × 4,191 B.I. = 41.9 fl oz./hr
Yearly consumption: 41.9 fl oz./hr
× 8,760 hr/yr × 1 gal/128 fl oz = 2,867 gal/yr

Cost of mineral oil:

$3.20/gal × 2,867 gal/yr = $9,175/yr

Preventive maintenance labor during 4 mo. of the year:

2 men × 800 hr/man × $20/hr = $32,000/yr

Anticipated yearly cost: = $41,175/yr

Switching to Year-Round Use of Synthetic Lubricant—

Cost of dibasic ester synthetic lubricant:

$8.40/gal × 2,867 gal/yr = $24,083/yr

For the sake of argument and to cover all possibilities, the actual analysis was expanded by calculating and crediting energy savings to the synthetic oil alternative. These credits accrue because rolling element bearings operating on synthetics do not have to overcome as high a frictional torque as would bearings operating on mineral oils [13].

For the mineral oil alternative, one also studied the cost effect of using only a single preventive maintenance technician and assuming that one incremental pump failure event would occur during each of the four winter months. However, in each case, the change to synthetic lube oils looked attractive.

Finally, it was decided to examine if synthetic and mineral oils would be compatible. Numerous compatibility tests were performed under varying operating conditions and full compatibility established. The mineral oil was drained from the main/storage tank feeding the entire plant (Figure 3-5) and also from the individual small reservoirs (Figure 3-6) inside each of the 17 oil-mist consoles. The main storage tank and all 17

Figure 3-5. Main storage tank for oil-mist lubricant at a major petrochemical plant. This plant is using dry-sump oil-mist lubrication for rolling element bearings almost exclusively. *(Source: Lubrication Systems Company.)*

Figure 3-6. Small reservoirs inside each of 17 oil-mist consoles were drained and refilled with dibasic ester lubricant when the plant converted from mineral oil. There are no compatibility concerns. *(Source: Lubrication Systems Company.)*

individual small console reservoirs were refilled without plant interruption or any other unusual event. It is important to note, though, that this highly successful switch to dibasic ester synthetic lubricant was made on a system with clean, rust-free distribution piping. If a detergent-action synthetic oil were to be introduced into piping with rust or similar corrosion products adhering to its walls, the debris might get dislodged and be transported into equipment bearings.

For a number of decades now, this 1978-vintage plant has done exceptionally well with the synthetic lubricant. No more than 8 working hours are expended each month by a single contract worker who services and reviews the 17 oil-mist systems in this modern plant.

Downtime statistics for this plant are equally impressive. There has been only one malfunction during a four-year period. A defective float valve allowed the oil reservoir level to drop below the minimum required volume and mist generation was temporarily interrupted. The annunciator feature (see page 29) was activated and the problem recognized and remedied long before bearing damage could occur.

Chapter 4

Components of a Plant-wide Oil-mist System

An oil-mist system consists of a mist lubricator unit that produces and controls the mist, distribution lines that convey the mist to bearing surfaces, application fittings that meter mist flow to each lubrication point (and, in many cases, also reclassify the "dry" mist to a wetter spray or large droplets), and vents that discharge spent air from bearings to the atmosphere. Such systems vary considerably in complexity as illustrated in Figures 4-1 and 4-2. (For a complete overview of recommended oil-mist systems, refer to the sample specification in Appendix A.)

The assembly shown in Figure 4-1 with its simple in-line mist lubricator unit includes only essential components. In sharp contrast, the system illustrated in Figure 4-2 has been provided with such accessories as solenoid valves for opening or closing the air supply line; heaters for warming both air and oil; and monitoring devices annunciating low air flow, high or low mist pressure, or low reservoir oil level.

Figure 4-1. Simple in-line oil-mist lubricator assembly used for preserving mothballed machinery. *(Source: Lubrication Systems Company.)*

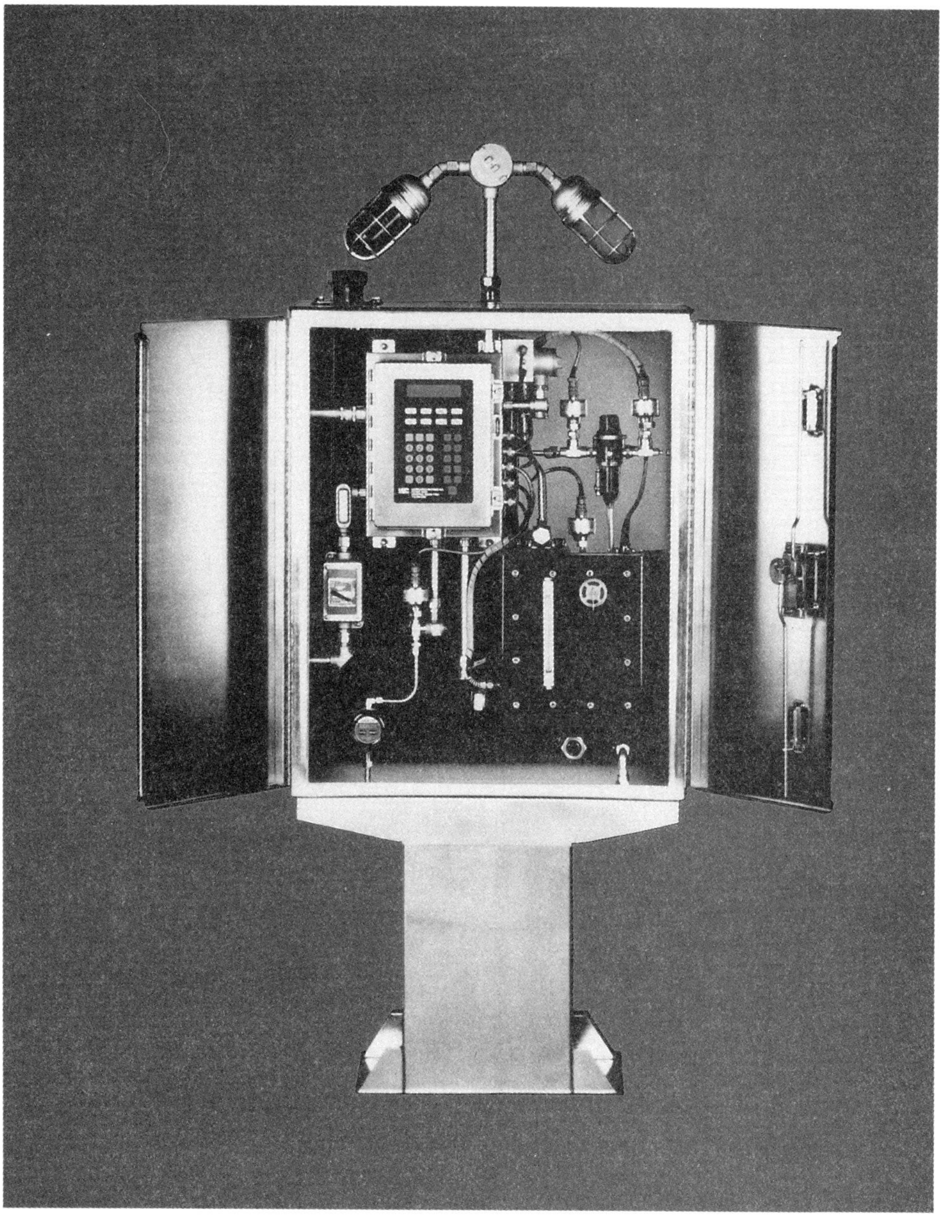

Figure 4-2. Fully instrumented, large-scale, oil-mist lubrication console. *(Source: Lubrication Systems Company.)*

Producing the Mist

Mist generator heads, as illustrated in Figure 4-3, are the key components in an oil-mist system. In the mist generator head the motive air makes contact with the lube oil to form mist.

Figure 4-3 shows how a vortex-type mist generator produces the mist. Air flows into the mist generator where it is guided by tangential slots to a vortex chamber. Here it flows in a whirling motion downward and out through an orifice at sonic velocity. Oil is drawn up the suction tube and through internal passages around the vortex chamber to a point just below the orifice where it mixes with the air to form mist. A deflector disc and baffle ring eliminate oil particles too large to be transported over long distances.

There is not much difference between the performance of vortex generator heads and venturi generator heads. However, any mist generator head must be sized to match its application. Often, generator heads are sized too large for the existing system because the specifying engineer wants to anticipate future expansion.

It is important to note that mist generators do not operate well below their rated

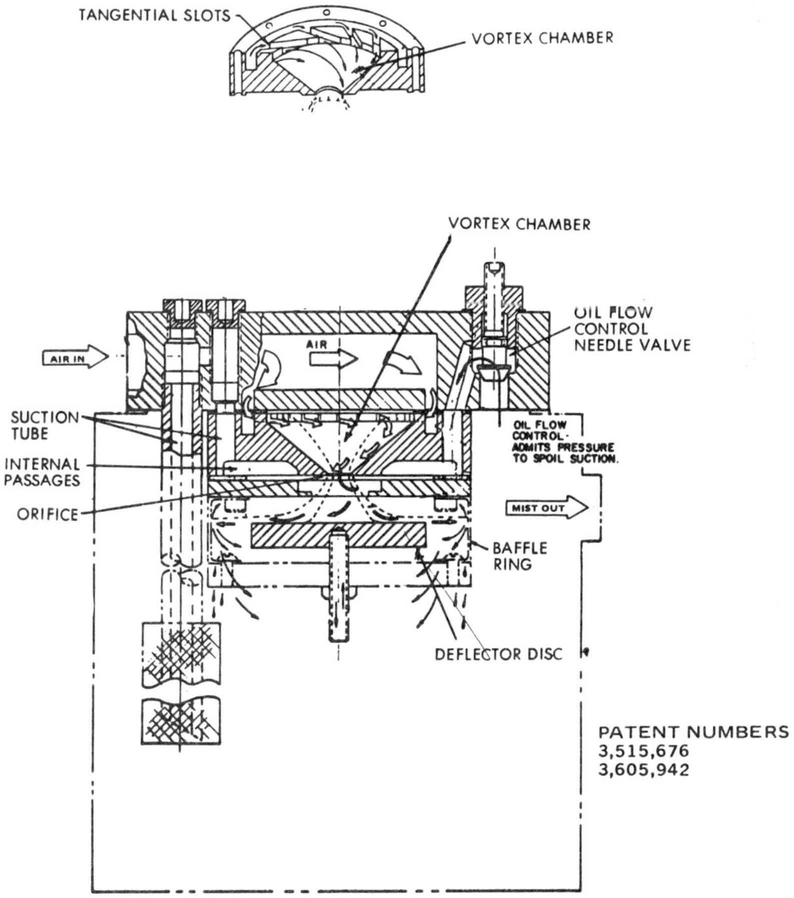

Figure 4-3. Vortex mist generator head. *(Source: Lubrication Systems Company.)*

capacity. When inlet air pressure is low, velocity through the dispersion orifice is also low. This results in insufficient vacuum to properly lift the oil from the reservoir and in turn causes inadequate dispersion of oil mist. Unnecessarily large reclassifiers are then used to increase air flow and allow the generator to operate in a more efficient range. This results in over-oiling, housekeeping problems, increased losses to the atmosphere, and wasted energy.

Typical oil-mist unit sizes can be selected in three very simple steps as will be shown later.

Controlling the Mist

Control and metering of oil mist is generally a very simple and straightforward procedure. This is demonstrated in Figures 4-4 and 4-5, which depict small lubricator units respectively utilizing the vortex mist and venturi mist principles. For each of these types, the differences between units of different generating capacities are essentially dimensional. Generally, construction features of either type are relatively unchanged across the range of capacities available.

The *air pressure regulator* is the basic control. It controls mist volume that is proportioned to the points of lubrication by the application fitting orifices. The regulator setting must be high enough for mist generation. That is, it must produce sufficient air flow through the mist generating head to reliably siphon oil from the lowest usable reservoir level to the mist head. The minimum setting is a function of the generating capacity of the head and of the height of the reservoir, and generally ranges between 5 to 20 psig (about 35 to 140 kPa). Further, the regulator is adjusted to produce the mist manifold pressure for which a given system was designed. Most frequently, systems are designed with mist pressures of 20 to 40 in. water column (5-10 kPa), although pressures as low as 2 in. water column (500 Pa) and as high as 80 in. water column (20 kPa) are sometimes used. In general, the lower pressures are used only with mist (metering only) fittings, and the higher pressures are used to improve the efficiency of reclassifying fittings (reduce stray mist) or to produce higher velocity outputs to penetrate air barriers around high speed elements.

The *oil flow valve* controls the mist density. Increasing oil flow to the mist head increases mist density (oil/air) ratio. The control is accomplished either by restricting oil flow or by reducing suction in the oil pickup tube (vacuum break). With the former, counterclockwise adjustment increases mist density, and with the latter, counterclockwise adjustment decreases mist density. It should be noted that the oil/air ratio or mist density depends on the characteristics of the oil and air delivered to the mist head. Oil output drops with temperature decrease.

The *air bypass valve*, included by some manufacturers, controls mist pressure without increasing oil output. However, the velocity of mist through the reclassifier fittings would increase as well. The same thing can be accomplished by designing the system for a higher manifold pressure, which will result in the selection of smaller orifice application fittings.

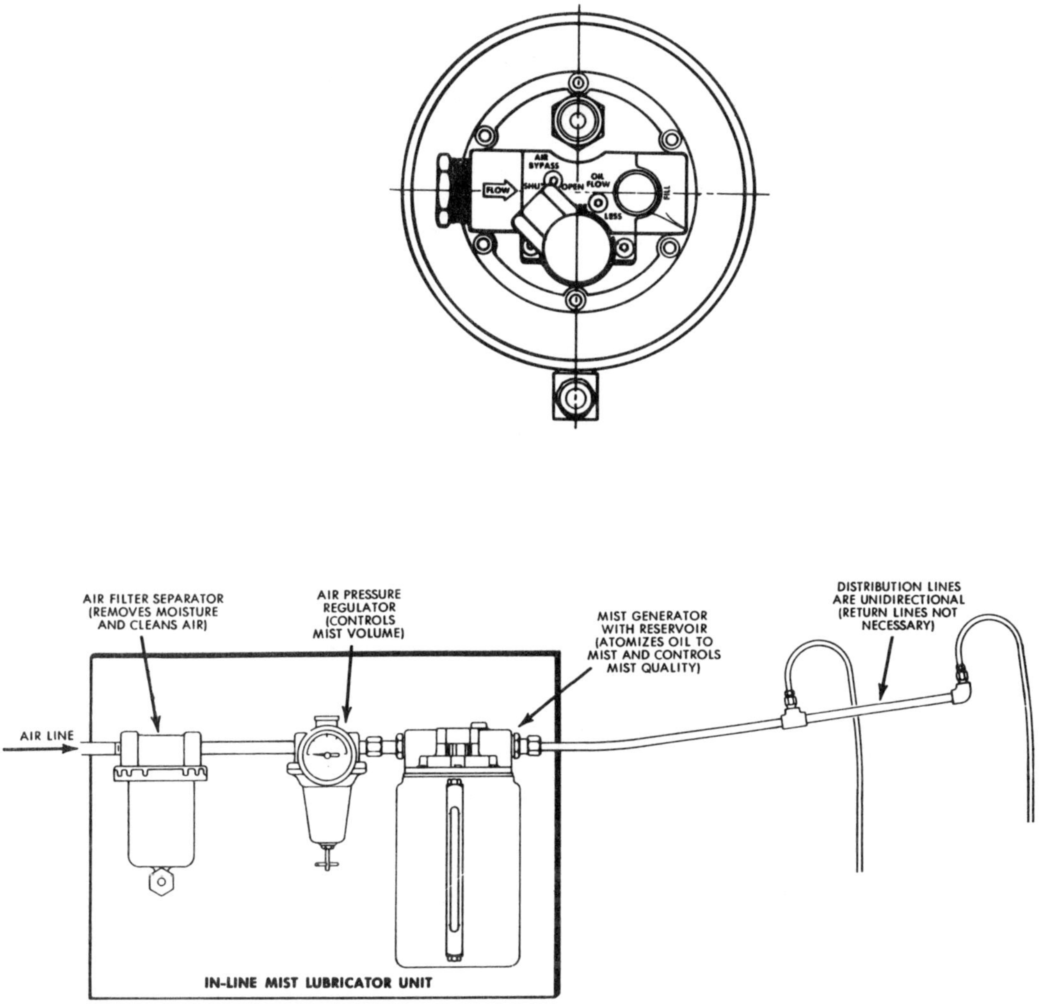

Figure 4-4. Small in-line oil mist lubricator unit using the vortex generator principle. Top view of mist generator showing controls for oil flow valve and air bypass valve. *(Source: Lubrication System Company.)*

In modern large-scale systems the mist generator is fully monitored and microprocessor controlled. Solid-state pressure and temperature transducers and level sensing devices have replaced the old-style electromechanical switches. Rather than gauges, all monitored variables are displayed on demand by an alpha numeric panel which not only shows typical gauge values but also provides messages describing the operating condition. In addition to improved reliability, the microprocessor control of the new mist generator units provides for customizing operating set points and alarm limits to exact user requirements [69]. See Figure 4-5a.

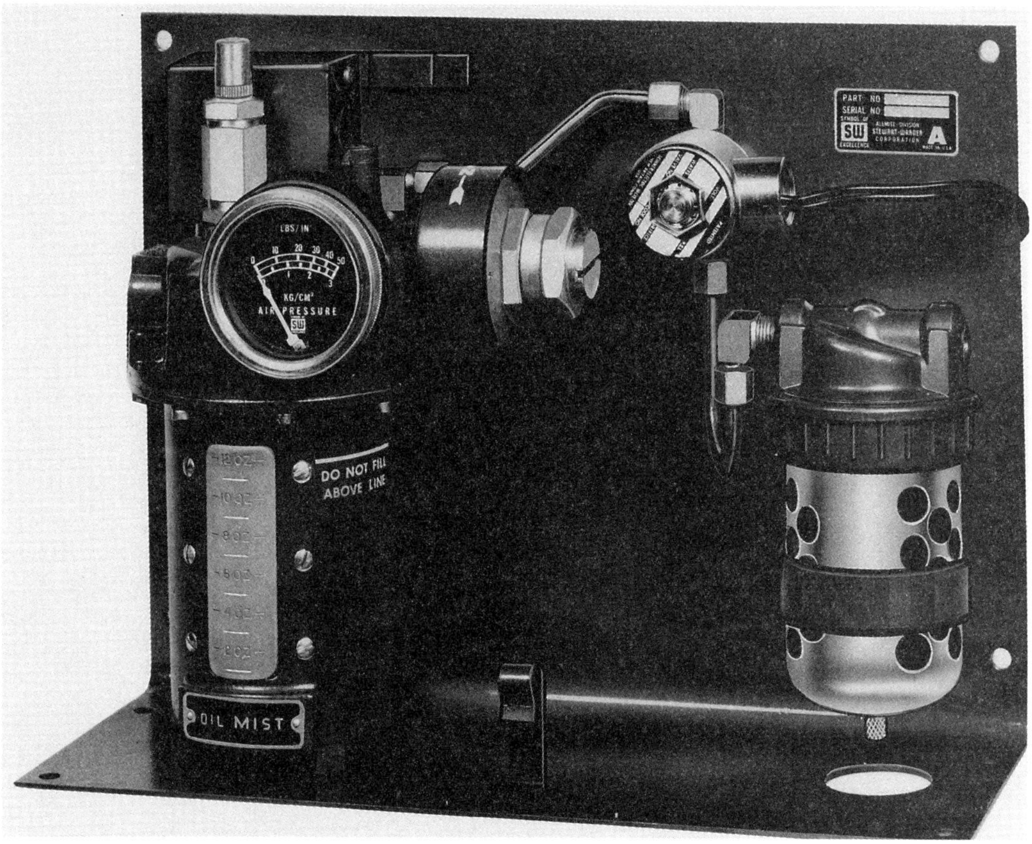

Figure 4-5. Simple, inexpensive oil-mist lubricator using venturi principle. *(Source: Alemite Division of Stewart-Warner Corporation.)*

Header System

Once generated, the oil mist is transported to the user equipment in the main header system. The header system should be sloped back toward the generator, but the distance the line is sloped will vary according to the length of the header. Figure 4-6 and Table 4-1 give some recommended slope percentages as a function of lube oil viscosity and temperature.

However, from a practical point of view, the user will find it easier to standardize on a fixed slope based on proven experience by petrochemical companies in the United States. For instance, if very long headers, say, 90 m (~300 ft), are used, at least the first 15 m (~50 ft) should be sloped toward the generator. If the lines are shorter, say 30-45 m (~100-150 ft), the entire line could be sloped back toward the generator. Vertical clearance in a pipe rack may govern the amount and direction of slope.

System drainage can also be through lubricated equipment, since neither pure nor purge mist permits overfilling.

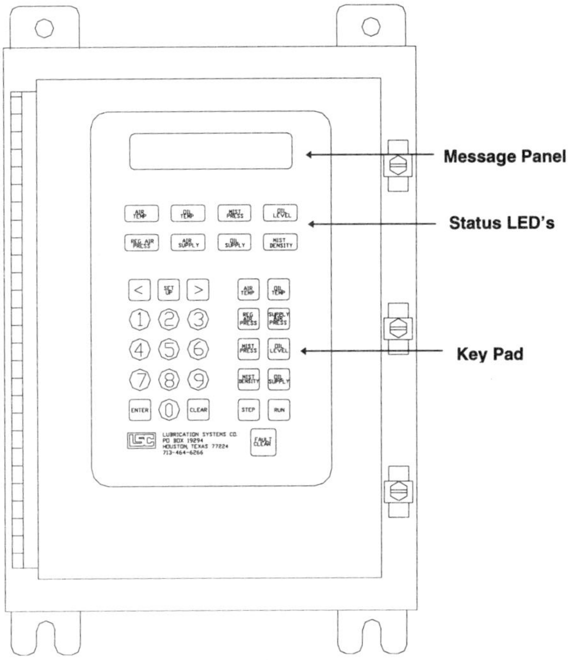

Figure 4-5a. Control panel for a large scale oil mist generation system. *(Source: Lubrication Systems Company, Houston, Texas).*

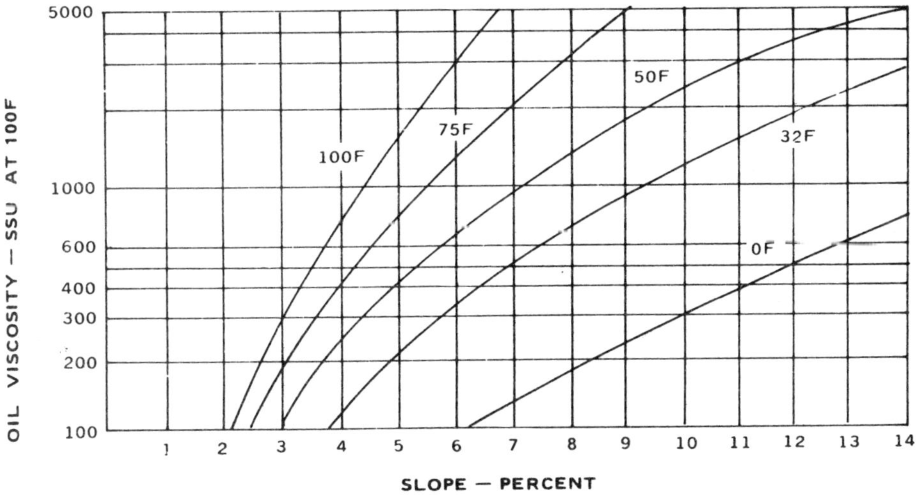

Figure 4-6. Minimum slope of mist distribution lines that slope toward a mist control unit. This chart applies only to lines that slope toward the mist control unit. It gives slope for continuous operation. If the system operates only part of each day (e.g. one or two shifts) divide by 2. Example: Line slope with 1,000 viscosity oil at 75°F would be 5-1/2% (5-1/2 in. drop every 100 in.).

Sloping is generally required because some mist particles collide with each other or strike the walls. This results in larger particles that are too heavy to remain airborne. These heavier particles fall to the bottom of the pipe and drain back to the reservoir or to a separate drain pipe. Most of this "condensation" occurs in the first 15 m (~50 ft). To reduce the severity of impact and time rate collisions, the velocity in the header is kept

Table 4-1.
Recommended Slope of Mist Manifold Toward Generator Source

Oil Viscosity (SSU @ 100°F)	Minimum Ambient or Manifold Temperature 0°F	32°F	50°F	75°F	100°F
100	5.3	3.7	3.0	2.4	2.1
180	8.8	5.4	4.1	3 0	2.5
300	10.5	6.1	4.6	3.7	2.9
500	12.2	7.2	5.5	4.4	3.5
800	18.0	8.5	6.5	5.1	4.0
1,500	—	11.0	8.8	6.1	4.9
2,500	—	15.0	10.4	7.1	5.4
5,000	—	—	14.4	9.0	6.7

***Percent slope of manifold**
2% slope equals 2" drop every 100" of manifold

Note:
1. Table is for manifold where condensed oil flow is opposite the mist flow
2. Table is for installations in continuous operation. For systems operating one or two shifts daily, divide slope by 2.
Source: Alemite Division of Stewart-Warner Corporation.

below 22 fps (7.3 m • s^{-1}). Laminar flow is thus ensured and premature or excessive condensation reduced.

Proper cleaning and commissioning the header system are important. Scale, debris, and moisture can present serious problems in a newly installed system. A suitable commissioning procedure is highlighted in the sample specification in Appendix A.

If major pipe branches or auxiliary headers are required, they should come off the top of the main header, as shown in Figure 4-7. The auxiliary header can be sloped either way (toward or away from main header), provided a drain leg is installed at the end of the auxiliary header if it is sloped away from the main header. A typical drain leg and plexiglass collecting container are shown in Figure 4-8. However, neither header nor branch lines should be installed with low spots since condensed oil could collect there. Pooled oil in low spots (traps) might not appreciably reduce flow but might locally increase turbulence and resultant oil wet-out, or even scrub oil out of the mist. It is good practice to review the piping installation before starting the mist system. Once commissioned, the system should be periodically checked to ensure that lines have not been bent or otherwise disturbed.

Drain Legs

Figure 4-8 depicts a drain leg and plexiglass collecting pot. Such an installation is necessary if long headers or routing over obstructions had to be used. Recent installations are automated, Figure 4-8a.

A drain leg is merely a vertical section of pipe that terminates with a valved collecting vessel or perhaps only a valved pipe cap. Condensed oil can then be drained and potentially troublesome oil accumulation is thus prevented.

As a general maintenance precaution, drain legs should be routinely emptied. It is also good practice to provide drain legs with an overflow orifice to prevent them from filling completely and thus risking oil accumulation in the main header. If the header is long enough, oil could seal off mist flow to the auxiliary header. Although not always feasible, a preferred design would eliminate the drain leg altogether and slope the header back toward the console.

Today's oil mist distribution systems utilize automated drain leg assemblies which do not require manual operation and can be fully integrated into closed-loop systems. These assemblies are equipped with an air activated level switch an pump. They collect the coalesced oil and automatically pump that oil overhead to a point in the distribution header that slopes back to the central generator.

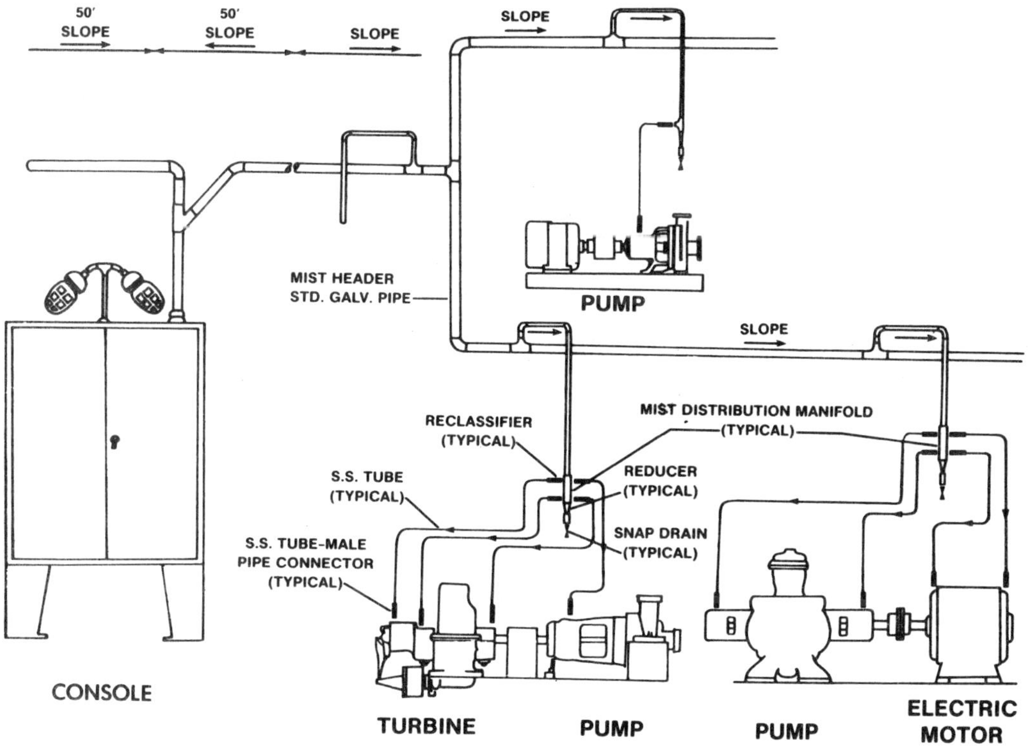

Figure 4-7. Mist distribution piping showing top take-off connections for pipe branches and auxiliary headers. This oil-mist system is serving general purpose equipment in a petrochemical plant. *(Source: Lubrication Systems Company.)*

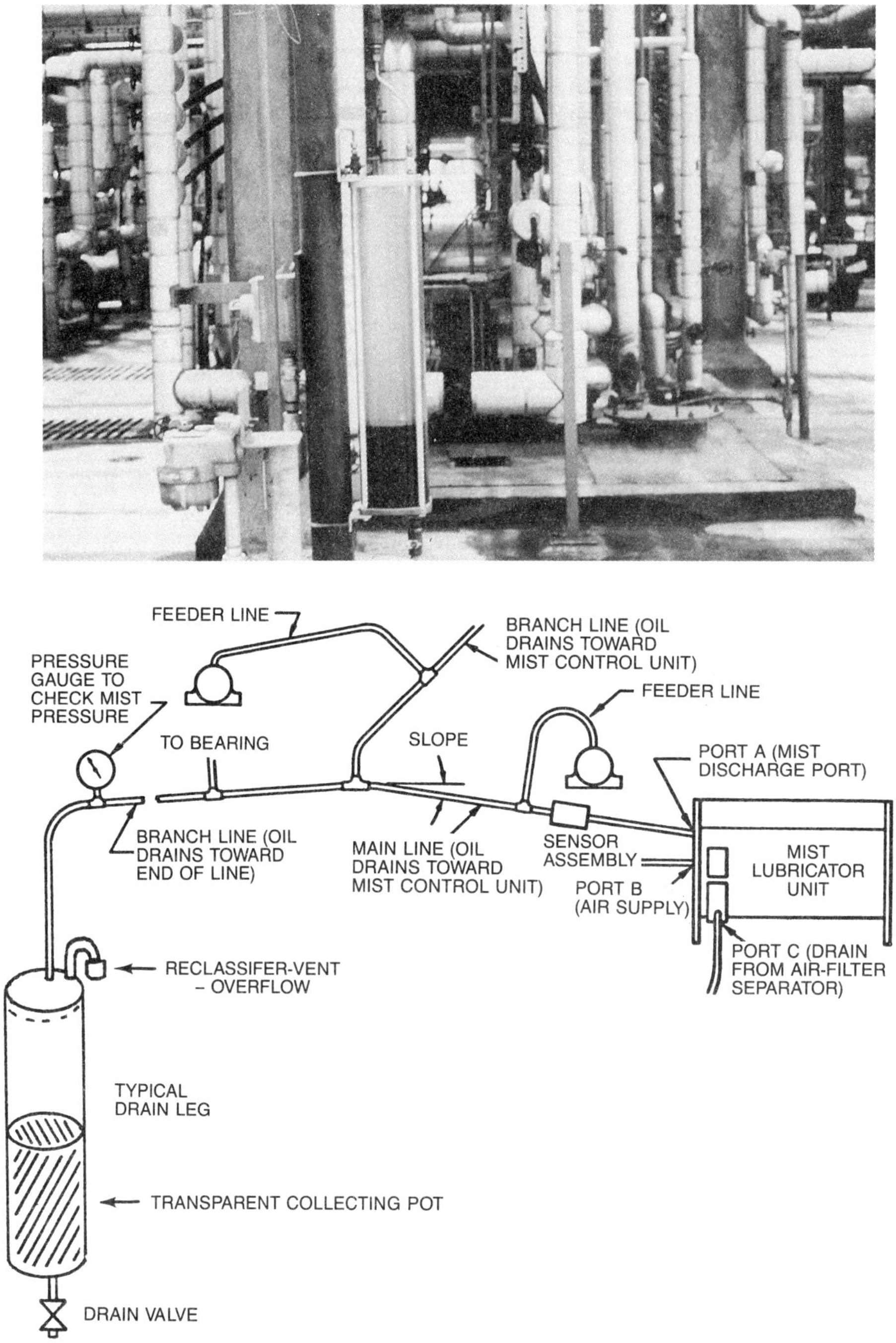

Figure 4-8. A drain leg assembly should be located at the end point location of an oil-mist header system.

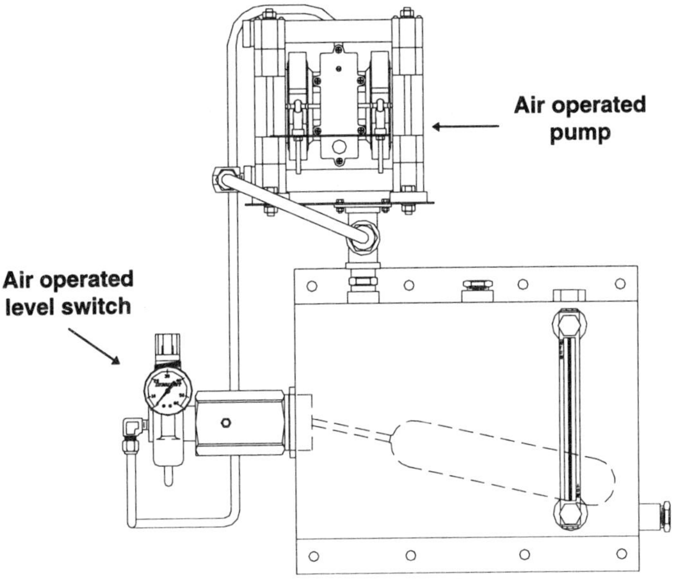

Figure 4-8a. Automated drain leg reservoir assembly [69]

Drop Points

Small oil-mist lines that originate either at the top or the bottom of the main header or pipe branch are called drop points. If an oil-mist line originates at the top of the header, the risk of getting condensed oil or other material into the reclassifiers is minimized. Two methods for terminating drop points are given in Figure 4-9. A third method is shown in Figure 4-10. It employs a multiport distribution block instead of the pipe-tee.

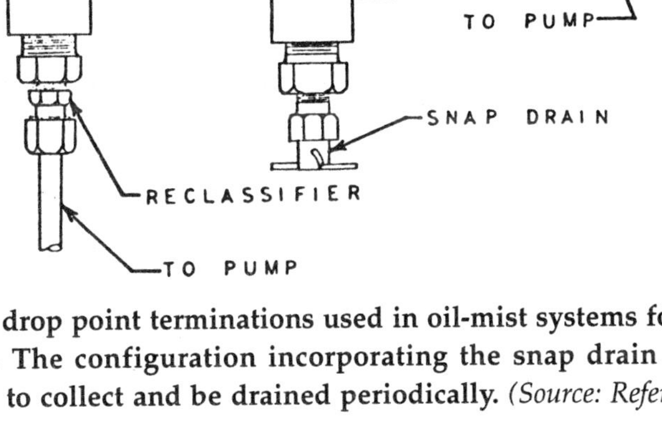

Figure 4-9. Typical drop point terminations used in oil-mist systems for pumps in petrochemical plants. The configuration incorporating the snap drain allows heavier-than-mist particles to collect and be drained periodically. *(Source: Reference 6.)*

As shown in Figures 4-7 and 4-9, reclassifiers are located at the drop point rather than on the equipment. In Figures 4-10 and 4-11, the reclassifiers are located on the equipment.

Experience has shown that if the reclassifiers are screwed into the bearing housing, they are sometimes taken to the shop when the pump is removed from the field. Depending on reclassifier configuration, shop personnel may mistake them for tubing fittings. Some get lost and others are inadvertently replaced by plain tubing fittings. Furthermore, with application fittings at drop points, equipment can be disconnected without disturbing the oil-mist system.

Nevertheless, locating the reclassifier at the equipment bearing housing is technically more advantageous because it defers the onset or formation of large oil droplets (condensation) until the mist arrives at the closest proximity to the point to be lubricated. Moreover, proper use of directed mist reclassifiers (Figure 4-12), sometimes used to better lubricate high-speed or heavily loaded bearings, requires that they be mounted at the bearing housing.

Operator and repair crew training should address these points. It should be pointed out to them that leaving tubing ends without a reclassifier fitting will result in reduced header pressure. In a properly instrumented system this deviation will trigger an alarm, and responsible personnel can correct the problem. If it continues unchecked, there is a risk of inadequate lubrication throughout the system. Moreover, a tubing line connected to a bearing without a reclassifier may cause extreme overlubrication, and a tubing line freely discharging oil mist to the atmosphere could represent an environmental nuisance.

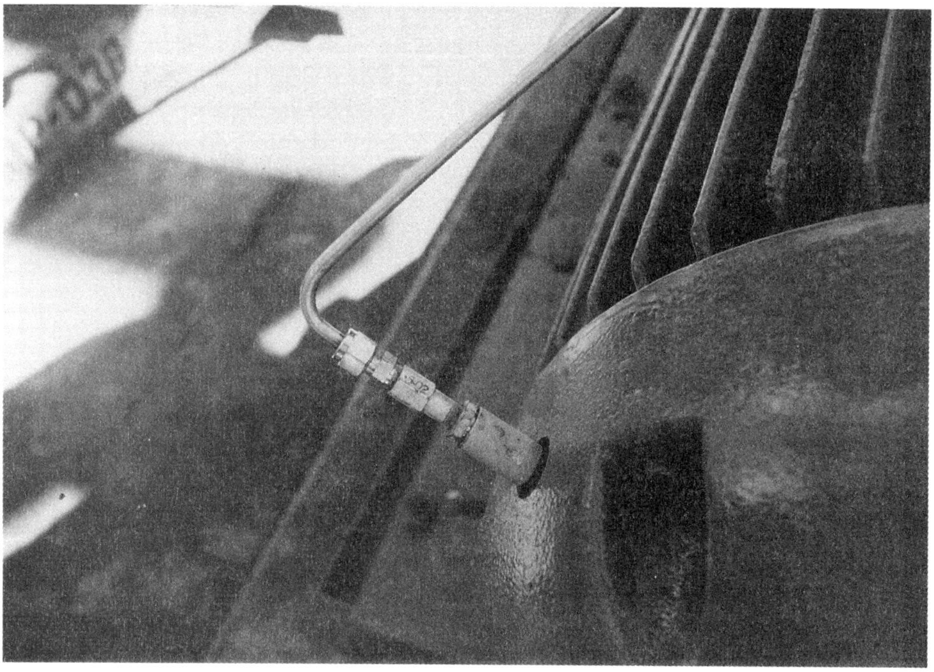

Figure 4-10. Drop-point termination employing multiport termination block. In this installation, the plant decided to locate the reclassifiers directly at the bearings to be lubricated.

Figure 4-11. Reclassifier mounted at electric motor bearing. This has the advantage of deferring the onset of oil-mist condensation until the mist arrives as close as possible to the point to be lubricated.

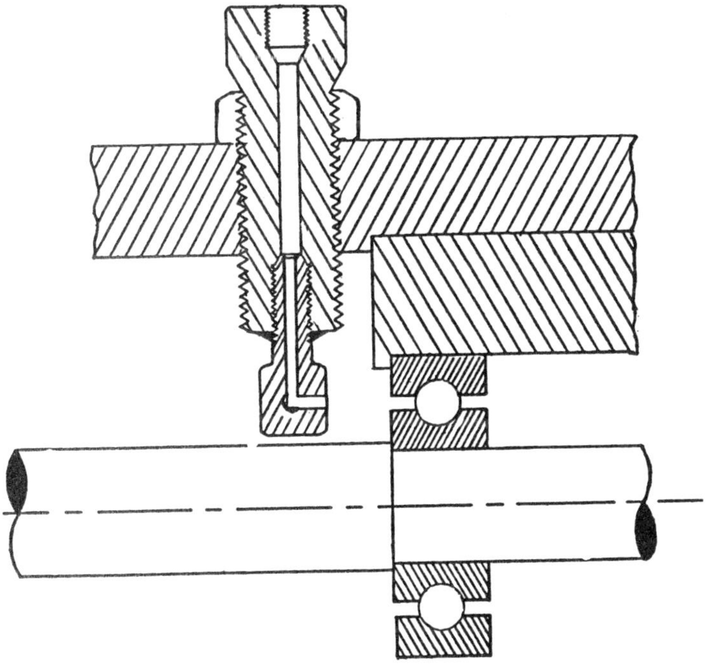

Figure 4-12. Directed mist reclassifier. The classifier outlet edge is close to the rotating elements. This helps to overcome windage created by high-speed bearings.

In either case, the cause of the pressure reduction should be found and remedied before making indiscriminate adjustments to the air supply pressure. For additional details, refer to Appendix B, "Oil-Mist System Troubleshooting Chart."

In large outdoor petrochemical plant installations, the line size for drop points is often 19-mm (3/4-in.) pipe. If the connected equipment is close to the main header, branch piping and drop points can be kept short. Where relatively long horizontal piping runs are needed, we must remember our velocity guidelines. With small diameter piping, mist velocity is increased. More mist particles will collide and condensation will be increased. Therefore, it is good practice to slope long piping runs so as to prevent excess oil from reaching the reclassifiers, although in more hydrocarbon processing industry applications excess oil at the reclassifiers would cause few, if any, problems. Some installations even drain headers through lubricated equipment. Speeds are not terribly high, and normal oil delivery rates are so small that even doubling or tripling them will not usually cause any great problems due to excessive lubrication.

The greater disadvantage of undersize piping is the loss of oil from the mist due to wet-out in the lines. Even if allowed to drain toward lubricated points, this oil will not be evenly distributed to all points, and, in an extreme case, many points might be oil starved even though the generator puts out an adequate amount of oil for all points.

Application Fittings

As mentioned earlier, application fittings are used to meter flow to the individual points lubricated by a mist system. Regardless of type, flow rates through these fittings are generally determined by the dimensions of small orifices (and, of course, by mist pressure).

There is little general agreement among either manufacturers or users of oil-mist systems regarding nomenclature of application fitting types. Some refer to all such fittings as "reclassifiers." Others reserve the term for fittings that in addition to metering flow change the dry mist to a wetter form and use "reclassification" to refer to this process. Still others call all application fittings "reclassifiers" but also refer to the process of changing the mist to wet spray or drops as "reclassification." This results in an apparent contradiction in terminology when referring to mist fittings that do not reclassify mist to another form.

Under conditions of laminar flow as approached in properly designed mist distribution systems, few oil particles collide with each other or with the passage walls with sufficient force to adhere. If mist flow becomes very turbulent, many such collisions will occur, producing larger particles or droplets of oil that will readily deposit on surfaces that they contact. This is the basic principle of operation of most reclassifying fittings.

Mist fittings (Figure 4-13) consist of small diameter, rather short orifices. Although turbulent flows occur in the orifices, they are not maintained long enough to produce appreciable reclassification. For practical purposes, mist fittings can be considered as only metering flow without changing the characteristics of the mist that passes through them. They should be used only to lubricate rolling element bearings operating at speeds

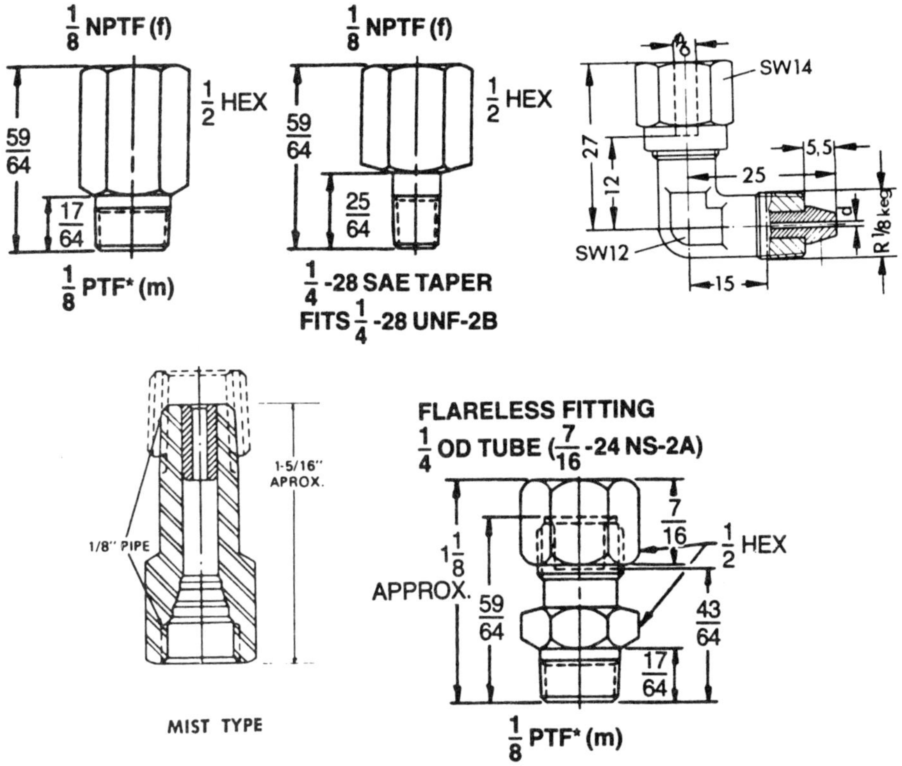

Figure 4-13. Typical mist reclassifiers are metering orifices that deliver mist, with minimum condensation, to machine elements. *(Sources: DeLimon-Fluhme; Alemite Division of Stewart-Warner Corporation; Lubrication Systems Company.)*

above one meter per second (200 lfm, or, linear feet per minute) at the mean diameter of the rolling elements. At such speeds, turbulence in and around the bearings will cause sufficient deposition of oil to adequately lubricate the rolling elements. However, at speeds down near 1 m • s^{-1} (200 lfm), reclassification might be incomplete enough to leave objectionable amounts of oil in the vented air. To minimize this "stray mist," mist fittings are often used only at speeds above about 5 m • s^{-1}, or 1,000 lfm.

Most fittings that reclassify the mist do so by inducing highly turbulent flow and maintaining it for sufficient distance to involve a majority of the oil particles in high-velocity collisions. Some fittings also incorporate baffling to increase the likelihood of such collisions. Spray fittings expel reclassified oil as fine, wet sprays. Generally, they reclassify by turbulent flow, either in a small passage that is very long (length at least 20 × diameter) or induced by a vane that produces a swirling motion some distance ahead of the output orifice (Figure 4-14).

Figure 4-12 shows a "directed mist" fitting, typically executed without a swirl vane. The diameter-length relationship of the orifice is such, however, that reclassification will occur, and "directed spray" would probably be a more accurate term. Such fittings are sometimes easier to install than a straight spray fitting where it is desired to aim the

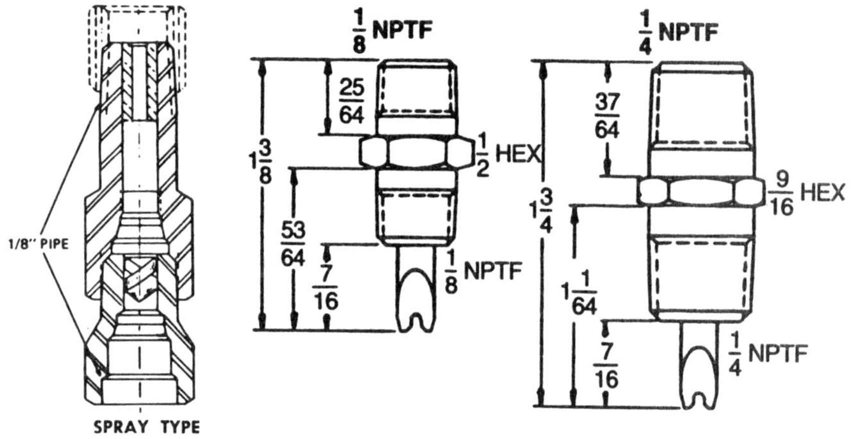

Figure 4-14. Spray application fittings are metering orifices that convert a high percentage of mist to an oil spray. Some models may incorporate a swirling vane. *(Sources: Alemite Division of Stewart-Warner Corporation; Lubrication Systems Company.)*

reclassifier output directly at the lubricated element. They are really just another configuration of spray fittings rather than a different type of fitting.

Condensing fittings (Figure 4-15) usually include some baffling or tortuous-path flow to increase reclassification efficiency. Some of these terminate in a small orifice through which oil is expelled as a wet spray and are sometimes referred to as condensed spray fittings. Others discharge reclassified oil through a large orifice with air velocity too low to carry the oil. From these fittings the oil just drips or runs down adjacent surfaces.

In general, the smallest reclassifier that will provide adequate bearing lubrication should be used. As explained later under "Lubricant Consumption," calculation techniques for air flow by volume, typically cubic feet per minute (cfm), or some other convenient method are provided by all oil-mist equipment suppliers. Bearing-inches are the shaft diameter in inches multiplied by the number of bearing rows. Volume flow is determined by multiplying bearing-inches by a constant service factor (either light, moderate, or heavy duty). In large petrochemical plants in the U.S. Gulf Coast area, experience has shown overwhelmingly that in all but the more severe applications (combining high speed, load, and temperature) the moderate-duty service factor is completely satisfactory. This will help prevent grossly oversizing the generator head, reduce oil consumption, and keep atmospheric emissions low. However, many users in the industry still use the heavy service formulas.

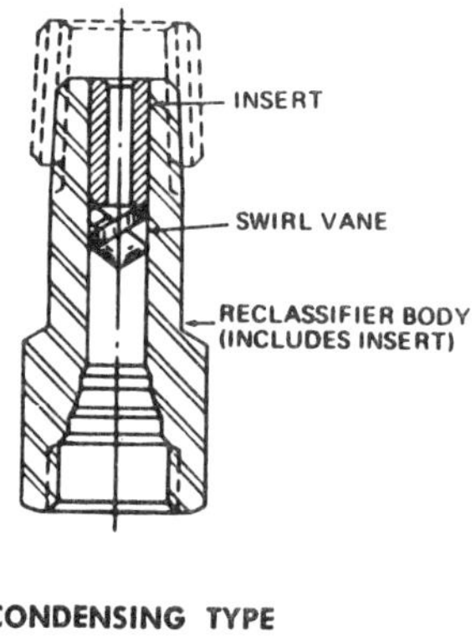

Figure 4-15. Condensing reclassifiers, also known as condensed-spray application fittings, are generally used in moderate-speed, light-load sleeve bearings and sliding mechanisms. *(Source: Lubrication Systems Company.)*

Connections at Rolling Element Bearings

Two different mist application methods, dry sump and wet sump, are available for pumps or, for that matter, any machine with rotating shafts requiring dependable lubrication. In the *dry* sump method, Figure 4-16, the bearing housing is drained of oil and all lubrication is accomplished by mist. A sight glass or other bottle-type device is installed at the bottom of the bearing housing to capture condensed oil. This catch pot is generally made of a transparent material to permit easy observation of water or other contaminants. Although it is not necessary to ensure *through*-flow of oil mist for single-row radial bearings, it is nevertheless good practice to do so. On thrust-loaded rolling element bearings, the oil mist should always be routed through the bearing rolling elements in order to ensure that complete lubrication and perhaps a small amount of additional cooling occur.

Venting of carrier air from closed housings is required to permit the continuous flow of oil mist into the housings. Relative placements of mist entries and vents are used to promote the movement of oil mist through certain assemblies. Proper venting of single-row bearings (pump radial bearings) equipped with dry mist lube is relatively simple. Leakage flow through a labyrinth or a small hole drilled in a lip seal usually amounts to acceptable venting. In some cases, where both sides of relatively high-speed (> 1 m • s^{-1}, or 200 fpm), moderate service, rolling element bearings can be freely exposed to mist, through-flow is not required. Windage generated by the rotating bearings promotes adequate circulation through them. With such applications, the housings must still be vented, and care must be taken to avoid venting much of the mist before it reaches the lubricated elements. Again, it is always prudent to review the bearing housing configuration so the correctness of venting or mist through-flow can be ascertained.

For multirow bearings, it is appropriate to vent the bearing housing to ensure equal flow through each row of bearings. A good sizing criterion for vents would be to design for at least twice the cross-sectional area of the oil-mist application fitting (reclassifier) or larger. If a machine has labyrinth shaft seals, additional vent provisions are seldom required.

Figure 4-17 shows a pump using the *wet-sump*, or purge oil-mist application method. This pump has its bearing housing filled to normal level with oil and is equipped with a constant level oiling device. In a wet sump lubricated machine, oil mist provides a positive pressure in the bearing housings to prevent the ingress of contaminants. It does not, however, counteract the possible harmful effects of recycling heavily oxidized oil or oil that has been contaminated by wear products from an oil ring. Periodic lube oil changes are still advisable when using wet sump, or purge oil-mist methods.

On wet sump installations using the constant level oiling device illustrated in Figure 4-17, it is customary to drill a 5-mm (3/16-in.) hole in the oiler body about 6 mm (1/4 in.) above the desired oil level. This hole allows excess oil produced by the addition of condensed oil mist to escape. (See Figure 4-18.)

Adequate venting is very important with purge mist. If venting is inadequate, oil can inadvertently be pressurized out of the bearing housing; this would leave the equipment operating on pure mist, dry-sump style. The bearings may remain unharmed if they

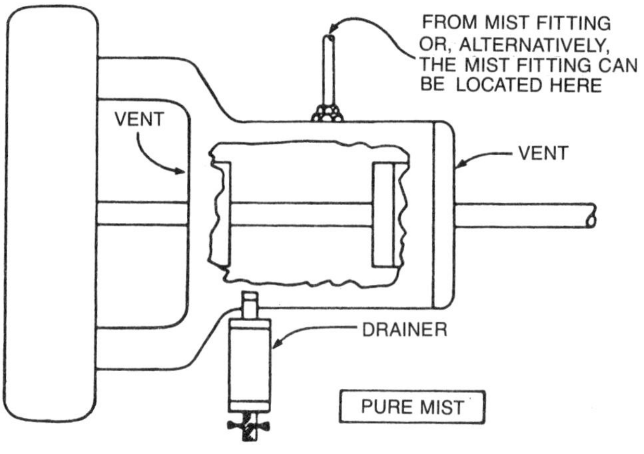

Figure 4-16. Schematic of centrifugal pump using dry-sump (pure mist) lubrication method. *(Source: Reference 6.)*

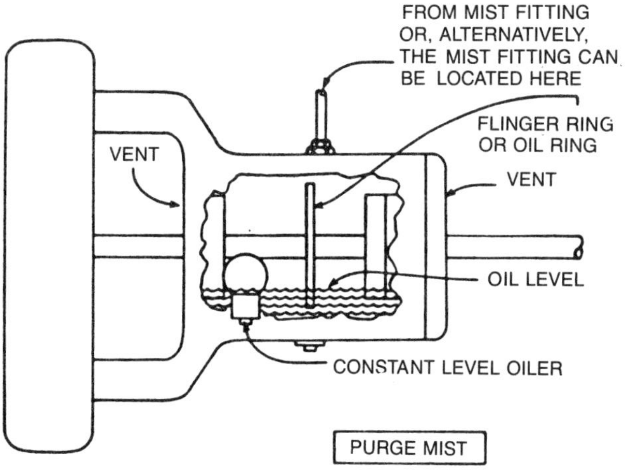

Figure 4-17. Schematic of centrifugal pump using wet-sump (purge mist) lubrication method. *(Source: Reference 6.)*

Figure 4-18. A vent hole should be drilled in body of constant level oilers used on purge-mist-lubricated equipment. *(Source: Lubrication Systems Company.)*

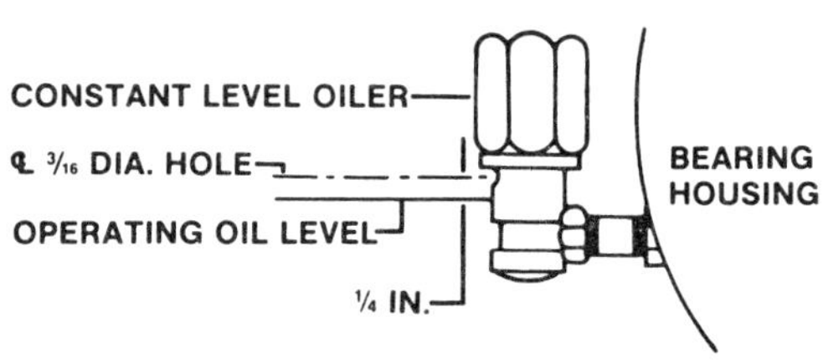

are only moderately loaded and if the oil-mist application fitting is a mist type of adequate size. Survival is less likely for heavily loaded rolling element bearings and sleeve bearings, or if the oil ring starts to wear and wear products find their way into the bearing.

Inadvertent overpressuring of purge misted (wet sump) bearing housings can be quite effectively avoided by using a balance line. This line connects the vapor space on top of the liquid oil in the bearing housing to the vapor space in the oil supply bottle, as shown in Figure 4-19. Here, the line labeled "air intake" performs the equalization function.

Controls and Alarms

Controls are provided to maintain oil and air temperatures, and in some cases, to maintain reservoir oil level (Figures 4-20 and 4-21). Depending on ambient temperature conditions and type of oil used, an oil-mist system might operate well without oil and air heaters. However, the heaters are often used to maintain mist density stability with variations in ambient temperatures. If not heated, oil in the reservoir might thicken at low temperatures and become more difficult to lift and atomize. So, the oil in the reservoir is heated primarily to ensure a reliable flow to the mist generating head. Air temperature affects the degree to which the oil that reaches the head will be atomized. With heavy oils, such as are used in rolling mill applications, air heaters are necessary for atomization. In outdoor applications, as in refineries and chemical plants, even the much lower viscosity oils commonly used might at some times require heated air to product mist. (See Chapter 12, "Specifications for Oil-Mist Systems," for more information on heater requirements based on oil viscosity and ambient temperature.)

When a mist unit is equipped with an automatic fill system, oil level controls consisting of a solenoid valve and a level control switch are used. Oil level is controlled over a narrow range, say 6 mm (1/4 in.) or less so that the bulk oil temperature in the reservoir is essentially constant.

Other controls include an air-pressure regulator for setting the pressure of air to the mist-generating head and an oil/air ratio screw, which is mounted on the generator head. Incoming air is adjusted until the desired header pressure, usually ~5 kPa (20-in. H_2O), is reached. The oil/air ratio is adjusted next. Some units are equipped with air bypasses that divert some air around the portion of the head where atomization occurs to permit broader adjustment of the oil/air ratio (Figure 4-4).

Generally, as air flow increases through the generator head, oil flow is also increased. At high air flows, oil flow can become excessive, so the oil/air ratio is adjusted to be leaner. In some units, it is possible at high air flow to run out of adjustment and still be providing more oil that necessary. This is where the bypass can be used. Regulated air pressure is reduced until the desired oil flow is produced at the generator head. The bypass is then opened to maintain header pressure at the desired level. Hence, header pressure is maintained while oil flow is reduced. If should be pointed out that the same thing can usually be accomplished by using a smaller application fitting, with the added benefit of a savings in air consumption.

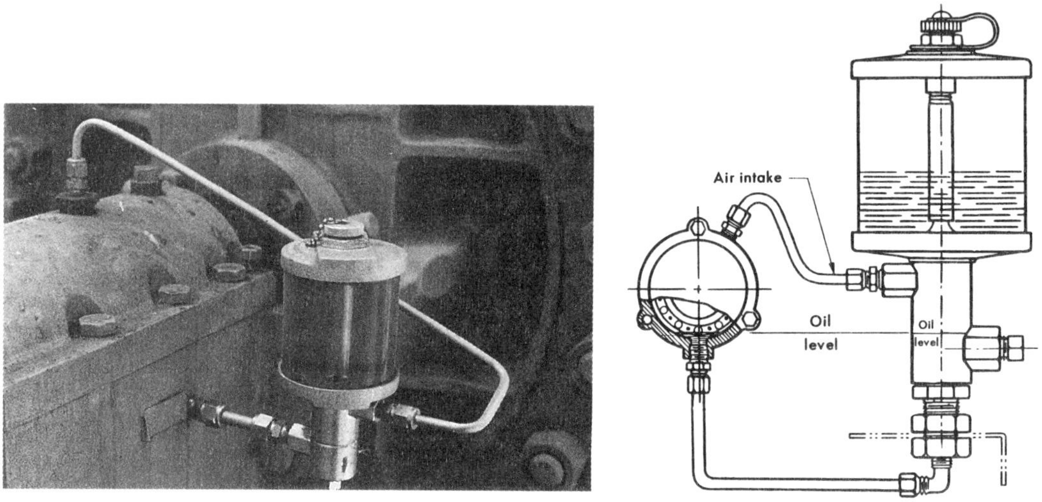

Figure 4-19. Constant level oil with equalizing tube prevents inadvertent lowering of oil level in bearing housing. *(Source: Oil-Rite Corporation, Manitowoc, Wisconsin.)*

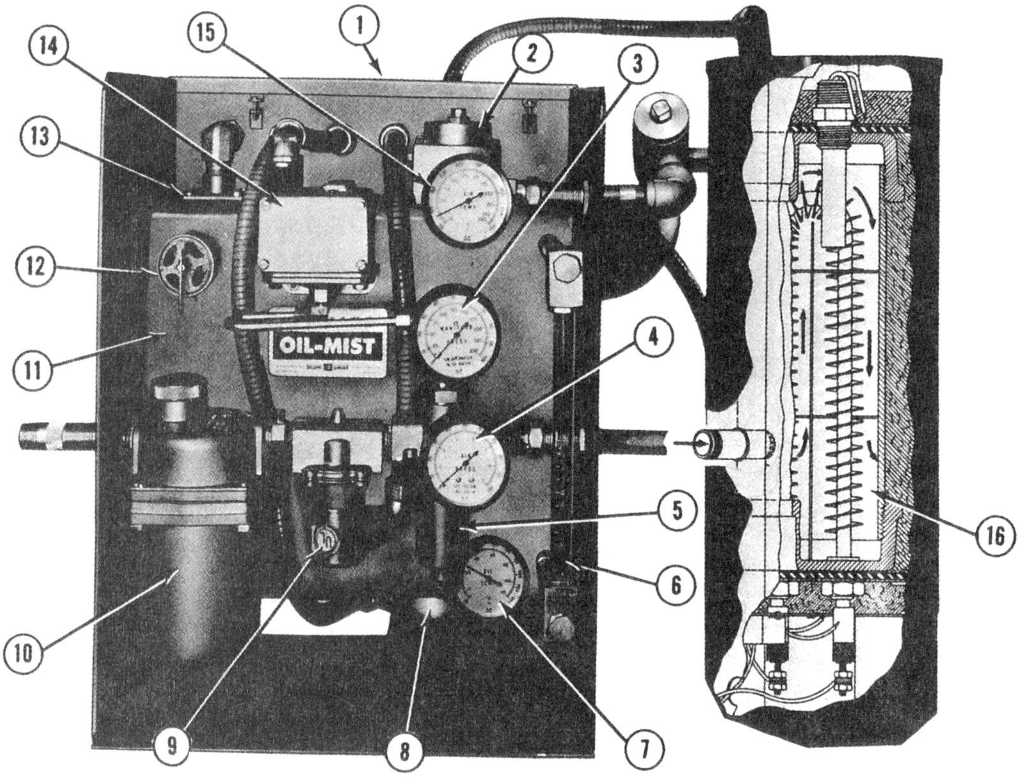

Figure 4-20. (Legend opposite) Large oil-mist lubrication unit equipped with heaters and various controls. *(Source: Alemite Division of Stewart-Warner Corporation.)*

Alarms can be provided for low and high mist pressure, air temperature, oil level, and oil temperature. Each alarm point can be adjusted separately according to user needs. Some units employ self-canceling alarms that clear automatically when rectified, while others must be manually reset. Mist density monitors that can indicate when mist density changes markedly from initial setting are available. These monitors use photoelectric sensors to indicate deviations from desired air/oil ratios. Figure 4-22 depicts a mist density monitoring unit that was introduced in 1984.

Refer to Appendix A for a concise overview of typical controls used in oil-mist systems.

1. **ELECTRICAL CONNECTIONS**—All enclosed in a large oil-tight terminal box with cover, for convenience and service.
2. **MIST-GENERATING HEAD**—Available in interchangeable CFM ratings for generating microscopic "airborne" mist.
3. **MIST PRESSURE GAUGE**—Indicates mist pressure in distribution system
4. **AIR PRESSURE GAUGE**—Indicates regulated air pressure to mist generating nozzle.
5. **AIR REGULATOR**—Provides accurate, regulated control of discharged air pressure.
6. **VISUAL OIL-LEVEL GAUGE**—Permits quick visual check of oil supply in reservoir.
7. **OIL THERMOMETER**—Indicates temperature of oil in reservoir.
8. **OIL HEATER**—Maintains oil at proper temperature for atomization, through automatic thermostat.
9. **SOLENOID AIR VALVE**—Automatically starts and stops the air supply.
10. **MOISTURE SEPARATOR**—Removes condensate from air. Automatic drain.
11. **OIL RESERVOIR**—Provides oil storage capacity.
12. **RESERVOIR SAFETY VALVE**—Protects oil reservoir from abnormal mist pressure.
13. **OIL-LEVEL SWITCH**—Closes or opens to energize warning signal, either visual and/or audible, when reservoir oil level is low.
14. **MIST PRESSURE SWITCH**—Two safety switches which respond to drop or rise in mist pressure. May be used to energize a warning signal and/or stop machine.
15. **AIR THERMOMETER**—Indicates air temperature entering mist head.
16. **THERMO-AIRE HEATER**—Preheats the incoming air entering the mist-generating head.

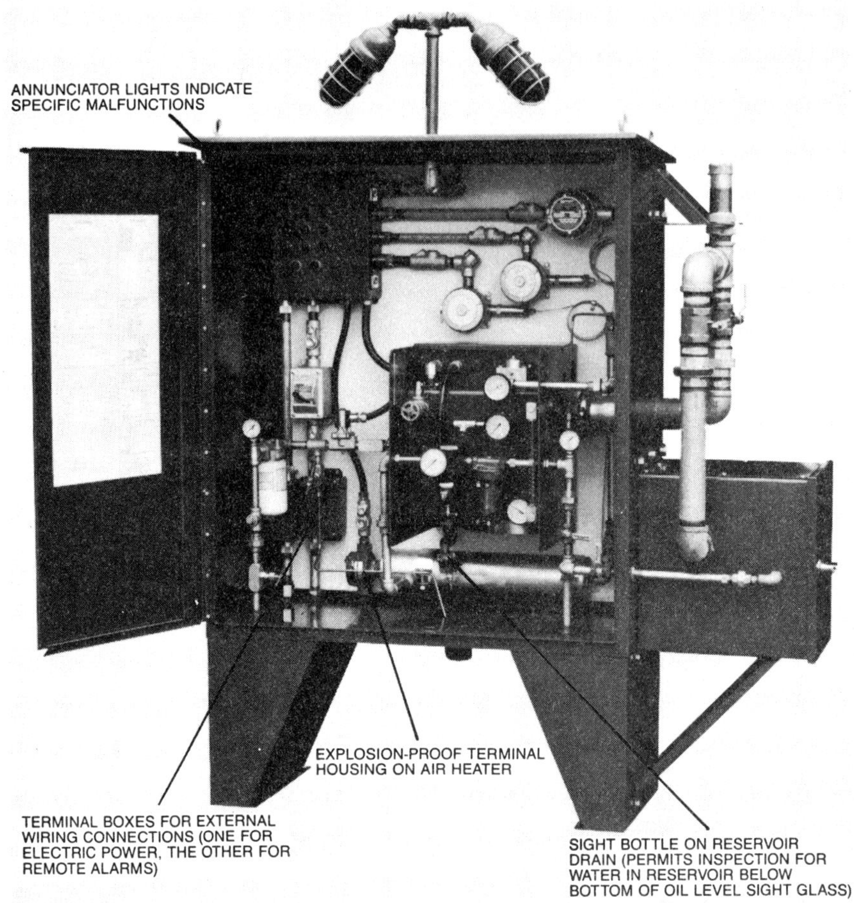

Figure 4-21. Oil-mist console assembly for hazardous area including controls and alarms. *(Source: Alemite Division of Stewart-Warner Corporation.)*

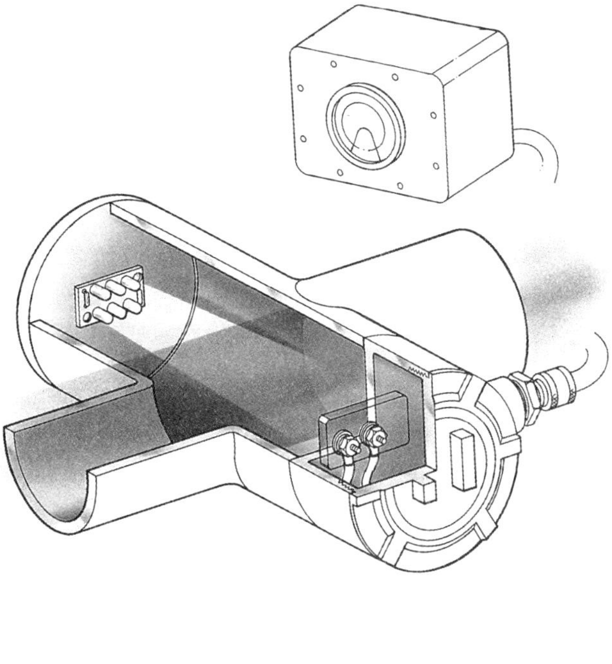

Figure 4-22. Oil-mist density monitor. Photoelectric sensors detect deviations from desired air/oil ratios. *(Source: Lubrication Systems Company.)*

Chapter 5

Oil-mist Application and Venting

As we examine how oil mist is applied, we should keep in mind the operating principles of mist lubrication. Recall that oil mist is a dispersion of extremely fine oil droplets in air arriving at or near the point to be lubricated. If the receiving location is a gear mesh, chain, slide, very-low-speed bearing or similar nonturbulent assembly, we must reclassify the oil *mist* into an oil *spray* or into large drops of oil. This type of reclassification is accomplished in an application fitting, orifice, or nozzle that is configured to produce collision of the fine mist droplets. The droplets are thus combined into larger drops or an oil spray.

If the receiving locating is a moderate- or high-speed bearing, turbulence is created by the motion of its component parts and only "dry," unreclassified mist—extremely fine droplets of oil suspended in air—need be applied at this point. The moving elements at this receiving location will do the reclassifying for us and a uniform coating of oil will establish itself on the machine components.

However, for dry, or unreclassified, oil mist to be supplied in sufficient quantity, through-flow is necessary. In other words, a pressure gradient must exist between the point of mist application and the point where spent oil mist—now almost entirely air—is vented. This is an important issue that must be understood to ensure the absolute adequacy and total success of an oil-mist installation. While there are thousands of pumps and electric motors whose rolling element bearings are properly lubricated without paying the slightest attention to vent locations and through-flow criteria, these considerations may nevertheless be quite important for heavily loaded and/or high-speed rolling element bearings and sometimes even for bearings under relatively moderate service conditions.

Bearing linear speed, load rating, and assembly method determine whether directed mist reclassifiers and special routing of oil mist through the bearing rotating elements are required. Bearings with two or more rows of rolling elements or thrust-loaded bearings require housing or mounting provisions ensuring that oil mist passes through, and not just past the bearing.

When rolling element bearings operate at inner race (bore) velocities in excess of 610 meters (approximately 2,000 feet) per minute, there exists the possibility of high windage preventing a sufficient quantity of oil mist from actually reaching the rotating elements. These bearings should be fitted with directed mist reclassifiers. Their discharge opening should be no more than 13 mm (approximately 1/2 in.) away from and directly opposite the rotating element. Only about 5% of the antifriction bearings commonly encountered in general purpose machinery require directed mist lubrication. In some cases, pumps may need minor modifications to accept the special directed mist reclassifier shown earlier in Figure 4-12.

Through-flow of oil mist can be achieved in various ways. Take pump bearing housings, for instance. Figure 5-1 depicts a typical pump bearing housing with oil mist applied to the center of the cavity. To pass from this cavity to the surrounding atmosphere, the oil mist must migrate through the bearing and shaft sealing areas. This is the natural flow path for oil mist supplied to bearing housings with labyrinth seals. If lip seals or other close-fitting housing enclosures are used, through-flow can be promoted by notching the lip seal or by drilling a vent hole in the bearing cap adjacent to the rolling elements. See Figure 5-2 for details.

Similarly, through-flow can be accomplished by allowing oil mist to enter the bearing from the atmospheric side at the housing end covers. One such execution is shown in

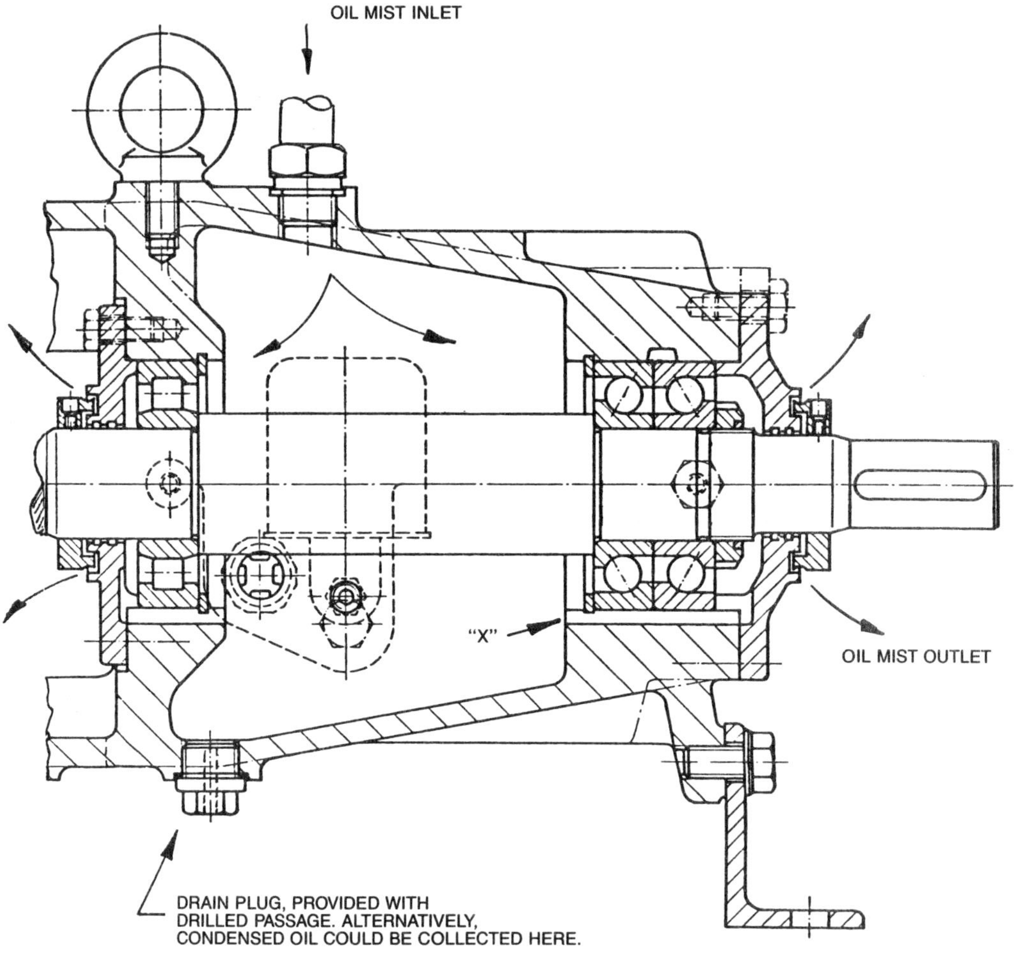

Figure 5-1. Dry-sump oil mist applied to center of pump bearing housing. Spent mist (stray mist) escapes along the shaft. The liquid oil lubricator (dotted) is no longer used. Oil rings have been removed. It is usually not necessary to plug the oil drain groove "X" when converting from conventional to oil-mist lubrication.

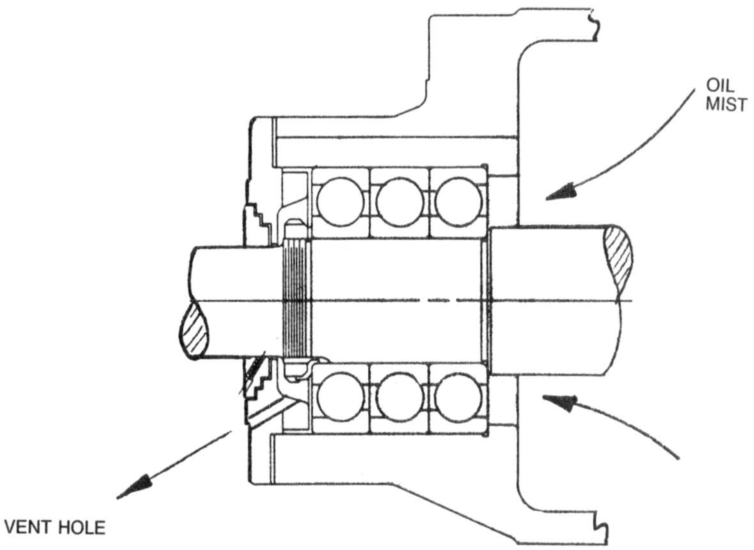

Figure 5-2. Bearing housing end cap with drilled vent hole passage. Through-flow from the center of the bearing housing towards the outside environment is thus facilitated. Note, however, that adequate lubrication for three rows of bearings may best be achieved by a "directed oil-mist fitting" close to the bearing farthest from the vent hole or by applying the oil mist from each end. *(Source: United Pump Company.)*

Figure 5-3 where the application fitting has been threaded into conventional end covers and venting is provided somewhere near the casing low point or at an overflow hole drilled into the condensed oil collection pot. It is not necessary to close the conventional drain groove, which is often provided in bearing housings as shown in Figure 5-3.

Arranging the oil-mist entry lines as shown in Figures 5-3 and 5-4 provides additional assurance that the oil-mist installation embodies the highest possible reliability standards:

- Windage or air flow induced by certain drive motor fan arrangements will not impede free venting.
- The oil-mist application fitting is mounted in such close proximity to the bearing that it constitutes a "directed" fitting, similar in function to the oil-mist fitting shown earlier in Figure 4-12. Directed fittings are recommended for bearings operating at bore velocities in excess of 610 m • min^{-1} (~2,000 fpm) since the fan effect of these bearings may tend to keep conventionally applied oil mist from liberally entering into the rolling element region.

See also Figure 5-5, which illustrates general guidelines for locating reclassifiers and vents.

It can be said that oil-mist application and venting are interrelated. Venting must be provided for the escape of carrier air from closed housings. It is customary to provide a

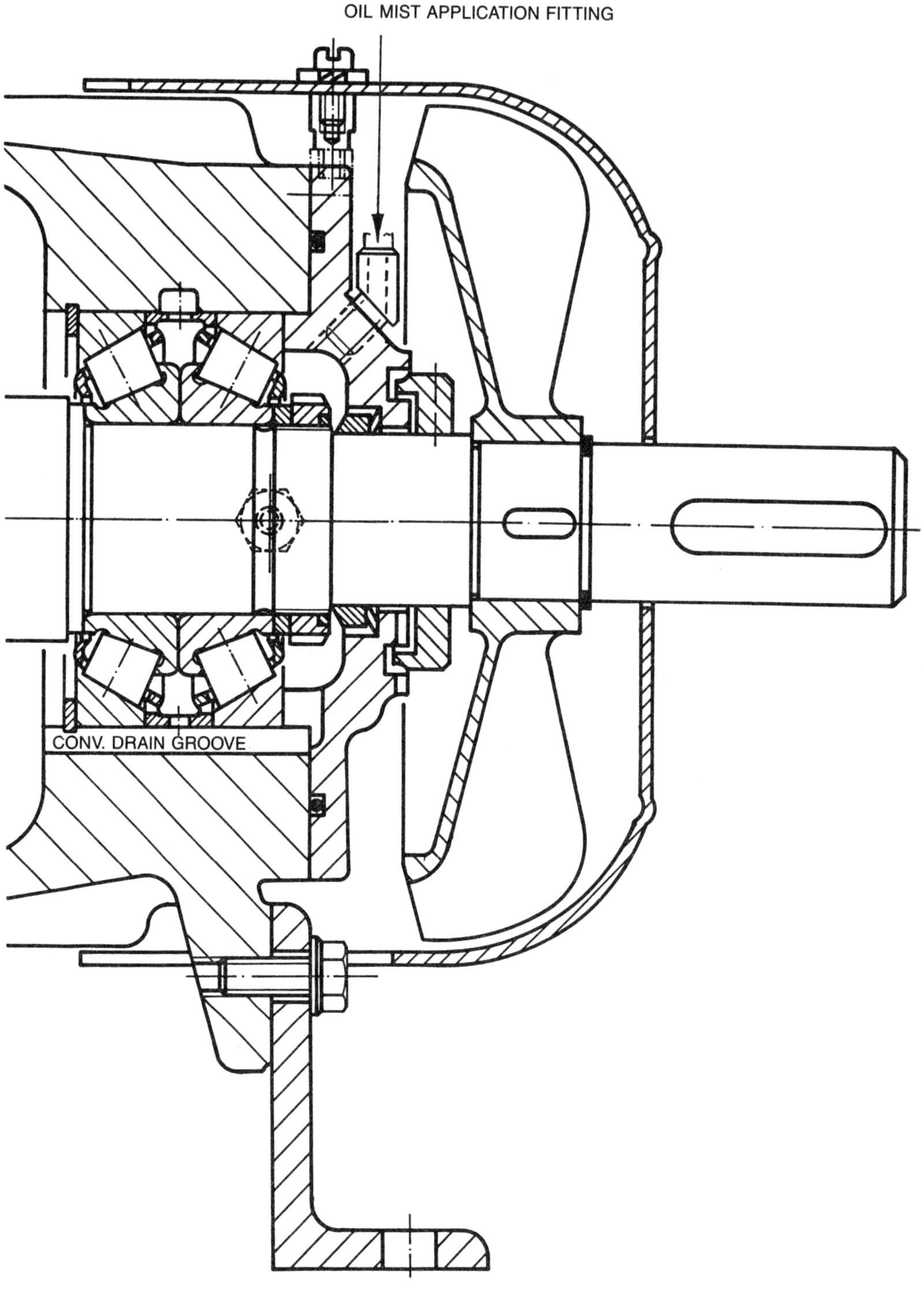

Figure 5-3. Fan-cooled equipment bearing housing with oil-mist application fitting threaded directly into cover for highest effectiveness.

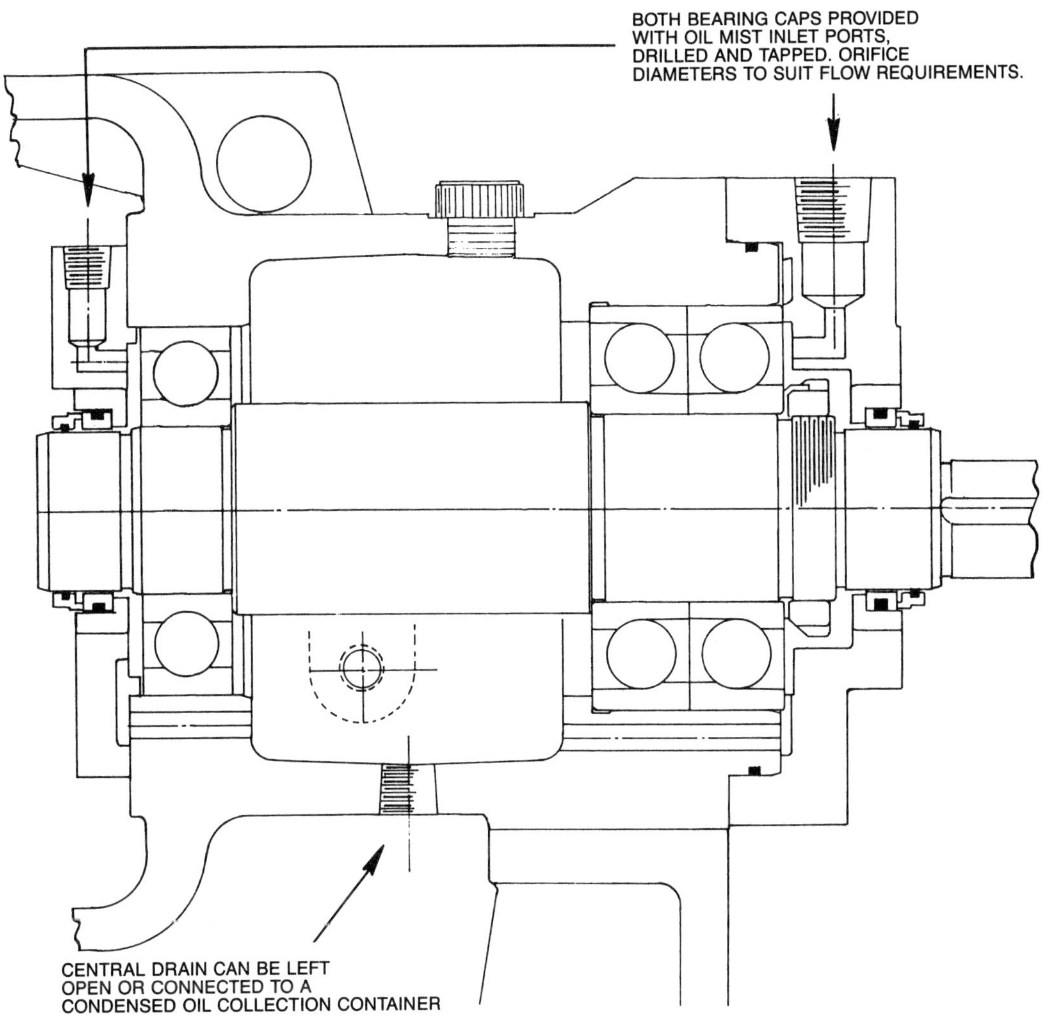

Figure 5-4. Optimized oil-mist path shows mist application through bearing housing end caps (directed oil mist) and venting and drainage at the bottom of bearing housing. *(Source: Carver Pump Company, Muscatine, Iowa.)*

minimum vent area equal to twice the total flow area of the application fittings supplying flow to that vent. Vent areas of this size will produce housing back-pressures equal to about 20% of manifold pressure.

When possible, relative positions of vents, applications fittings, and lubricated elements should promote flow from application fittings to and through lubricated surfaces.

Venting can be by means of appropriately located drilled holes, or frequently, by existing ports in the housing. Labyrinth seals will usually provide adequate venting, although a small one might have insufficient clearance for this purpose and require the addition of a drilled hole. Contact seals can be notched to provide venting, but this is not recommended because of the likelihood that notching will not be provided when seals are replaced.

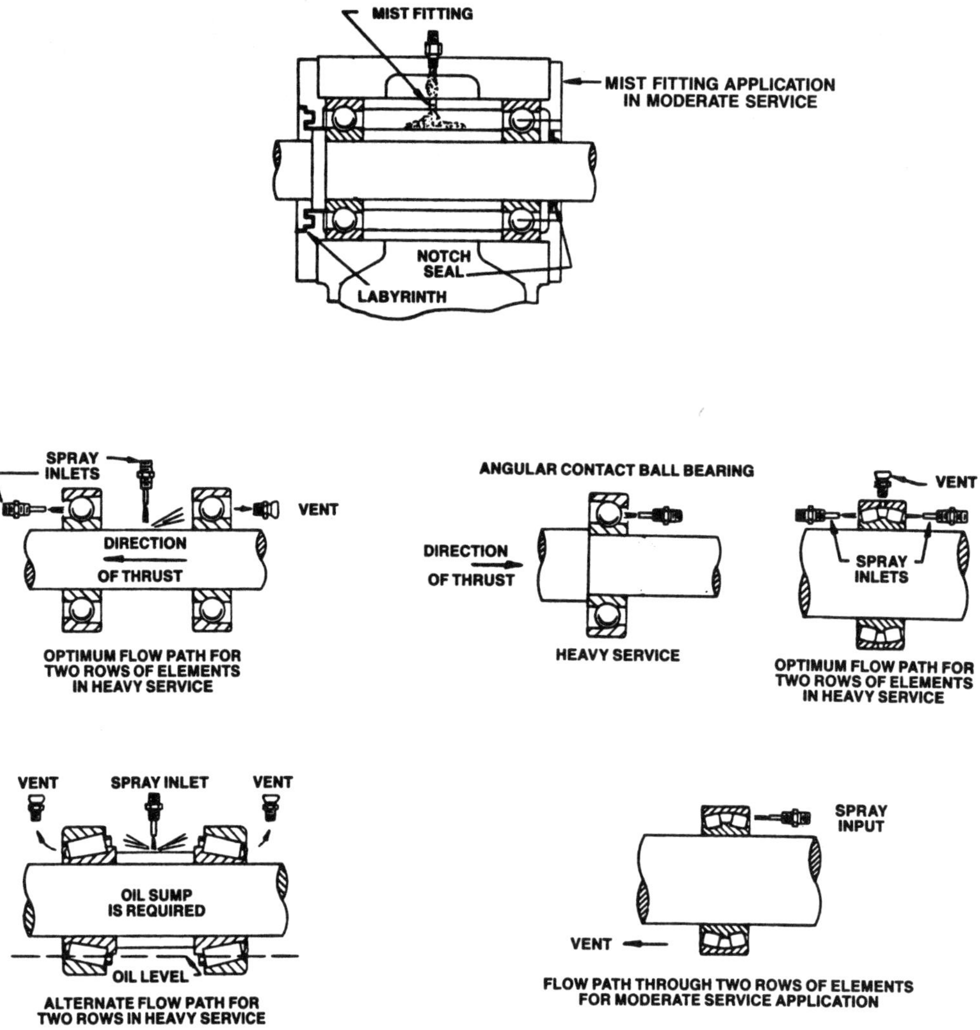

Figure 5-5. Generalized guidelines for locating reclassifiers and vents in rolling element bearing housings. *(Source: Alemite Division of Stewart-Warner Corporation.)*

Vent ports can often serve as oil overflow or drain ports. In a wet sump, or purge mist application the vent can be placed just above the normal sump oil level to provide an overflow path for any excess oil delivered by the oil-mist system. Such vents should be located so that liquid oil will not splash out through the port. For a dry sump application the vent can be placed at the bottom of the housing to drain all liquids. (See Figure 5-4.)

Vents should generally be protected from outside contaminants. Holes in the side of housings should slope downward to the outside. Vent ports in the tops of housings should have shielded vent fittings installed. Figures 5-6 and 5-7 show drilled vent holes and shielded vent fittings, respectively.

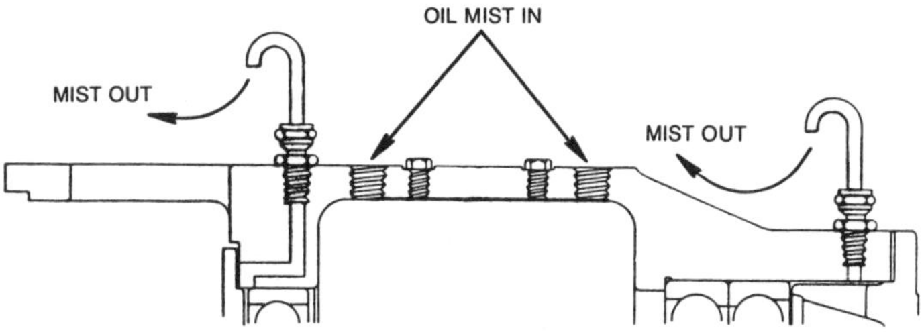

Figure 5-6. Vent holes drilled and tapped at top of pump bearing housing are fitted with tubes bent to prevent ingress of airborne contaminants. *(Source: United Pump Company.)*

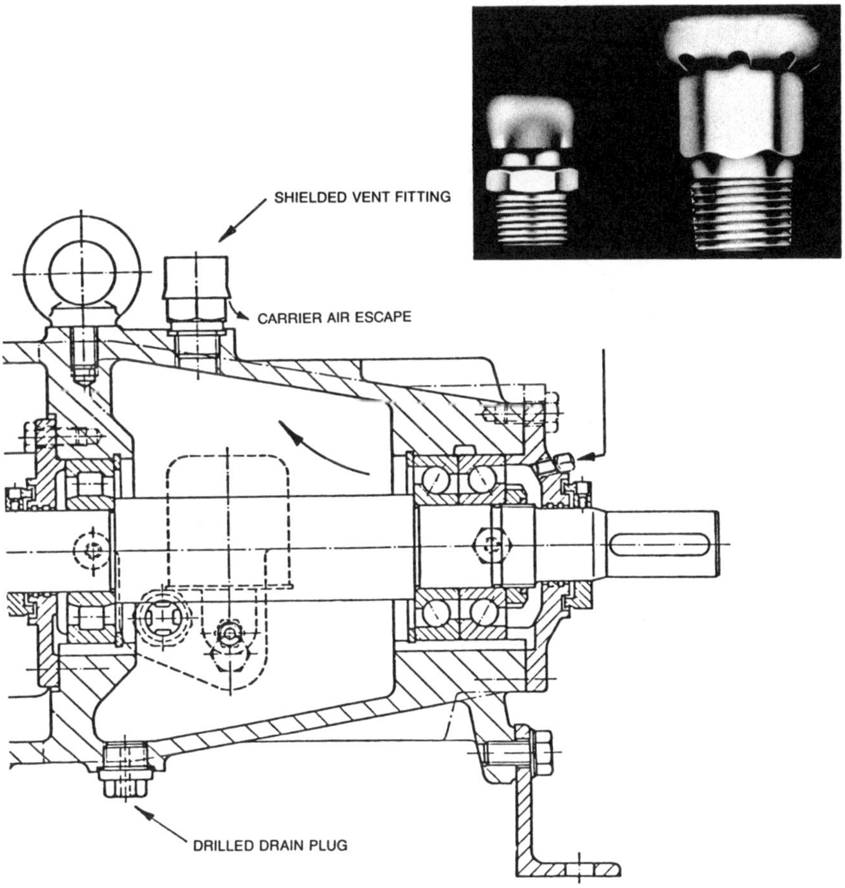

Figure 5-7. Dry-sump oil mist applied to pump through bearing housing end cap. The stray mist (spent air) can escape through either the shielded vent fitting or the drilled drain plug.

Generalized guidelines for locating reclassifiers and vents in rolling element bearing housings are illustrated in Figure 5-5.

Of course, plain bearings must also be vented. Manufacturing tolerances are usually large enough to allow air to escape. If normal clearance is insufficient for venting, then additional venting must be provided.

A vent hole should be located in the same radial plane as the reclassifier entry hole and connected to it by a radial groove. This vent hold must be located with respect to shaft rotation as shown in Figure 5-8. For plain bearings, generalized vent location guidelines are also illustrated in Figure 5-9.

Finally, mist can be made to circulate through the rolling elements of bearings by windage created by the rotating parts. This method is limited to single row, moderate service ball bearings operating at a surface speed greater than 200 linear fpm (~1 m • s^{-1}) and with a shaft diameter less than 4 in. (~100 mm). Figure 5-10 illustrates how certain machine assemblies can make use of windage from rotating parts to circulate mist through rolling element bearings and gears.

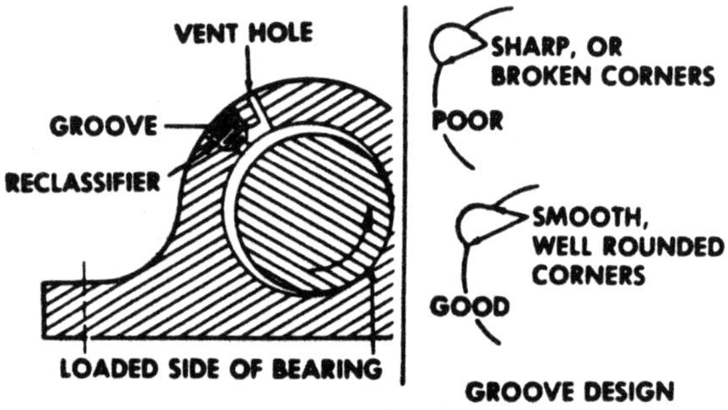

Figure 5-8. Vent holes in plain bearings should be located in the same radial plane as the reclassifier entry hole and connected to it by a radial groove. Direction of rotation must be observed. *(Source: C. A. Norgren Company.)*

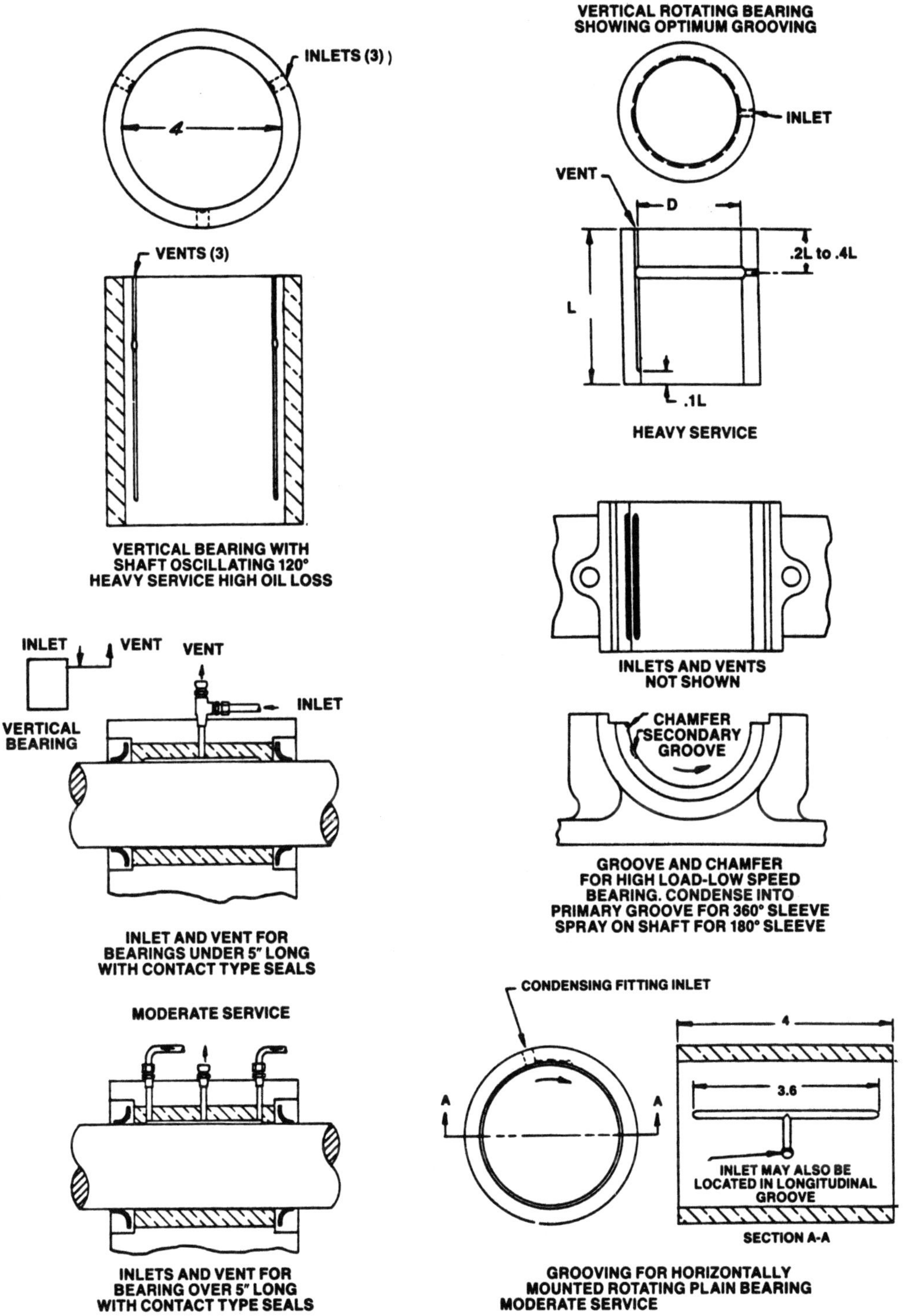

Figure 5-9. Generalized groove and vent location guidelines for plain bearings. *(Source: Alemite Division of Stewart-Warner Corporation.)*

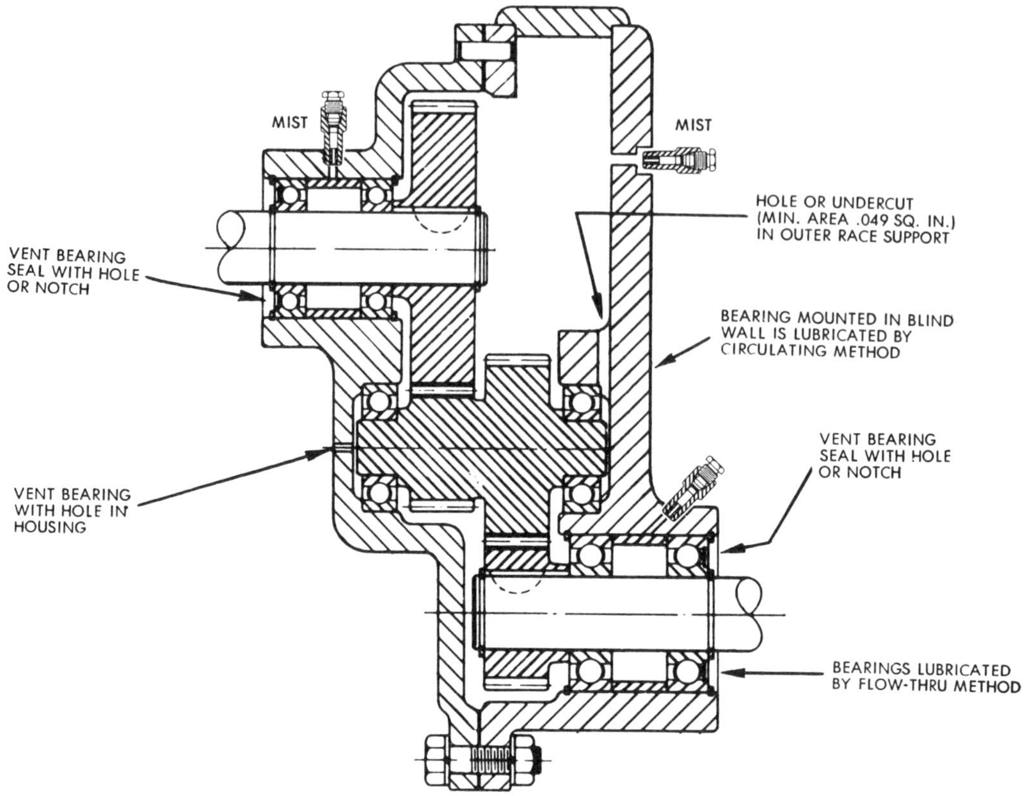

Figure 5-10. Windage from rotating parts can circulate mist through rolling element bearings and gears. *(Source: Lubrication Systems Company.)*

Chapter 6

Lubricant Collection

By now, the reader should understand how oil mist is generated and transported to the machine components to be lubricated, and how turbulence in the reclassifier (application fitting) or bearing itself causes the oil droplets to recombine into liquid lube oil. Any oil mist that does not revert to liquid oil will escape through the nearest vent. If this is objectionable or perhaps uneconomic due to the value of a premium synthetic oil, the oil mist that is about to escape to the environment can be drawn off by applying a slight vacuum. This mist can then be condensed by leading it either through a centrifugal separator or through an electrostatic precipitator.

First, however, we should consider how liquid oil can be collected from bearing housings of pumps, electric motors, and similar equipment. The liquid oil exists in these bearing housings either because it was introduced as an oil spray or as droplets downstream of a suitable reclassifier, or because the turbulent action in the immediate vicinity of a rolling element bearing caused condensation and subsequent overflow of oil from the bearing into the surrounding housing or sump.

Condensed oil drains from the equipment bearing housing and is collected in a transparent plastic container or sight glass, as shown in Figures 6-1 and 6-2. As oil accumulates, the excess can be drained through a suitable line to a separate collection vessel or common header. This arrangement helps reduce housekeeping problems because oil does not drip on the baseplate or around machinery installed outdoors, such as pumps. A transparent container allows the condition of the oil to be observed. The small line leading to the collection header can also be bent into a U-tube shape to provide a suitable back pressure on the bearing housing—generally about 0.2-0.5 in. H_2O (50-125 Pa). If water or solid contaminants should enter the bearing housing, the contamination can be spotted in the transparent container. Sight glasses should be drained periodically.

An observant viewer may notice that the stainless steel drain tubing lines illustrated in Figure 6-2 are actually entering the collecting pot at the bottom. As condensed oil starts to accumulate, it will cover the tubing entry points. Thus, the individual vent tube associated with a given drain tubing will show stray mist only if a particular bearing is actually receiving oil mist. If the drain tubing from several bearings entered at the top of the collecting pot, it would theoretically be possible for a certain bearing mist supply orifice to be plugged, and yet each vent would show evidence of escaping stray mist.

Sight glasses are also used for the bearing housings of small general purpose turbines. Turbine bearing housings often use purge mist because they have sleeve bearings. Sight glasses of the type shown in Figure 6-3 are generally screwed directly into the bottom of the bearing housing. Since small steam turbines are often troubled by steam leaks

Figure 6-1. Oil mist enters the bearing housing at the top via stainless steel tubing and reclassifier. Condensed oil is collected in the small sight glass at the bottom of the bearing housing *(Source: Reference 12.)*

Figure 6-2. Oil-mist condensate from pump bearings (opposite motor) and from two motor bearings is collected in a single transparent container. Note small branch tubing serving as vents.

through the labyrinths and subsequent moisture condensation in the bearing housings, sight glasses allow water to be readily observed and drained before damage occurs. The line connecting the bearing housing and sight glass must be sized to take the surface tension of water into account. With very small lines, the surface tension could prevent water from dropping into the sight glass.

A number of large-scale oil mist installations in the U.S. Gulf Coast region are using plastic milk bottles or similar transparent containers in lieu of small screw-in collection containers. Figures 6-4 and 6-5 show these inexpensive bottles which, due to their relatively large capacity, require only infrequent emptying.

Large oil mist collectors with transparent or heavy metallic construction with multiple inlet connections are also offered by several manufacturers as shown in Figure 6-5a.

Mist Draw-off

Oil-mist draw-off is a method of controlled removal or capture of stray mist, or mist that would otherwise escape to the environment. Typical controlled draw-off means are shown in Figures 6-6 through 6-9.

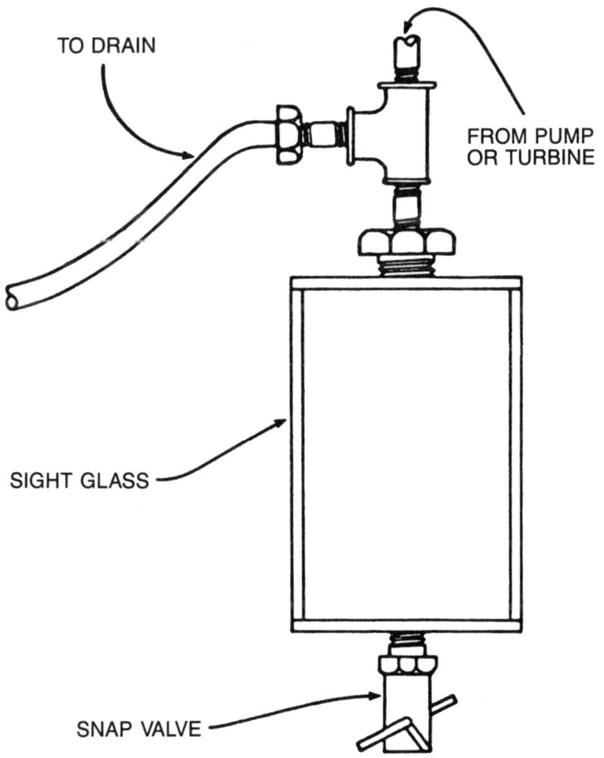

Figure 6-3. Sight glasses are used to collect condensed oil mist in dry sump systems, as shown. *(Source: Reference 6.)*

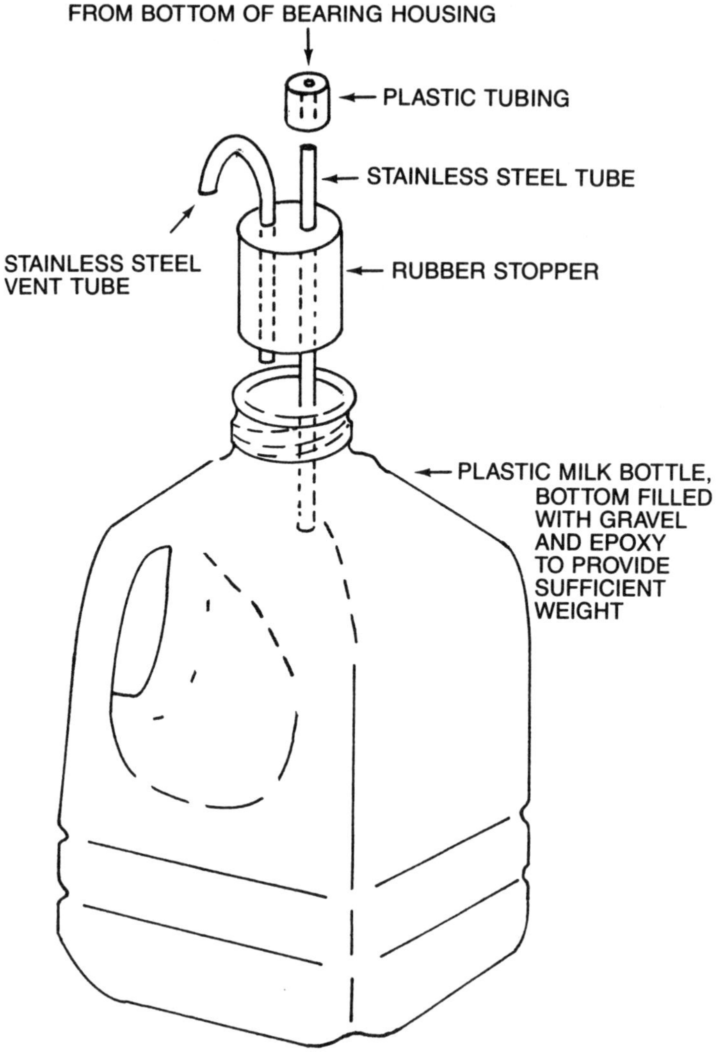

Figure 6-4. Plastic milk bottle modified to function as large-capacity condensed-oil collecting container.

Figures 6-6 and 6-7 depict noncontacting rotor-stator seals (bearing isolators), which can serve as bearing housing closures. They are equipped with an expulsion port to which tubing under slight vacuum can be connected. More tightly fitting closure seals are known as lip seals (Figure 6-8) or magnetic face seals (Figure 6-9). If these tight-fitting enclosures are used, a separate draw-off port should be provided. This draw-off port would be so located as to first allow the oil mist to pass through the bearing or region to be lubricated. Next, the oil mist would either impinge on a centrifugal separator—essentially a high-speed blower that creates extreme turbulence—or it would enter into an electrostatic precipitator.

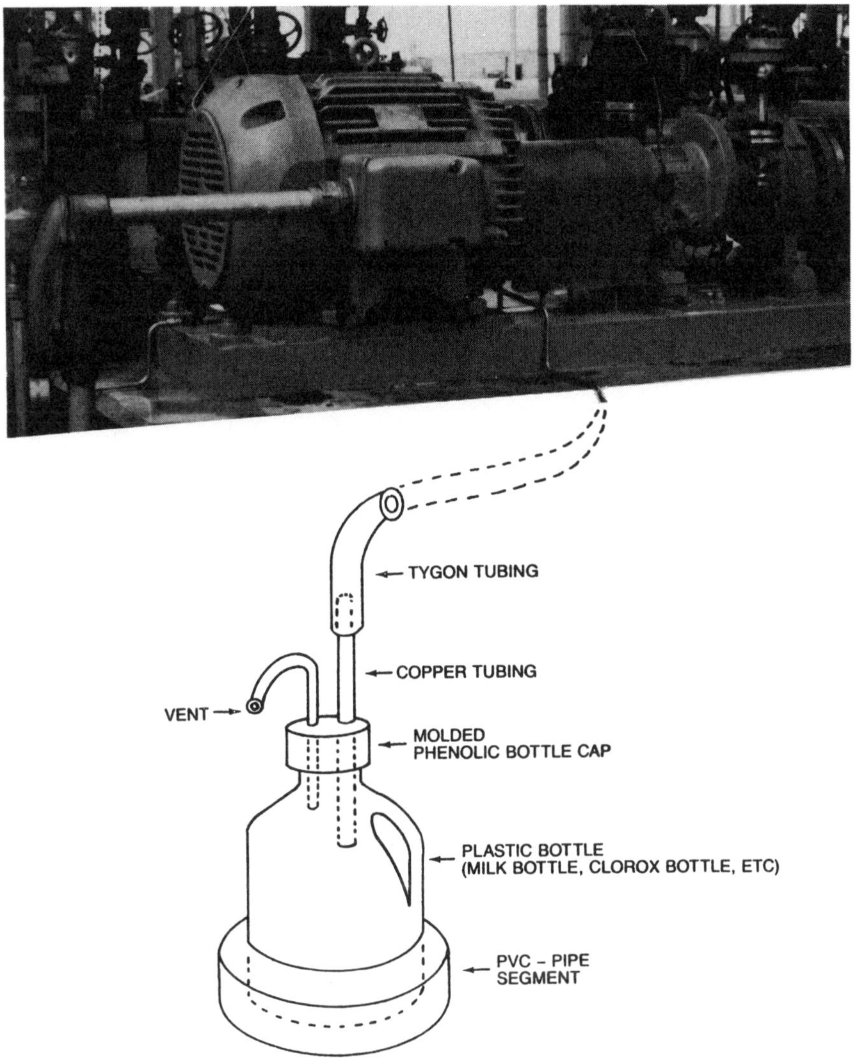

Figure 6-5. Condensed oil mist can be collected in this type of transparent container.

Forced Condensation

Electrostatic precipitators, shown later in Figure 10-3, allow smoke and mist-laden air to pass through an ionizing section, where all the particles are given a strong electrical charge. The air then passes through a collecting cell, where the charged particles are attracted to and collected on oppositely charged plates. Cleaned air passes through an outlet port. The collected oil, now in liquid form, runs off the collecting cell plates and can be captured for disposal or reuse.

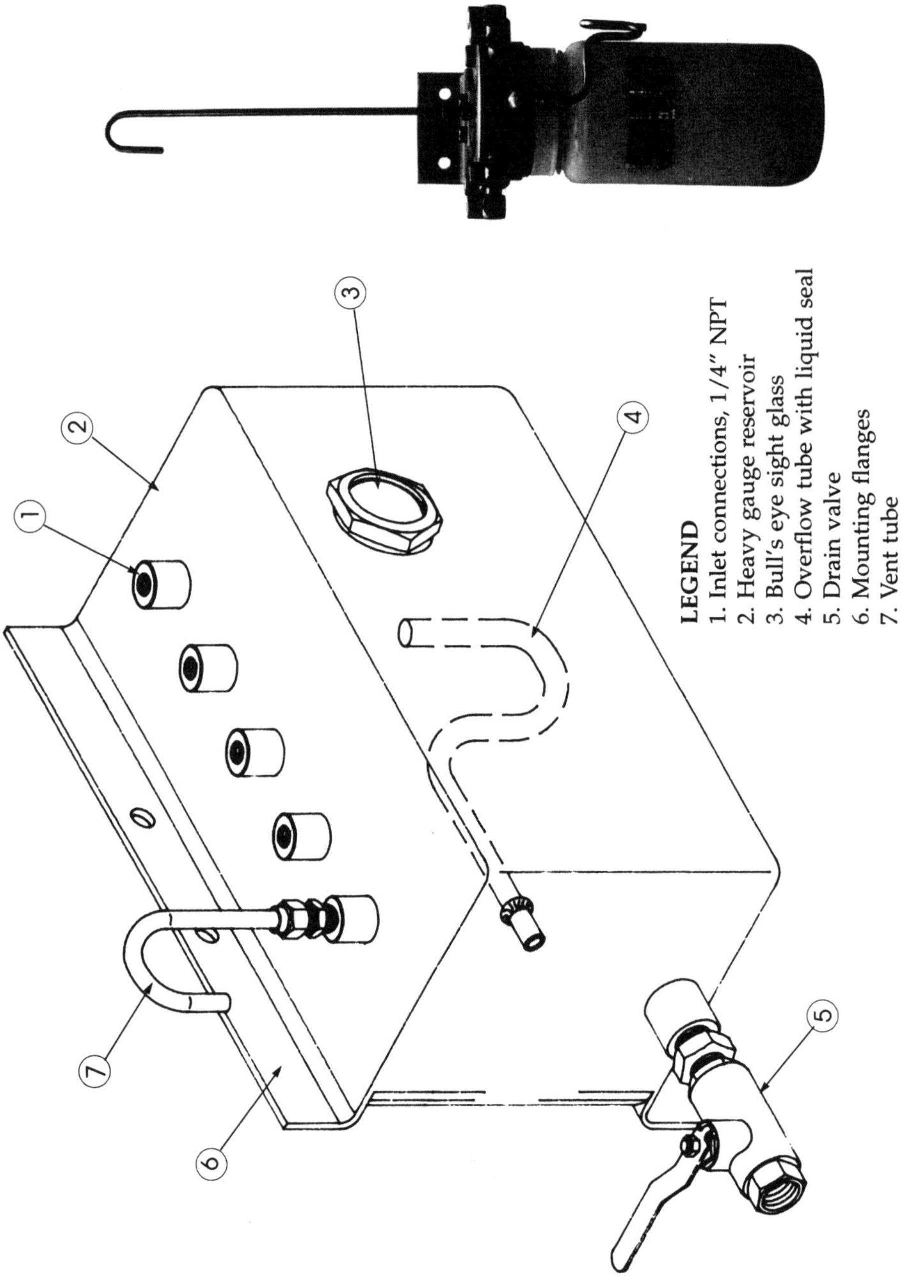

Figure 6-5a. Large oil mist collectors with transparent or heavy metallic constructions. *(Source: Bill Spitzer & Associates, Houston, Texas [Transparent type] and Lubrication Systems Company, Houston, Texas [Metallic type])*

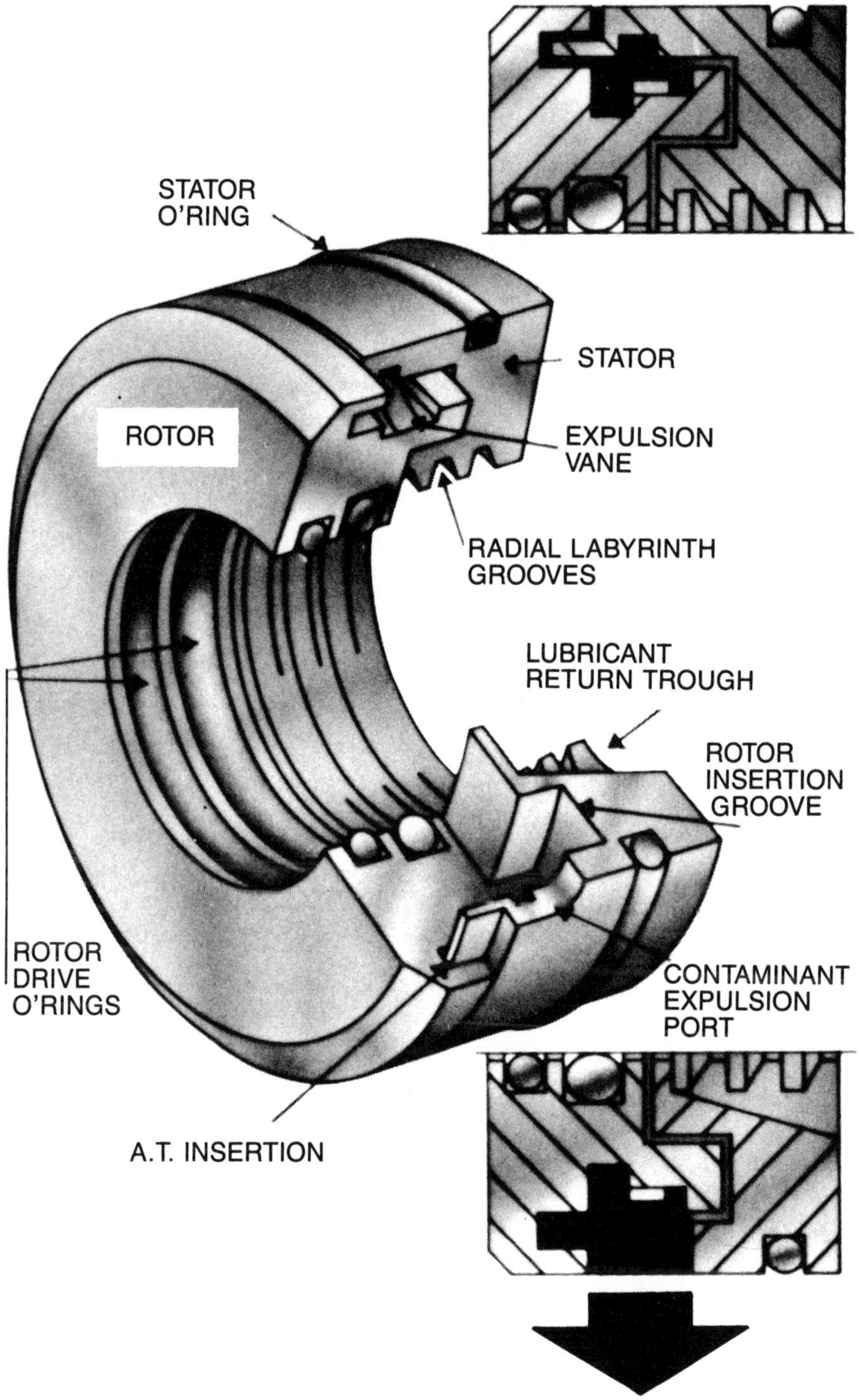

Figure 6-6. Noncontacing rotor-stator seals ("bearing isolators") can serve as bearing housing closures. Connecting a vacuum line to the contaminant expulsion port will allow controlled removal of stray oil mist. *(Source: INPRO/Seal.)*

Figure 6-7. Noncontacting rotor-stator seal ("bearing isolator") installed on a centrifugal pump bearing housing. Note vacuum line connected to the contaminant expulsion port. *(Source: INPRO/Seal.)*

Figure 6-8. Vent port drilled below conventional lip seal. *(Source: Goulds Pumps.)*

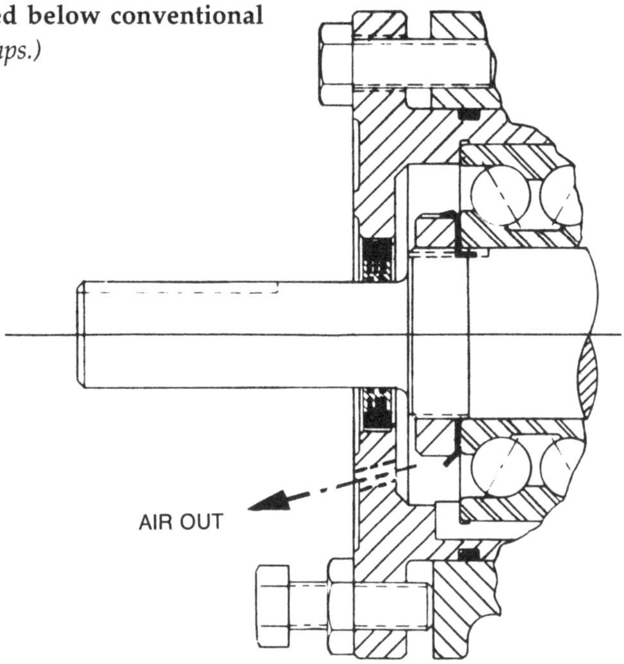

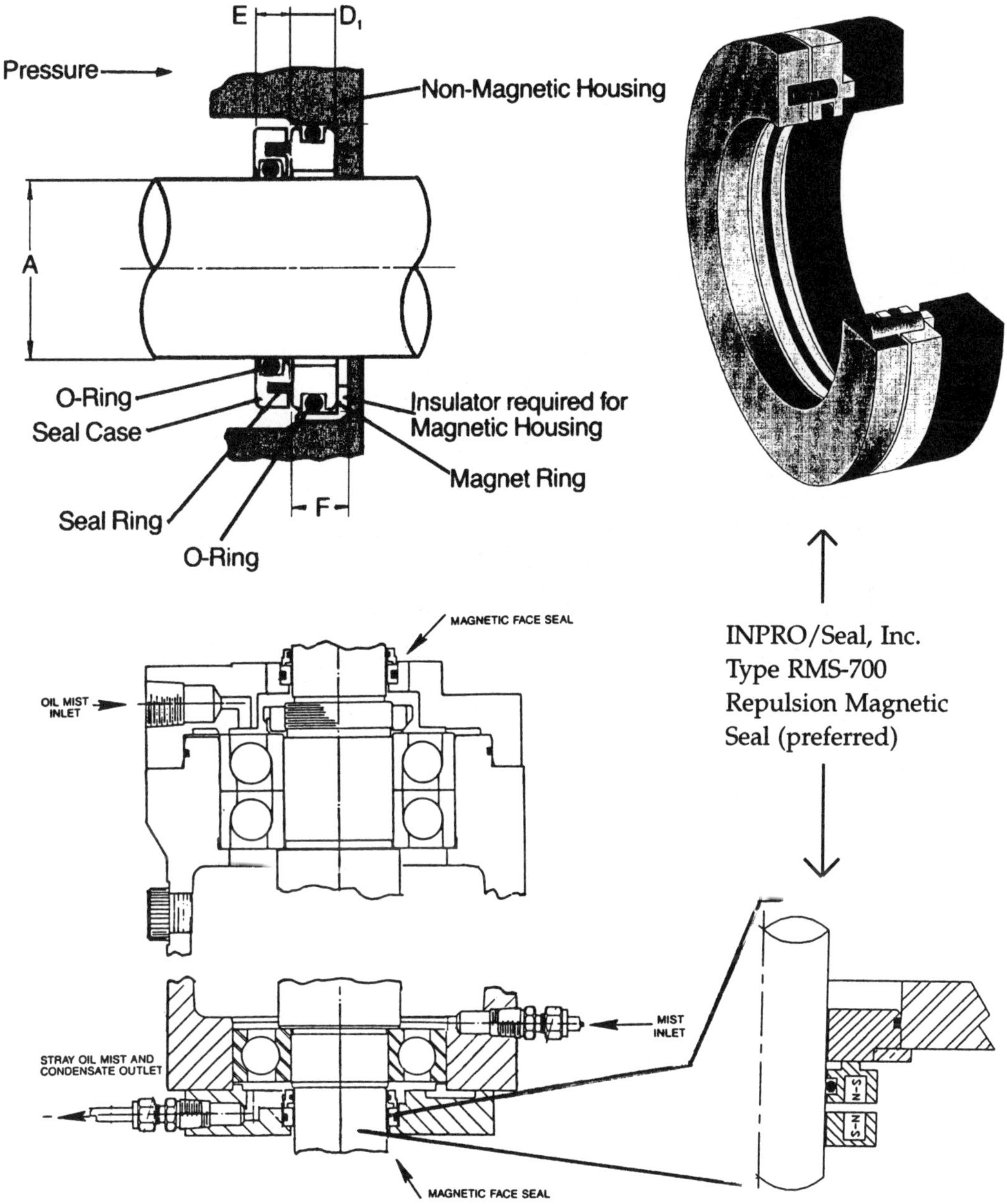

Figure 6-9. The escape of stray oil mist into the environment can be totally avoided by using magnetic face seals and stray mist outlet passages piped to centrifugal separators or electrostatic precipitators. *(Source: Carver Pump Company, Magnetic Seal Company, and INPRO/Seal, Inc., Rock Island, Illinois)*

Chapter 7

Selecting the Application Fittings

Conventional Application Fittings

Again, the term "application fitting" is used as the collective term including mist, spray, condensing, and even pressure jet fittings. These components were discussed in an earlier chapter dealing with plant-wide oil-mist systems.

Reclassifiers are nozzle-like application fittings that reclassify the dry oil mist or fog into a wet usable oil. They should be used at each application point. Reclassifiers also proportion the oil mist to the various points of application in accordance with the lubrication requirement.

Each major manufacturer of oil-mist systems usually selects a certain size range of reclassifiers. One company standardizes on 1, 3, 6, 10, 15, and 20 bearing-inches.

Basic application fitting or reclassifier ratings for another vendor are 1, 2, 4, 8, 14, 20, and 40 bearing-inches. When calculating the requirements of machine elements, choose the next highest rated fitting or reclassifier whenever calculations give a result between any two available ratings.

Reclassifiers are rated according to the amount of oil they will deliver. An eight-bearing-inch reclassifier will deliver approximately four times as much oil as a two-bearing-inch reclassifier.

One manufacturer's reclassifier (Lubrication Systems Company) performance is shown in Figures 7-1 and 7-2. Although the two tables list only six reclassifier sizes and their respective flow ratings at systems of pressures of 20 and 30 in. H_2O (~5 and 8.7 kPa), interpolation allows rough estimates of the performance of similar reclassifiers from other manufacturers. Another manufacturer's reclassifier (Stewart-Warner Corporation, Alemite and Instrument Division) performance is shown in Figures 7-1a through 7-1e. For final systems sizing, the user would apply specific data available from the manufacturer who is ultimately selected to implement a system.

When it is not possible to install fitting-type reclassifiers due to space limitations, it is usually possible to drill appropriately sized nozzles into the housing or bearing spacers to permit mist impingement on the bearing surface. Figure 7-3 shows how this can be accomplished in principle. However, detailed guidelines on drill diameter and length of drilled passage should be requested from the oil-mist system manufacturer.

For some gear and chain applications, pressure-jet reclassifiers are very helpful. These should be selected from special rating charts available from the manufacturer. Pressure-jet reclassifiers incorporate the standard type of reclassifier with an auxiliary source of air jetting along the reclassifier axis. The result is the delivery of lubricant with sufficient force to penetrate the boundary-air layer common to high-speed parts. They require an auxiliary supply of filtered air at a pressure of 10 to 12 psi (~69 to 83 kPa) and may be connected as shown in Figure 7-4.

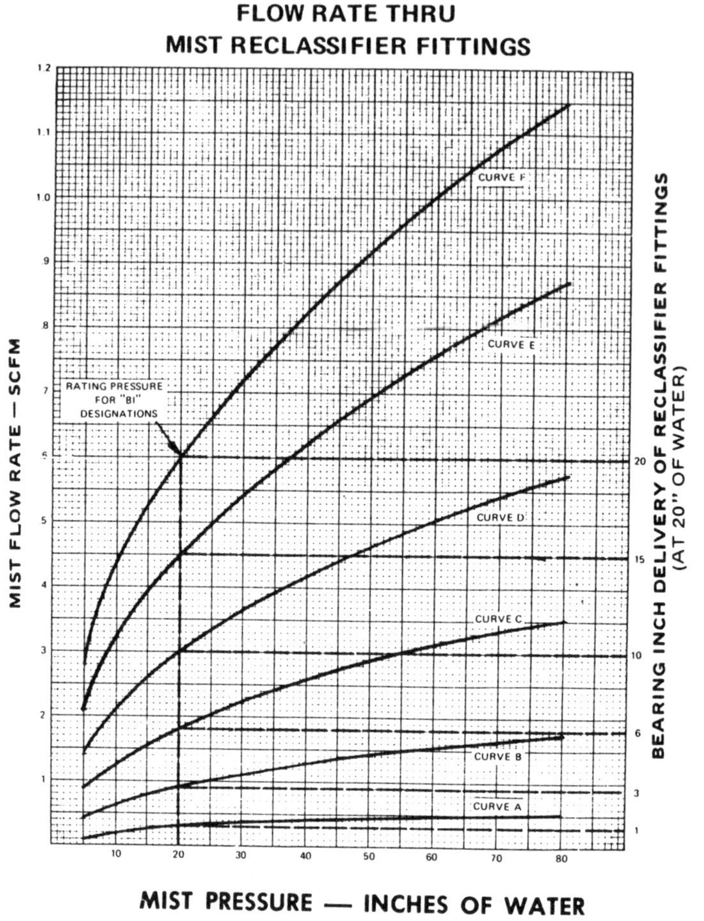

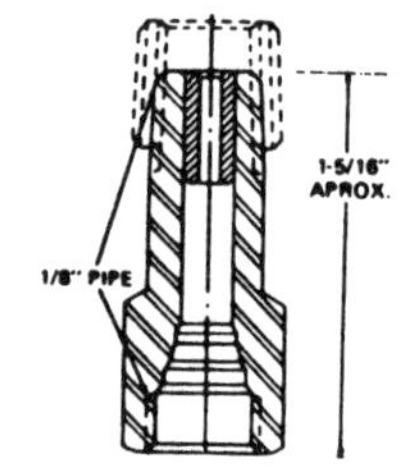

MIST TYPE

SCFM RATING	BI RATING	RECLASSIFIER CURVE SYMBOLS		BORE SIZE			MIN. VENT DIA.
		SYM.	MIST TYPE	DIA.	MIN. LENGTH	AREA	
.03	1	A	77-800-500	.018	13/64	.00025	.025
.09	3	B	-501	.032	1/4	.00080	.045
.18	6	C	-502	.047	1/4	.00173	.066
.30	10	D	-503	.060	3/8	.00278	.084
.45	15	E	-504	.073	7/16	.00419	.103
.60	20	F	-505	.086	1/2	.00581	.122

Figure 7-1. Relationship between mist flow rate, mist reclassifier size and mist pressure. *(Source: Lubrication Systems Company.)*

MIST APPLICATION FITTINGS

Mist fittings are metering orifices which deliver mist, with minimum condensation, to machine elements.

The mist must be converted to liquid oil by the machine element.

The fittings are made of brass. Color—black.

The air flow graph shows average output.

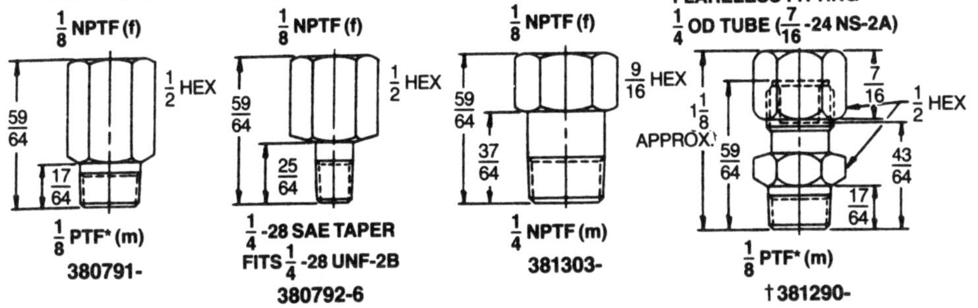

‡INLET ⅛ NPTF (FEMALE)			‡INLET ¼ OD TUBE FTG	ORIFICE		
OUTLET ⅛ PTF*(m)	OUTLET ¼-28 SAE TPR	OUTLET ¼ NPTF(m)	OUTLET ⅛ PTF*(m)	DIAMETER	LENGTH	FLOW AREA
380791-2			381290-2	.032	.44	.00078
380791-4		381303-4	381290-4	.045	.44	.00159
380791-6	380792-6	381303-6	381290-6	.055	.44	.00237
		381303-10	381290-10	.078	.44	.00484
		381303-16	381290-16	.089	.44	.00622

*SAE Special Short.
‡PREFERRED ARRANGEMENT. INLET AND OUTLET CAN BE INTERCHANGED WHERE NECESSARY.

†PART NO. INCLUDES:
381289-2,-4,-6,-10, OR -16 FTG
328301-4 COMP. NUT
328302-4 COMP. SLEEVE

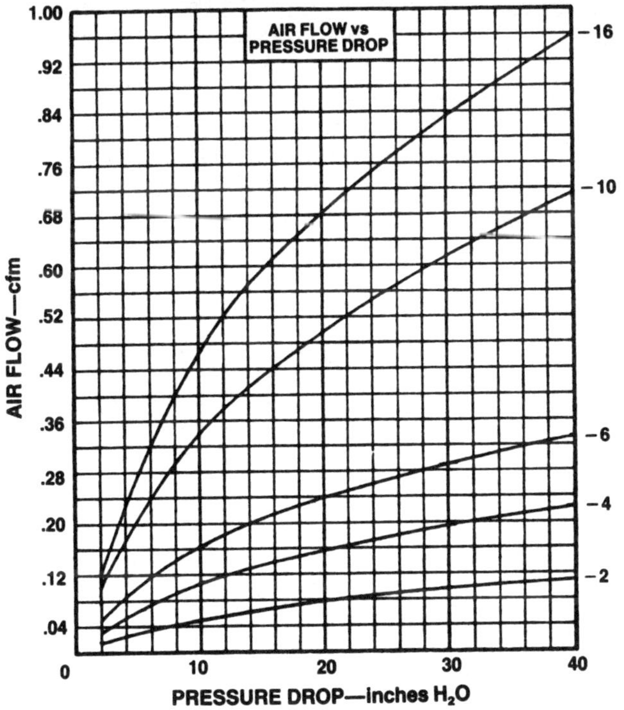

Figure 7-1a. Design and performance characteristics of mist type application fittings. *(Source: Stewart-Warner Corporation, Alemite and Instrument Division.)*

CONDENSING APPLICATION FITTINGS

Condensing fittings are metering elements which convert mist to liquid oil within the fitting. The condensed oil flows by gravity to the machine element.

The fittings are made of brass. Color—silver.

The air flow graph shows average output.

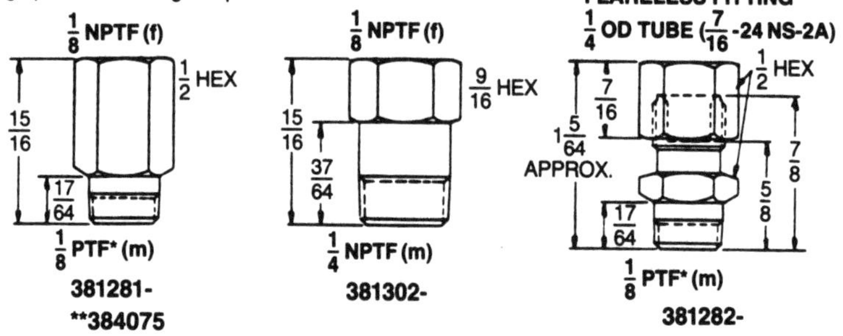

**VENTED FITTING 384075 HAS A MILLED SLOT IN THE MALE THREAD OF THE FITTING. THE MALE END MUST BE USED AS THE OUTLET, AND THE SLOT SERVES AS A VENT.

INLET 1/8 NPTF (f)		‡INLET 1/4 OD TUBE FTG	ORIFICE (BAFFLED)		
OUTLET 1/8 PTF*(m)	OUTLET 1/4 NPTF(m)	OUTLET 1/8 PTF*(m)	SIZE	LENGTH	FLOW AREA
381281-05			SLOT .014 R	.38	.0003
381281-1	381302-1	381282-1	SLOT .020 R	.38	.0006
381281-2	381302-2	381282-2	2 SLOTS .020 R	.38	.0012
381281-3	381302-3	381282-3	3 SLOTS .020 R	.38	.0018
381281-4	381302-4	381282-4	5 SLOTS .020 R	.38	.0030
381281-5	381302-5	381282-5	6 SLOTS .020 R	.38	.0036
384075	VENTED FITTING		SLOT .020 R NO BAFFLE	.38	.0006

*SAE Special Short.

‡PREFERRED ARRANGEMENT.

INLET AND OUTLET CAN BE INTERCHANGED WHERE NECESSARY.

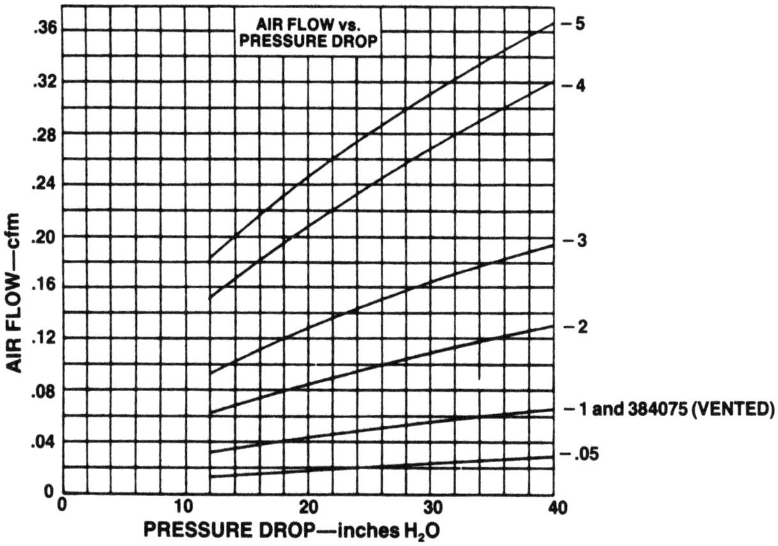

Figure 7-1b. Design and performance characteristics of condensing type application fittings. (Source: Stewart-Warner Corporation, Alemite and Instrument Division).

SPRAY AND CONDENSED-SPRAY APPLICATION FITTINGS

Spray fittings are metering orifices which convert a high percentage of mist to an oil spray. The condensed-spray fitting is a metering element which converts nearly all of the mist to an oil spray.

The fittings are made of brass. Color—olive drab.

The air flow graph shows average output.

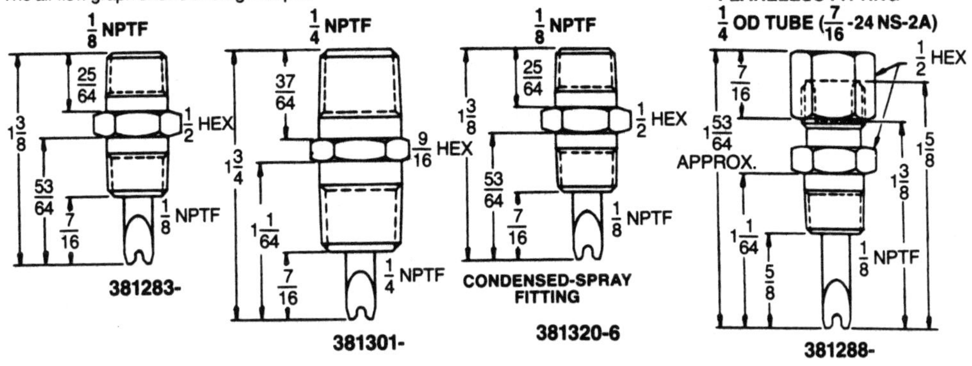

INLET ⅛ NPTF OUTLET ⅛ NPTF	INLET ¼ OD TUBE FTG. OUTLET ⅛ NPTF	INLET ¼ NPTF OUTLET ¼ NPTF	ORIFICE DIAMETER	 LENGTH	 FLOW AREA
381283-1	381288-1		.024	1.38	.000452
381283-2	381288-2		.035	1.38	.000962
381283-3	381288-3	381301-3	.042	1.38	.00138
381283-5	381288-5	381301-5	.055	1.38	.00237
381283-8	381288-8	381301-8	.067	1.38	.00352
381320-6	CONDENSED-SPRAY FITTING		.067 AND 6 SLOTS .020 R	.97 .38	.00352 .0036

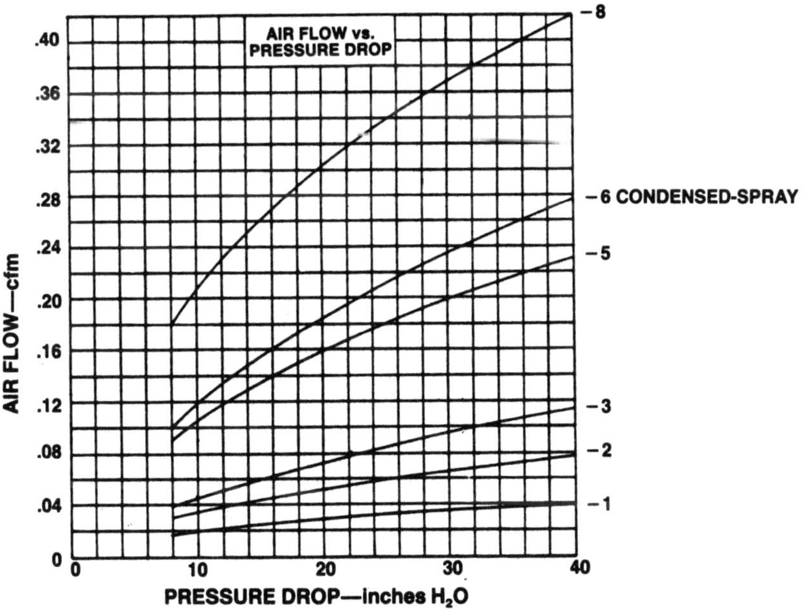

Figure 7-1c. Design and performance characteristics of spray and condensed-spray type application fittings. *(Source: Stewart-Warner Corporation, Alemite and Instrument Division).*

SPRAY NOZZLE APPLICATION FITTINGS

Spray nozzles contain one or more metering orifices which convert a high percentage of mist to an oil spray. The capacity of each metering orifice is nearly equal to that of one – 8 spray fitting.

The nozzles are made of brass.

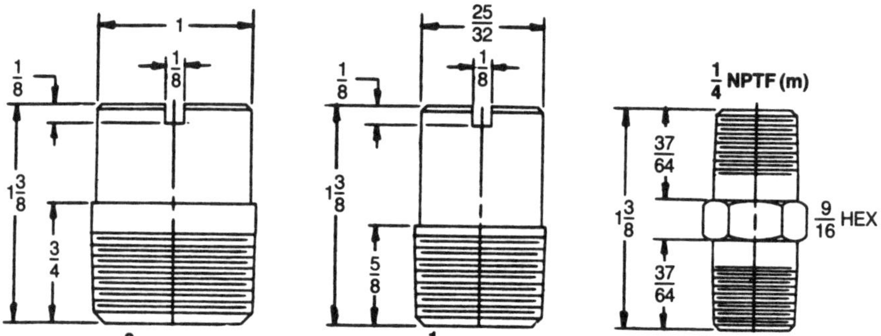

3/4 NPTF (m) 383617-

1/2 NPTF (m) 326370-

1/4 NPTF (m) 383588-

ONE END THREADED		BOTH ENDS THREADED		NO. OF ORIFICES .067 DIA. 1.38 LG.	TOTAL FLOW AREA SQ. IN.
½ NPTF(m)	¾ NPTF(m)	¼ NPTF(m)	½ PTF*(m)		
326370-1		383588-1	384280-1	1	.00352
326370-2		383588-2	384280-2	2	.00704
326370-3		383588-3	384280-3	3	.01056
326370-4		383588-4	384280-4	4	.01408
326370-5		383588-5	384280-5	5	.01760
326370-6		383588-6	384280-6	6	.02112
326370-7		383588-7	384280-7	7	.02464
326370-8			384280-8	8	.02816
326370-9			384280-9	9	.03168
326370-10	383617-10		384280-10	10	.03520
	383617-11			11	.03872
	383617-12			12	.04224
	383617-13			13	.04576
	383617-14			14	.04928

*SAE Short

1/2 PTF*(m)

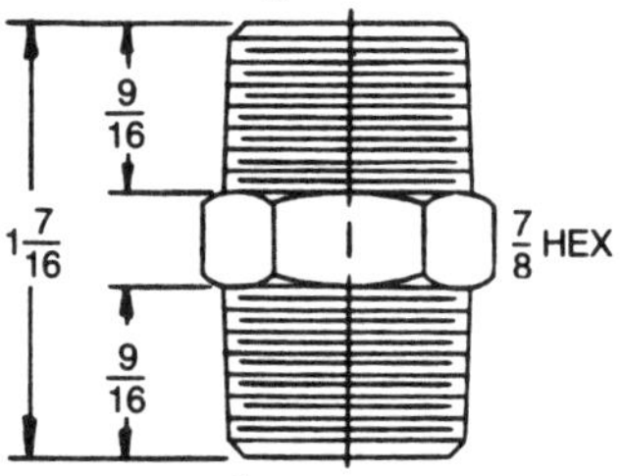

1/2 PTF*(m)

384280-

Figure 7-1d. Design and performance characteristics of spray nozzle type application fittings. *(Source: Stewart-Warner Corporation, Alemite and Instrument Division).*

SPRAY NOZZLE APPLICATION FITTINGS (continued)

Dash number for each curve indicates number of .067 diameter orifices in nozzle and is used as suffix of nozzle part number.

The flow through one .067 diameter orifice is .56 cfm at a pressure drop of 60 inches water column, and .68 cfm at 80 inches water column.

The nozzle air flow graph shows average output.

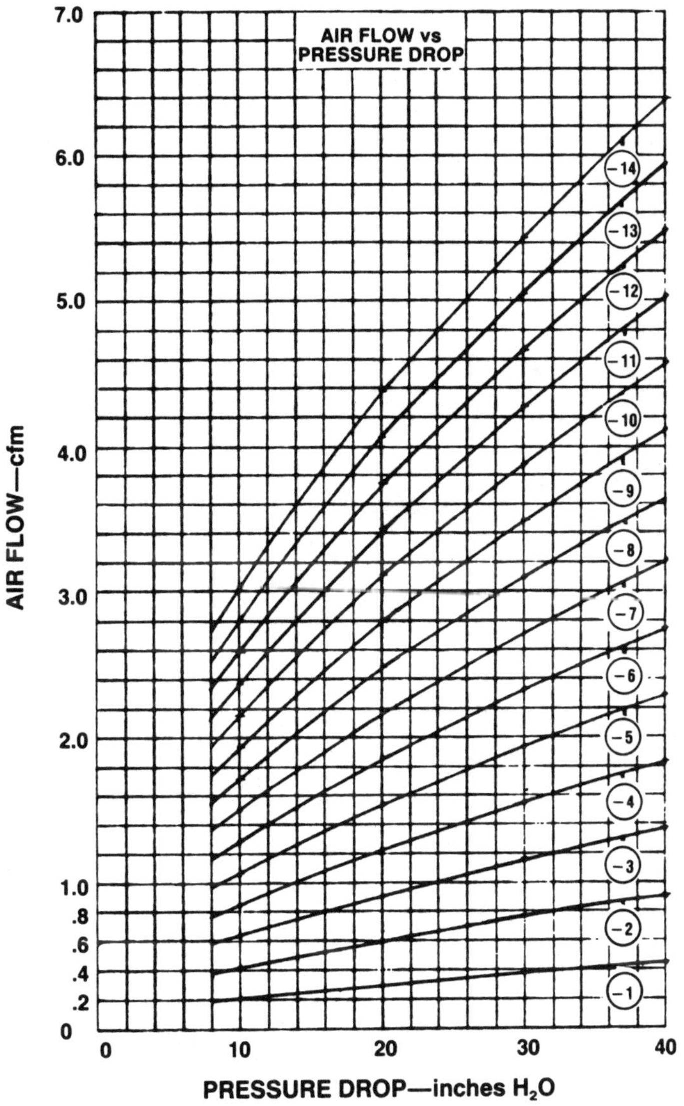

Figure 7-1e. Design and performance characteristics of spray nozzle type application fittings (continued from Figure 7-1d). *(Source: Stewart-Warner Corporation, Alemite and Instrument Division).*

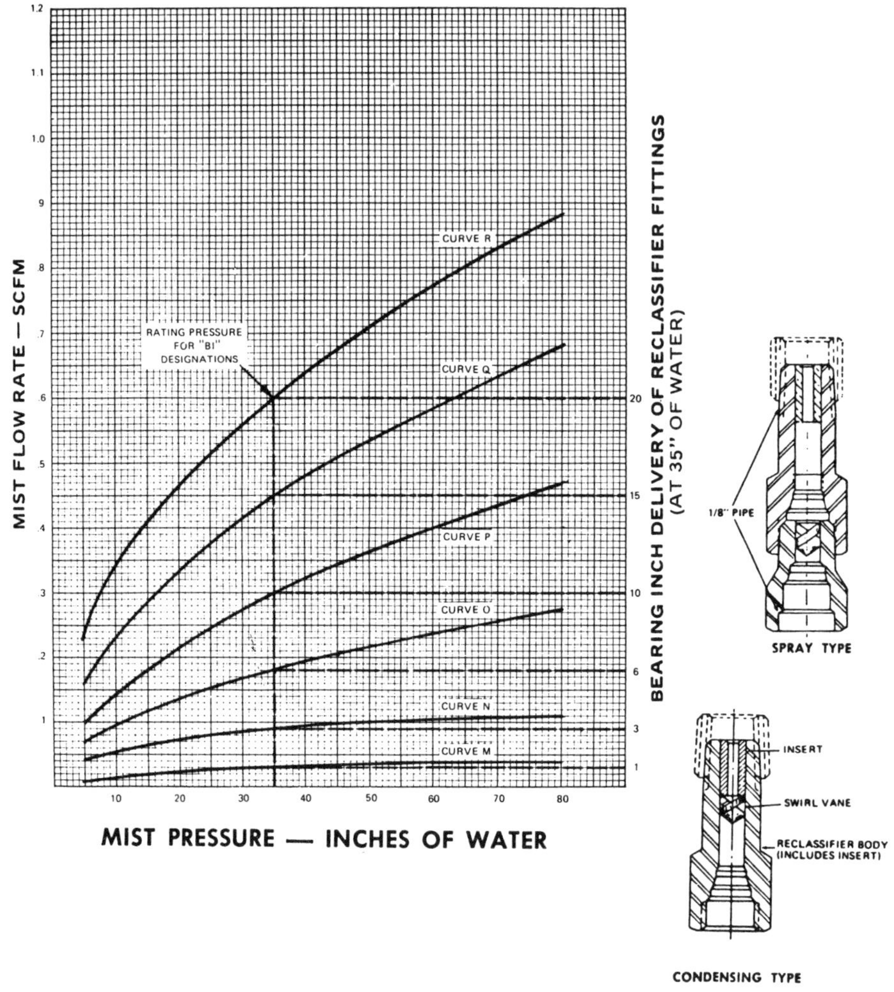

SCFM RATING	BI RATING	RECLASSIFIER FITTING NUMBER & CURVE SYMBOLS				BORE SIZE			MIN. VENT DIA.
		SYM.	COND TYPE	SYM.	SPRAY TYPE	DIA.	MIN. LENGTH	AREA	
.03	1	M	77-800-520	M	77-800-528	.018	13/64	.00025	.025
.09	3	N	-521	N	-529	.032	1/4	.00080	.045
.18	6	O	-522	O	-530	.047	1/4	.00173	.066
.30	10	P	-523	P	-531	.060	3/8	.00278	.084
.45	15	Q	-524	Q	-532	.073	7/16	.00419	.103
.60	20	R	-525	R	-533	.086	1/2	.00581	.122

Figure 7-2. Relationship between mist flow rate, condensing or spray reclassifier size, and mist pressure. *(Source: Lubrication Systems Company.)*

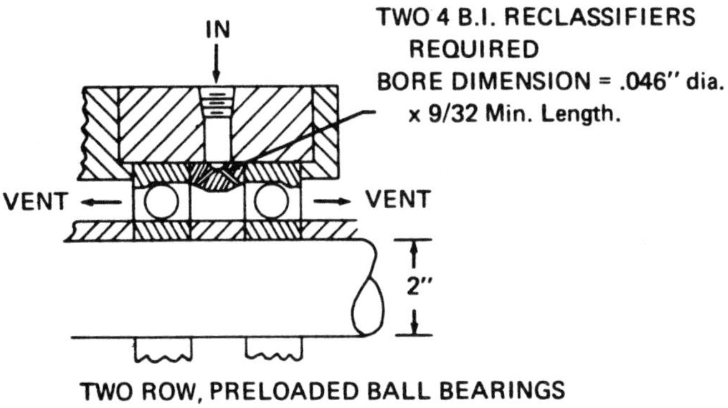

Figure 7-3. When space limitations make it difficult to install commercial reclassifiers, it is usually possible to drill appropriately sized nozzles into the machine component. *(Source: C.A. Norgren Company.)*

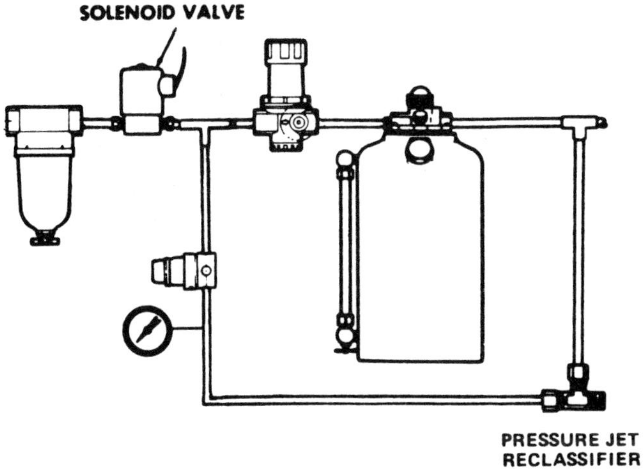

Figure 7-4. Pressure jet reclassifiers deliver the lubricant with sufficient force to reach certain high-speed parts. They require an auxiliary supply of filtered air at pressures typically in the range of 10 psi (~69 kPa). *(Source C.A. Norgren Company.)*

When using pressure-jet reclassifiers, a connection can also be made in the line between the filter and regulator to supply the required air. A pressure regulator and pressure gauge should be used in the auxiliary line that feeds this filtered air to the reclassifier.

As will be noted in Figures 7-1 and 7-2, each size reclassifier has its characteristic bore and minimum length of bore. If it is preferred, the reclassifiers may be integrated into the machine element by locating orifices of these dimensions at the lubrication points (see Figure 7-3). It may be convenient to use small-bore tubing as a reclassifier, particularly to inaccessible bearing locations. The small-bore tubing should have an I.D. and length similar to the reclassifier bore. The use of such tubing frequently simplifies installation at some points.

High-efficiency Reclassifiers

Just as do standard reclassifiers, high-efficiency reclassifiers serve three principal purposes:

1. To create wet oil mist by coalescing extremely fine oil droplets into larger droplets that can then coat the components to be lubricated.

2. To meter the oil-mist flow and to allow a given quantity to migrate towards the part to be lubricated.

3. To maintain a desired back pressure in the oil-mist supply line.

High-efficiency reclassifiers as shown in Figure 7-5 are offered by DeLimon Fluhme GmbH & Company (P.O. Box 5209, Dusseldorf, Germany). They incorporate a packing of extremely small metal pellets that cause the airborne mist to adopt a "tortuous path" flow pattern through the restriction encountered. In the process, the oil mist forms a coating on the pellets, which then gets picked up by the air and is hurled against the bearing surfaces as an oil spray.

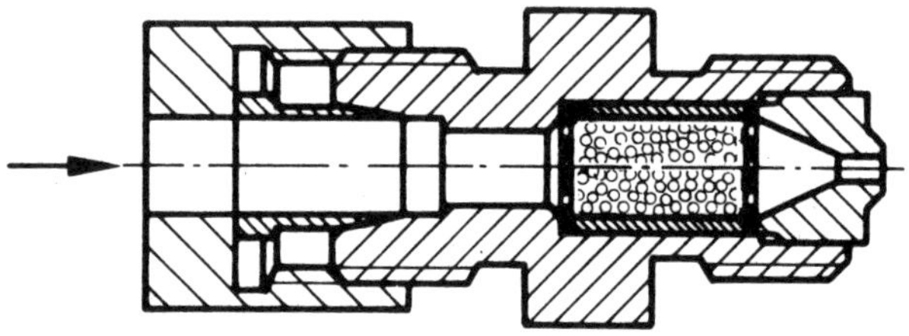

Figure 7-5. High-efficiency reclassifiers produce coalescence by forcing the mist through a pellet-type packing. They virtually eliminate stray mist. *(Source: DeLimon Fluhme, GmbH & Company.)*

The manufacturer claims that close to 99% of the oil mist is thus reclassified into droplets, and stray mist a virtually eliminated. Aside from the obvious benefits to the environment, the high degree of oil separation can result in substantial systems capacity upgrading. An increase of 40% has been reported in the number of bearing points compared to the use of conventional reclassifiers. Air consumption will decrease quite considerably.

The operating pressure in a system with high-efficiency reclassifiers ranges between 30 and 65 in. H_2O (~7.5-16.3 kPa). Although this is higher than the more common operating pressures of mist systems with conventional application fittings, it is usually possible to use conventional fittings and high-efficiency reclassifiers in the same oil-mist system. The selection of conventional fittings sizes is simply based on the higher mist pressure required by the high-efficiency reclassifiers.

De Limon Fluhme has also supplied reclassifier nozzles with different pellet sizes designed to discharge both very large particles and a residual oil mist.

Chapter 8

Rating Individual Machine Elements

Rolling Element Bearings—Ball, Roller, and Needle Bearings

The lubrication requirements of rolling element bearings are calculated by multiplying the shaft diameter by the number of rows and applying a load factor, LF or Y.

$$B.I. = D_1 \times R \times LF \text{ (bearing-inch method)} \tag{8-1}$$

$$L.U. = D_2 \times R \times LF/25 \text{ (lubrication unit method)} \tag{8-2}$$

$$cfm = D_1 \times R/Y \text{ (scfm method)} \tag{8-3}$$

where D_1 = shaft diameter, in.

D_2 = shaft diameter, mm

R = number of rows or balls, rollers, or needle bearings

LF, Y = load factor governed primarily by nature of loading and characteristics of mist generator

The factors and the criteria for their selection vary among the manufacturers of oil-mist systems. (For example, where Norgren uses an LF of 1 to calculate B.I., Lubrication Systems Company would, if they used B.I. formulas, have to use LF = .733 to make their B.I. and cfm ratings consistent. Where Alemite uses Y = 40, Lubrication Systems Company uses Y - 45.45, etc.)

Some manufacturers' criteria for load factor selection are:

LF = 3 for—

- Spherical, straight, and tapered roller bearings with preload.

LF = 2 for—

- Spherical roller bearings without preload.
- Ball bearings with initial preloading.

LF = 1 for—

- Ball, straight, and tapered roller bearings without preload.
- Needle bearings.

Y = 14 for—

- Work roll and backup roll bearings in rolling mills.

Y = 20 for—

- Constantly thrust-loaded bearings.
- Preloaded bearings.
- Bearings on shafts transmitting more than 30 kW (40 hp).
- Bearings subjected to high inertial loads, either by frequent hard starting and stopping or by unbalanced shaft designs.

Y = 40 for—

- Any not included in the preceding service definitions.

Assuming a load factor of 1, a single antifriction bearing running on a 1-in. shaft requires a 1 B.I. reclassifier. A 4-in. shaft mounting a 4-row antifriction bearing would require 16 B.I. of reclassifier rating (4 × 4 × 1 = 16). A 75-mm shaft mounting a 3-row antifriction bearing would require a reclassifier with a 9 L.U. rating (75/25) × 3 × 1 = 9). A 2-in. shaft mounting a 2-row antifriction bearing would require a mist flow of (2 × 2)/40 = 0.1 scfm.

Normally, the speed of the bearing need not be considered for the purpose of these calculations. However, if the bearing bore operates at a speed in excess of 610 m/min (~2,000 fpm), there is some possibility that windage (fan effects) impede the mist flow. In these high-speed cases the user may elect to install directed mist fittings of the type shown earlier in Figure 4-12.

It is important to note that bearings may well be of different types but due to different load factors they may end up using the same reclassifier size. Or, in applying Equations 8-1, 8-2, or 8-3 to shaft diameters of fractional inch, or certain metric sizes, the resulting B.I., L.U., of scfm number may place the resulting reclassifier requirement between two of these available from a given manufacturer. In those cases the next larger reclassifier offered should be used. A few examples will illustrate this convention. (Also, the reader may wish to refer to Figures 7-1 and 7-2, which are typical of the sizing information published by major manufacturers of oil-mist systems.

Example:

Shaft diameter - 1.187 in.
Bearing = single-row, tapered, without preload. Using Equation 8-1,

B.I. = 1 × 1.187 × 1 = 1.187

Recommended: 2-B.I. rating reclassifier (next available size).

Example:

Shaft diameter = 40 mm
Bearing = needle type. Using Equation 8-2,

L.U. = (1 × 40 × 1)/25 = 1.6

Recommended: 2-L.U. rating reclassifier (next available size).

Example:

Shaft diameter = 2.125 in.

Bearing = three-row, straight roller. Using Equation 8-3,

$$\text{cfm} = (3 \times 2.125 \times 1)/40 = 0.16$$

Recommended: 0.18-scfm reclassifier (next available size).

Of course, in actually designing a mist system, one would use the manufacturers' procedures and data provided by those being considered as potential suppliers. For our examples we will use rating formulas of two different manufacturers and simplified lists of application fittings (Figures 7-1, 7-2, and Table 8-1) that we will assume to be available.

Table 8-1. Available Application Fittings

Bearing Inch		scfm	
1	9	.03	.27
2	10	.06	.30
3	12	.09	.36
4	15	.12	.45
5	20	.15	.60
6	25	.18	.75
7	30	.21	.90
8	40	.24	1.20

Example:

Shaft diameter = 7.75 in.

Bearing = double row ball, without preload.

$$\begin{aligned} \text{B.I.} &= 2 \times 7.75 \times 1 \\ &= 15.5 \\ \text{scfm} &= 2 \times 7.75/40 \\ &= .39 \end{aligned}$$

Recommended: Use a 20-B.I. reclassifier or a .45-scfm fitting.

Tapered Roller Bearings without Preload

On tapered roller bearings without preload, the reclassifier should be positioned to apply the lubricant on the small end of the rollers because of the natural pumping action of the rollers. The reclassifier should be located a minimum of ~3 mm (1/8 in.) to a maximum of ~25 mm (1 in.) from the bearing surfaces (see Figure 8-1).

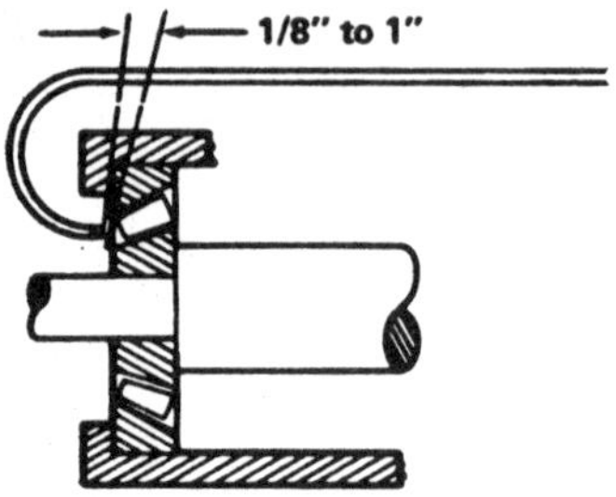

Figure 8-1. Oil mist applied to tapered roller bearing without preload. *(Source: C.A. Norgren Company.)*

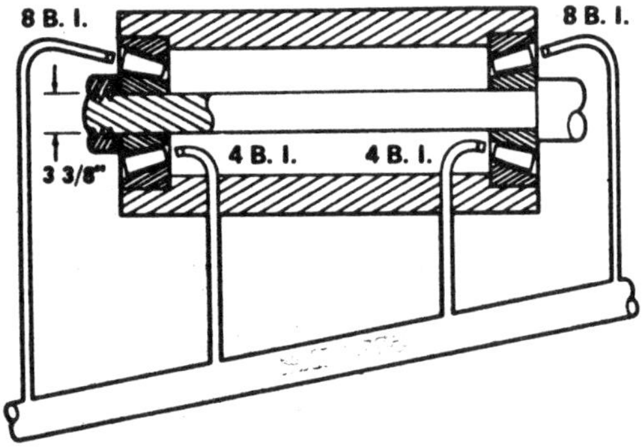

Figure 8-2. Two tapered roller bearings with preload require six times the oil flow of a single tapered roller bearing of the same size without preload. *(Source: C.A. Norgren Company.)*

Tapered Roller Bearings with Preload

Tapered roller bearings with an initial preload require two to three times the lubrication of a nonpreloading bearing. This is applied by using two reclassifiers so that 1/3 of the lubricant is applied to the small end and 2/3 to the large end of the bearing (see Figure 8-2).

Example:

Shaft diameter = 3.375 in.
Bearing = single row, tapered roller, preloaded

B.I. = 3.375 × 1 × 3
= 10.125
scfm = 3.375 × 1/20
= .17

Recommended: Use a 7-B.I. or a .12-scfm application fitting at the large end of the bearing and a 4-B.I. or .06-scfm fitting at the small end.

Heavily preloaded bearings require a small oil sump. The oil should contact the lowest rolling elements. Since a preload can squeeze lubricant out of bearings during idle periods, the sump is required to provide lubrication during the starting revolutions.

Recirculating Ball Nuts

The bearing-inch rating of recirculating ball nuts is equivalent to the pitch diameter of the screw plus 10% for each row of balls additional to the first. The scfm formulas are similarly based. The reclassifier should be directed at the approximate center of the loaded portion. No additional venting is necessary.

$$\text{B.I.} = d + [.1(R - 1)] \quad (8\text{-}4)$$
$$\text{scfm} = d/30 + dR/300 \quad (8\text{-}5)$$

where d = pitch diameter of screw, in.
R = number of rows of balls

Plain Bearings

Oil-mist lubrication of plain bearings is entirely feasible as long as the direction of loading is constant and properly defined. This is generally the case in certain machine tools, but cannot always be assured in pumps and motors in petrochemical plants. Consequently, plain bearings in refinery machinery continue to depend on conventional lubrication methods and use oil mist as a purge only.

Lubrication requirement calculations for plain bearings are based on projected areas of the bearing surfaces. The bearing-inch rating is determined by multiplying the bearing length by the shaft diameter, the load factor, and by a constant factor of 0.125. The scfm calculation divides the product of bearing length and shaft diameter by a load factor.

$$\text{B.I.} = (D \times L \times LF)(0.125) \quad (8\text{-}6)$$
$$\text{scfm} = DL/Y \quad (8\text{-}7)$$

where D = shaft diameter, in.
L = bearing length, in.
LF, Y = load factor

(Projected area = D × L)

Load factors (LF), used by one manufacturer, are functions of the static loading on the projected area, as shown in Table 8-2. This table implies that loads up to 35 kg/cm^2 ($500lb/in.^2$) can be lubricated using oil-mist technology. Higher bearing loads have been accommodated, but this manufacturer recommends that the factory be consulted for such applications.

Load factors (Y), used by another manufacturer, are functions of oil losses from the bearings. For example, bearings mounted in any position where the oil is retained in the bearing by contact-type seals would be considered as moderate service. Heavy service, high oil-loss bearings would be, for example, large bearings without seals that are mounted on vertical shafts.

Example: (Refer to Figure 8-3)
Shaft diameter = 2 in.

Table 8-2. Load Factors as a Function of Static Loading on Projected Area of Plain Bearings

LF	Static Loading Projected Area lbs/in.^{2}
1	Under 100
2	101 to 200
4	201 to 400
8	401 to 500

Bearing length = 2-3/4 in.
Static loading - 150 lb/in.^{2}
Axis horizontal, no seals, therefore Y = 100

$$\text{B.I.} = (2 \times 2.75 \times 2)(0.125) = 1.375$$

$$\text{scfm} = 2 \times 2.75/100 = .055$$

Recommended: Use a 2-B.I. reclassifier or a .06-scfm fitting.

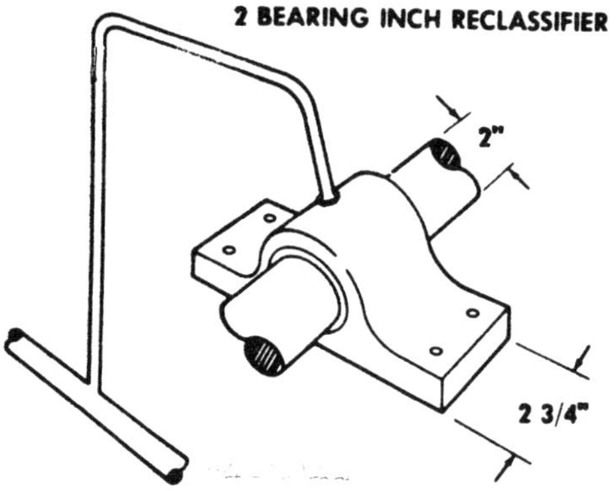

Figure 8-3. Plain bearings can be lubricated with oil mist as long as the direction of loading is constant and properly defined. *(Source: C.A. Norgren Company.)*

The reclassifier should be located to deliver oil to a longitudinal groove in the unloaded portion of the bearing. This groove should be approximately 90% of the length of the bearing cap. To make the groove the full length of the bearing cap would increase the end losses and defeat the distribution of oil along the length of the bearing (see Figure 8-4 and also refer back to Figure 5-9).

The groove location should be ahead of the load area as shown earlier in Figure 5-8. This location is also satisfactory where the heavy load is at the top of the bearing on the working stroke and at the bottom on the return stroke. The groove edges should be smoothly rounded to avoid a scraping action.

The optimum distance between the reclassifier and the shaft is 1/4 in. The minimum is 1/8 in. and the maximum is 1 in. (see Figure 8-5).

Each 5-6 in. of bearing length or fraction thereof (depending on the oil-mist system manufacturer) requires a reclassifier (see Figure 8-5).

Example:

Shaft diameter = 4 in.
Bearing length = 8 in.
LF = 2
Y = 100

The 8-in length requires two reclassifiers.

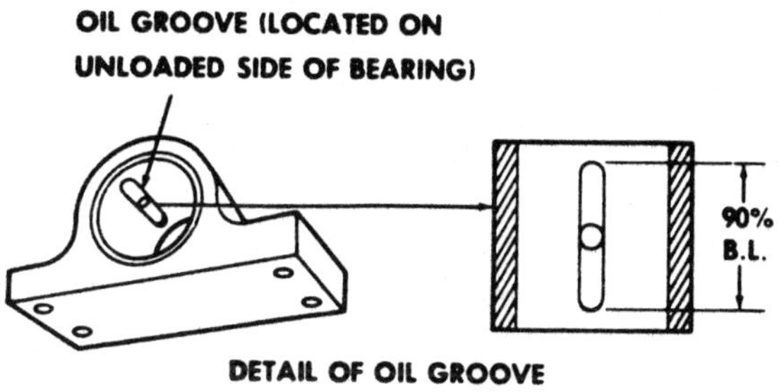

Figure 8-4. Oil must be delivered to longitudinal groove in the unloaded portion of the bearing *(Source: C.A. Norgren Company.)*

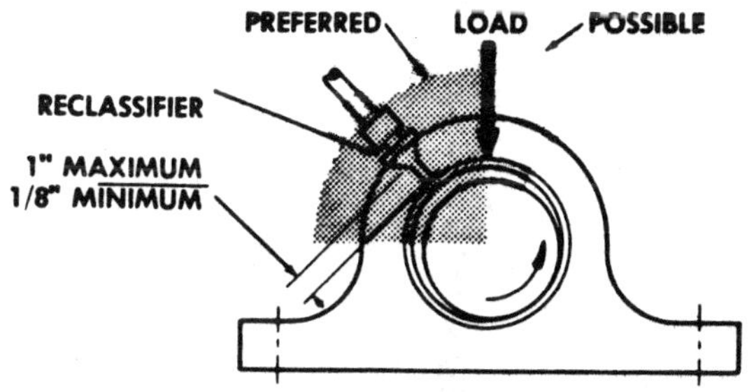

Figure 8-5. Reclassifier and groove locations for plain bearings. *(Source: C.A. Norgren Company.)*

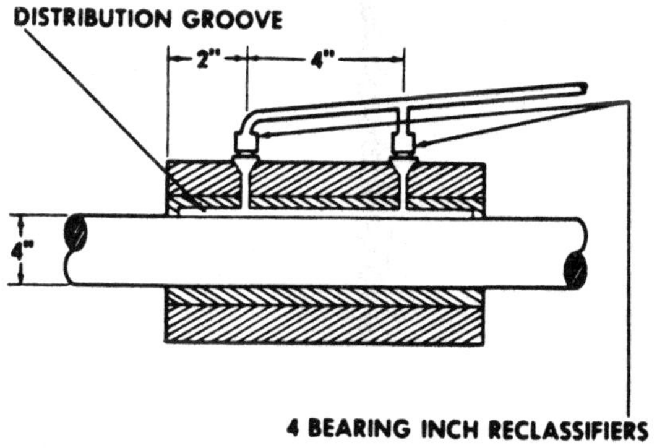

Figure 8-6. Each six inches of bearing length or fraction thereof requires a reclassifier. *(Source: C.A. Norgren Company.)*

$$
\begin{aligned}
\text{B.I.} &= (4 \times 8 \times 2)(0.125) \\
&= 8 \\
\text{scfm} &= 4 \times 8/100 \\
&= .32
\end{aligned}
$$

Recommended: Use a 4-B.I. reclassifier 2 in. from each end of the bearing, or two .18-scfm fittings similarly placed.

Grease-lubricated bearings are frequently found to have a "figure 8" or "X" groove in the loaded portion of the bearing (see Figure 8-7). These grooves will interrupt the formation of an oil film and should be eliminated before oil-mist lubrication is applied.

Oscillating Bearings

The bearing-inch calculation of a lightly loaded oscillating bearing is the same as a plain bearing. The number of reclassifiers required is dependent on shaft diameter and width. For shaft diameters of 1 in. or less, two reclassifiers are used diametrically opposed. For larger shafts, a minimum of two reclassifiers is required with the maximum number dependent on locating reclassifiers along the circumference no more than 3 in. apart. Reclassifiers should be equally spaced (see Figure 8-8).

For moderately loaded oscillating bearings, one experienced vendor (Alemite) prefers to base number and placement of fittings on angle of rotation: three fittings around circumference for 120° rotation, four for 90°, six for 60°, etc.

For horizontal bearings, each 6 in. of bearing length or fraction thereof requires a reclassifier.

Figure 8-7. Grease grooves in loaded area of plain bearings may interfere with the formation of an oil film if the bearing is converted to oil-mist lubrication. *(Source: C.A. Norgren Company.)*

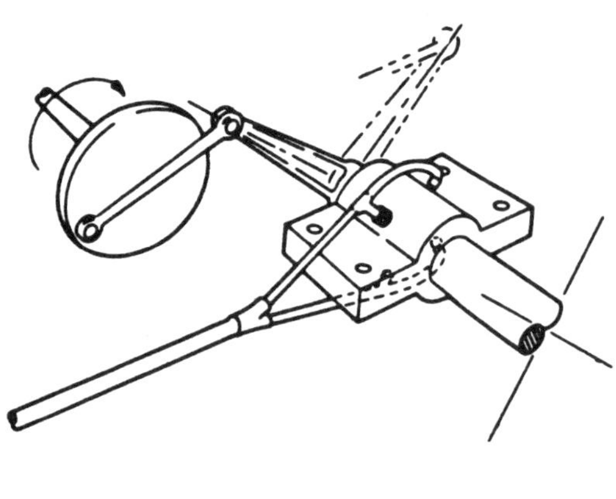

Figure 8-8. Lubrication guidelines for oscillating sleeve bearings differ with shaft size and bearing length. *(Source: C.A. Norgren Company.)*

For vertical bearings, the reclassifier should be set to deliver oil to a circumferential groove in the upper 1/3 of the bearing.

Gear Lubrication

Reclassifier ratings of gear pairs or trains are determined by adding the pitch diameters, multiplying this sum by the face width, and applying an adjustment or load factor.

$$\text{B.I.} = F\,(P_1 + P_2 + P_3 + ...)(0.25) \qquad (8\text{-}8)$$

$$\text{scfm} = F\,(P_1 + P_2 + P_3 + ...)/160 \qquad (8\text{-}9)$$

where F = face width of gear, in.
P_1 = pitch diameter of small gear, in.
P_2 = pitch diameter of large gear, in.
P_3 ... = pitch diameters of additional gears, in.

Example: (Refer to Figure 8-9)
Small gear = 4-in. pitch diameter, 2-in. face
Large gear = 7-3/4-in. pitch diameter, 2-in face

$$\text{B.I.} = 2(4 + 7.75)(0.25) = 5.87$$

$$\text{scfm} = 2(4 + 7.75)/160 = .147$$

Recommended: Use a 6-B.I. or a .15-scfm reclassifier.

A reclassifier should be used for each 2-in. of face width (see Figure 8-10).

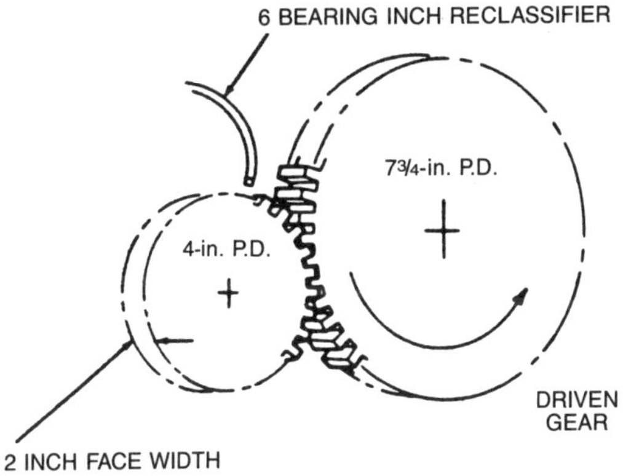

Figure 8-9. Gear lubrication with lube application in the incoming mesh of a gear set. *(Source: C.A. Norgren Company.)*

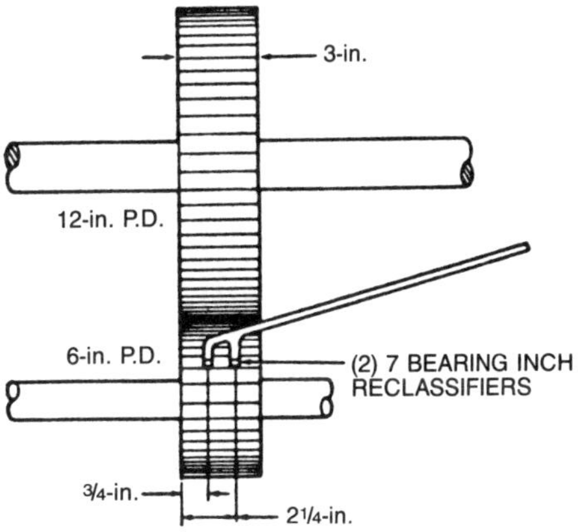

Figure 8-10. Each two inches of gear face width or fraction thereof requires a reclassifier. *(Source: C.A. Norgren Company.)*

Example:

Small gear = 6-in. pitch diameter, 3-in. face
Large gear = 12-in. pitch diameter, 3-in. face

$$\text{B.I.} = 3(6 + 12)(0.25) = 13.5$$

$$\text{scfm} = 3(6 + 12)/160 = .34$$

Recommended: Two 7-B.I. or two .18-scfm reclassifiers, located at the 1/4 points of the face width.

The foregoing procedures are applicable on plain, spur, beveled, helical or herringbone gears operating at surface speeds up to 2,000 fpm when using standard reclassifiers. From 2,000 to 3,000 fpm, pressure jet reclassifiers might be used to advantage. Information on pressure jet reclassifiers was given earlier. Speeds greater than 3,000 fpm have been achieved by experienced oil-mist systems manufacturers. The reader may wish to seek their advice, as applicable.

Large-ratio Gearing

Figure 8-11 shows gearing with a pitch diameter ratio exceeding 2:1. If the pitch diameter of any gear is more than twice that of the small gear (P_1), use $2P_1$ for the pitch diameter of the large gear. The formulas then become:

$$\text{B.I.} = F(3P_1)(0.25) \qquad (8\text{-}10)$$

$$\text{scfm} = F(3P_1)/160 \qquad (8\text{-}11)$$

An example from industry is illustrated in Figures 8-12 through 8-14. The rotary kilns shown in Figure 8-12 are installed in a cement plant. Each kiln is equipped with an

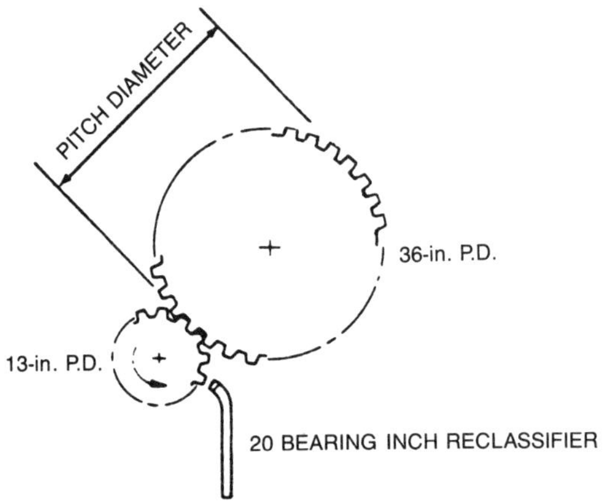

Figure 8-11. Gearing with pitch diameter ratio exceeding 2-to-1 requires oil flow per Equation 8-10 or 8-11.

Figure 8-12. Rotary kilns in a cement plant. *(Source: Alemite Division of Stewart-Warner Corporation.)*

open girth gear having a face width of 18 in. and a pitch diameter of 84 in., much more than twice the 12-in. diameter of its pinion.

$$\begin{aligned} \text{B.I.} &= 18(3 \times 12)(0.25) \\ &= 162 \\ \text{scfm} &= 18(3 \times 12)/160 \\ &= 4.05 \end{aligned}$$

Recommended: Use nine 20-B.I. or .45-scfm reclassifiers, starting 1 in. in from the edge of the pinion, as illustrated in Figures 8-13 and 8-14.

NOTE: Figure 8-13 is a photograph of the actual installation with real application fittings rather than those from our hypothetical list. Since the fitting configuration selected was only available in ratings up to 0.3 scfm (@ 20-in. H_2O), it took fourteen of them to provide the necessary flow.

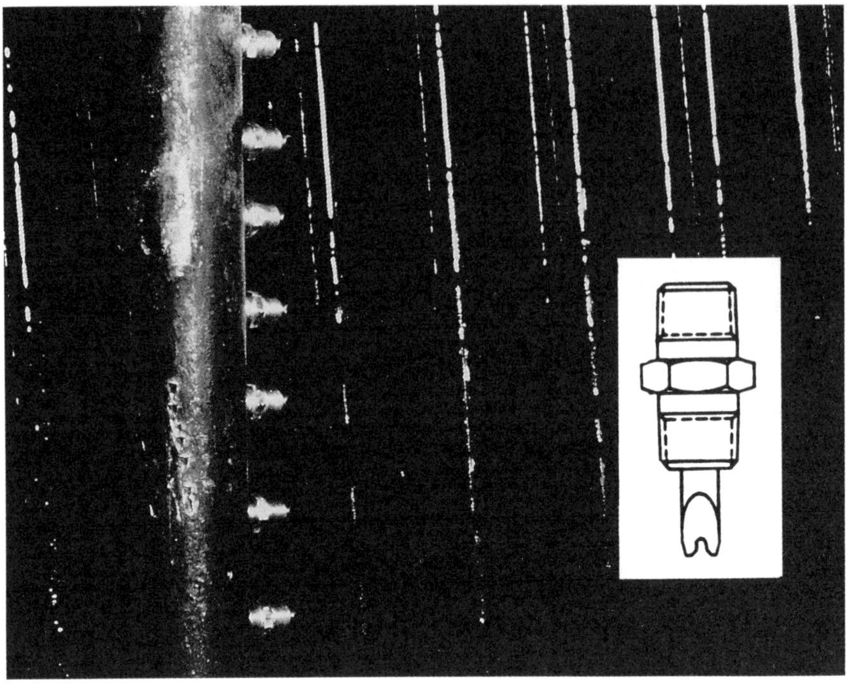

Figure 8-13. Spray-type oil mist reclassifiers installed on open gear train. *(Source: Alemite Division of Stewart-Warner Corporation.)*

Figures 8-15 and 8-16 illustrate the placement of oil-mist application fittings on gear trains.

Reversing Gears

Reversing gears require more lubrication than those in non-reversing service, because both sides of each tooth must be lubricated. (Figure 8-17).

$$\text{B.I.} = 2F\,(P_1 + P_2 + P_3 + ...)(0.25)$$

$$= F\,(P_1 + P_2 + P_3 + ...)(0.5) \qquad (8\text{-}12)$$

$$\text{scfm} = F\,(P_1 + P_2 + P_3 + ...)/110 \qquad (8\text{-}13)$$

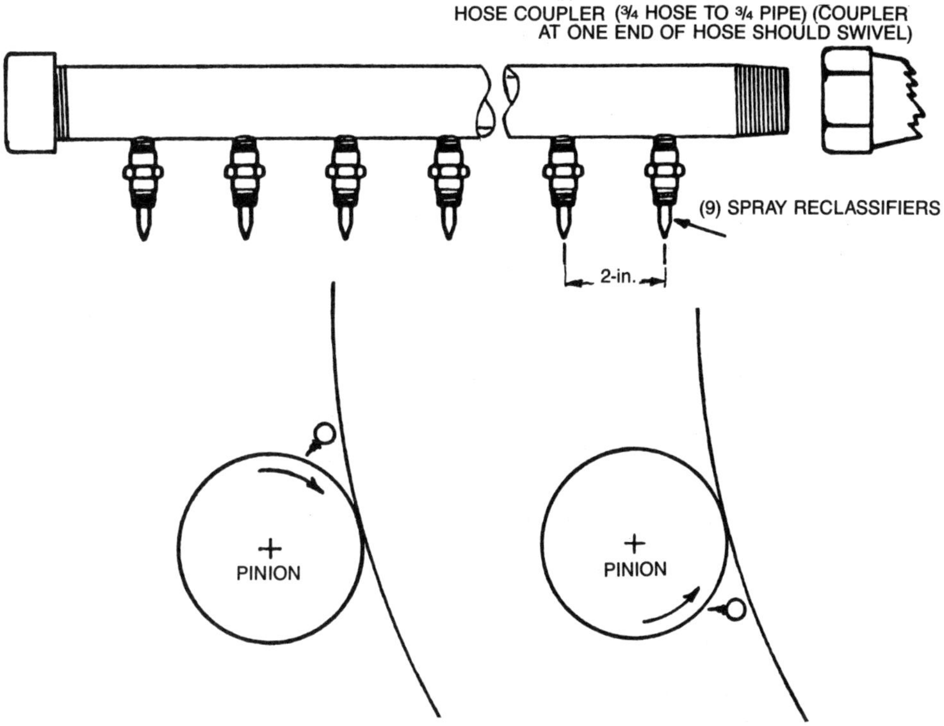

Figure 8-14. Oil-mist manifold for open gear lubrication of 18-in. wide kiln drive. *(Source: Alemite Division of Stewart-Warner Corporation.)*

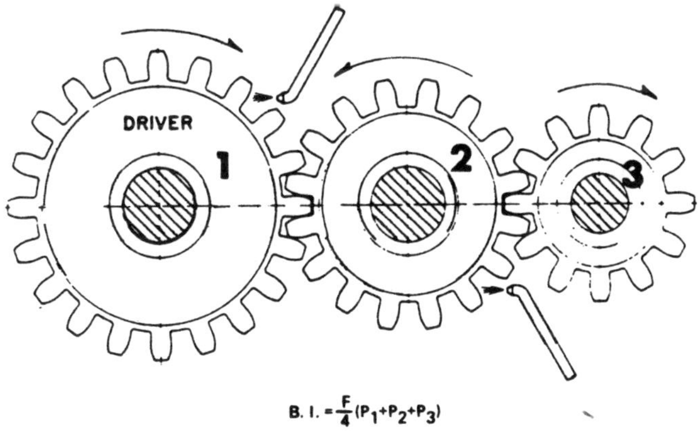

Figure 8-15. Oil mist applied to simple gear train. *(Source: C.A. Norgren Company.)*

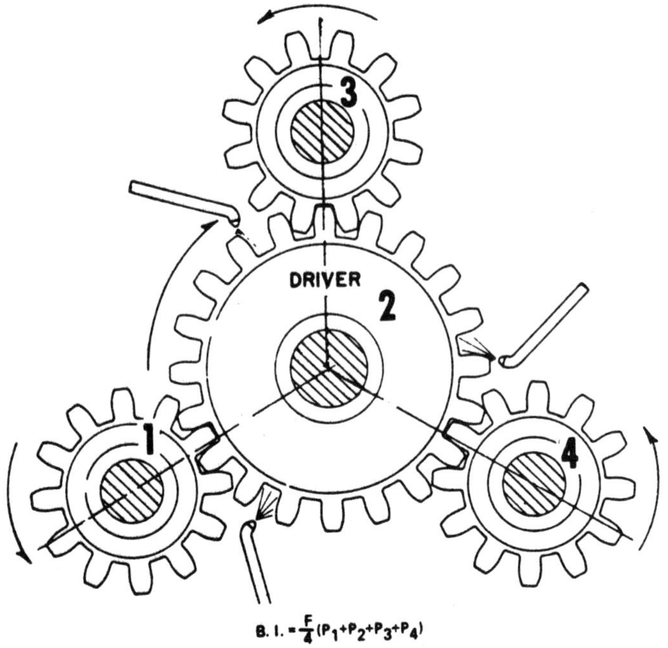

Figure 8-16. Four-gear assembly with oil-mist lubrication. *(Source: C.A. Norgren Company.)*

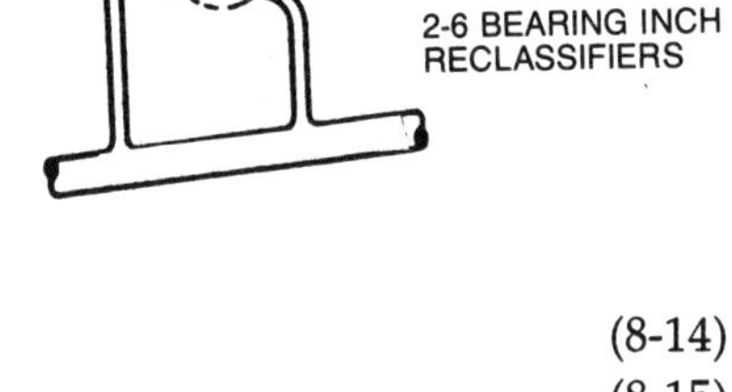

Figure 8-17. Reversing gears require twice the normal amount of lube oil. *(Source: C.A. Norgren Company.)*

Worm Gearing

The bearing-inch and scfm systems that we are using for illustration use different worm and gear dimensions to calculate lubrication requirements.

$$\text{B.I.} = 0.25(L_W \times P_1 + F \times P_2) \tag{8-14}$$
$$\text{scfm} = F\,(2P_1 + P_2)/160 \tag{8-15}$$

where L_W = length of worm, in.

P_1 = pitch diameter of worm, in.
P_2 = pitch diameter of gear, in.
F = face width of gear, in.

Example: (Refer to Figure 8-18)
Worm length - 2 in.

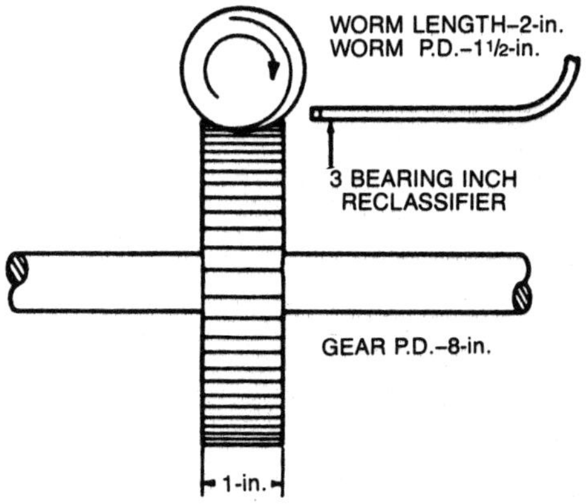

Figure 8-18. Worm gearing shown with oil spray applied to incoming mesh. *(Source: C.A. Norgren Company.)*

Worm pitch diameter = 1.5 in.
Gear pitch diameter = 8 in.
Gear face width = 1 in.

$$\begin{aligned} \text{B.I.} &= .25(2 \times 1.5 + 8 \times 1) \\ &= .25(3 + 8) \\ &= 2.75 \\ \text{scfm} &= 1(2 \times 1.5 + 8)/160 \\ &= .069 \end{aligned}$$

Recommended: Use a 3-B.I. or a .09-scfm reclassifier.

Worm gears should have the reclassifiers directed toward the loaded side of the tooth of either the worm or the gear. Reversing worm gears require more lubricant than non-reversing worm gears, since both sides of the tooth need to be lubricated:

$$\text{B.I.} = 0.5(L_W \times P_1 + F \times P_2) \quad (8\text{-}16)$$

$$\text{scfm} = F\,(2P_1 + P_2)/110 \quad (8\text{-}17)$$

Rack and Pinion

In the particular manufacturers' systems that we are using for illustration, the bearing-inch total for a rack and pinion is 1/2 the projected area of the pinion while the scfm calculation uses the reversing gear formula with the active length of the rack divided by 3 (for pi) for the pitch diameter of the second gear.

$$\text{B.I.} = (F \times P)(0.5) \quad (8\text{-}18)$$

$$\text{scfm} = F\,(P + L/3)/110 \quad (8\text{-}19)$$

where F = face width of pinion, in.
P = pitch diameter of pinion, in.
L = length of rack, in.

Reclassifier Location for Gears

Reclassifier discharge should be between 1/8 in. and 1 in. from the outside of the gear teeth and directed toward the loaded side of the teeth. On high ratio gearing it is better to lubricate the small gear.

The preferred point of lubricant application is on the loaded side of the driving tooth, approximately 90° to 120° from the point of mesh (Figure 8-19).

To lubricate both sides of the teeth of reversing gears, one manufacturer recommends the use of separate reclassifiers for each side while another recommends directing all spray fitting discharges toward the gear axis.

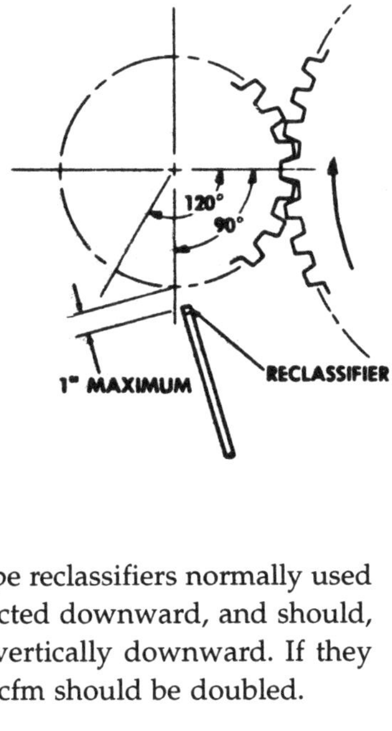

Figure 8-19. For gears, the preferred point of lubricant application is on the loaded side of the driving tooth. *(Source: C.A. Norgren Company.)*

Another consideration is the operation of the spray type reclassifiers normally used to lubricate gears. They perform most efficiently when directed downward, and should, wherever possible, be directed between horizontally and vertically downward. If they must be directed upward the calculated bearing inches or scfm should be doubled.

Cams

The bearing-inch rating for cams is determined by multiplying the face width of the cam by the maximum cam diameter and dividing this product by an adjustment factor (Figure 8-20).

Each 2 in. of cam width or fraction thereof requires a reclassifier that should be located at an optimum distance of 1/4 in. from the cam surface, and not more than 1 in. or less than 1/8 in. away.

$$\text{B.I.} = (F \times D_m)/10 \quad (8\text{-}20)$$
$$\text{scfm} = (F \times D_m)/400 \quad (8\text{-}21)$$

where F = face width of cam, in.
D_m = maximum diameter of cam, in.

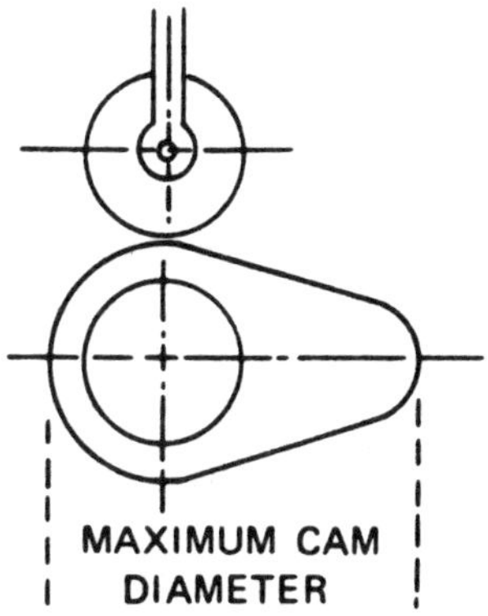

Figure 8-20. The bearing-inch rating for cams is determined by multiplying the face width of the cam by the maximum cam diameter and a factor of 0.1. *(Source: C.A. Norgren Company.)*

Slides and Ways

Generally, oil-mist calculations are based on the area of contact between slide and way.

$$\text{B.I.} = (L \times W)(0.05) \quad (8\text{-}22)$$
$$\text{scfm} = (L \times W)/800 \quad (8\text{-}23)$$

where L = length of slide, in.
W = width of contact, in.

For other than rectangular contact shapes, use appropriate area formulas in place of (L × W). Other considerations such as the physical size of the traveling member or the attitude of the member will also influence the total lubrication requirement.

Application techniques for slides and ways are relatively simple. The reclassifiers should discharge into grooves across the contact surface perpendicular to the direction of motion. The grooves should be similar to those described earlier under plain bearings. Reclassifiers should enter the grooving so that there is sufficient air flowing for impingement and be positioned to give an impingement distance of from 1/8 in. minimum to 1 in. maximum.

When the slides and ways are nearly horizontal, the slides should have a reclassifier every 4 in. of length or fraction thereof with the end reclassifier fitted within 1 in. of the leading and trailing edges. Every 6 in. of slide width (or contact width) will require a reclassifier.

Sliding members under 4 in. in length require only one reclassifier.

Example:

Slide length = 5 in.
Contact width = 5 in.

$$\text{B.I.} = (5 \times 5)(0.05) = 1.25$$
$$\text{scfm} = (5 \times 5)/800 = .03$$

Since the length exceeds 4 in., two reclassifiers are required. Since the width is less than 6 in., no additional reclassifiers are required.

Recommended: Use two 1-B.I. or .03-scfm reclassifiers, one 1 in. from leading edge and the other 1 in. from trailing edge, on the slide center line (Figure 8-21).

Example:

Slide length = 10 in.
Contact width = 8 in.

Slide length exceeds 4 in. Because 10/4 = 2-1/2, three reclassifiers are required for distribution over the length. Width exceeds 6 in. 8/6 = 1-1/3 so two rows of reclassifiers are required.

Recommended: Six 1-B.I. reclassifiers, spaced as shown in Figure 8-22.

Figure 8-23 illustrates one method of grooving the slide and for providing oil mist access to the bearing surfaces. The same procedure for applying reclassifiers to horizontal surfaces can be applied to inclined or vertical slides.

Vertical Slides

Advantage can be taken of gravity by placing the reclassifiers near the top of the slide and allowing gravity plus grooving to distribute the oil. Every 6 in. of width or part thereof should have its reclassifier.

These reclassifiers can be located at the top of the sliding portion and allow gravity

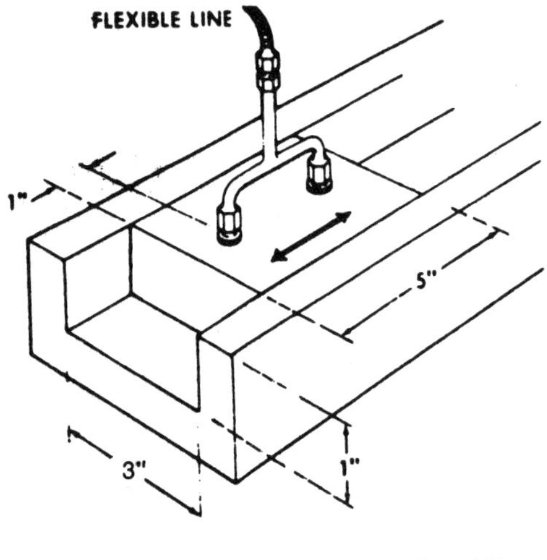

Figure 8-21. Two one-B.I. reclassifiers would be required for a 5-in. × 5-in. slide. Note that width is 1 + 3 + 1 in. *(Source: C.A. Norgren Company.)*

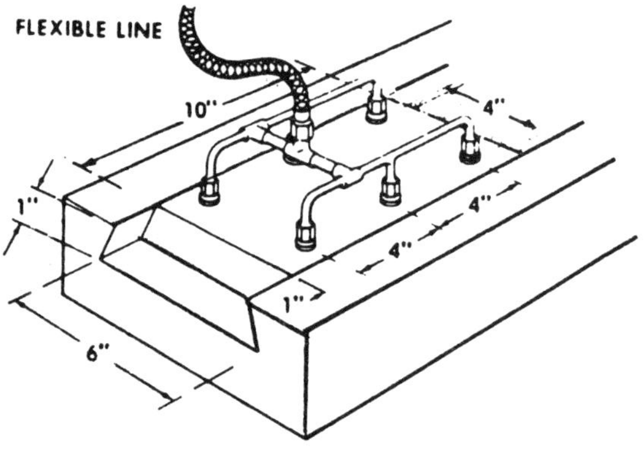

Figure 8-22. Six one-B.I. reclassifiers would be required for a 10-in. × 8 in. slide. Note that width is 1 + 6 + 1 in. *(Source: C.A. Norgren Company.)*

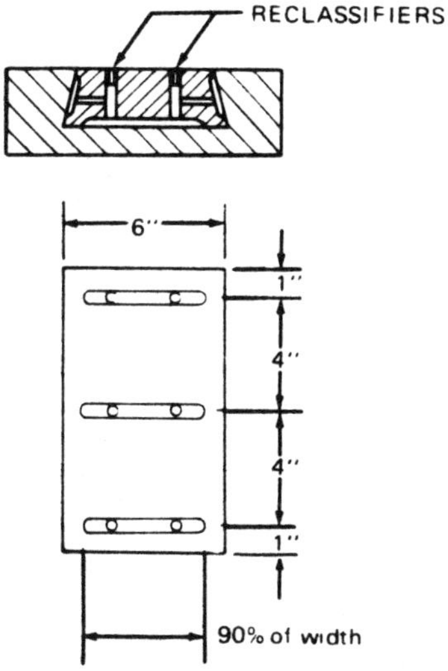

Figure 8-23. Slide grooving details. *(Source: C.A. Norgren Company.)*

to distribute the oil the length of the slide. Reclassifiers are sized by taking the contact area in square inches and multiplying it by an adjustment factor.

Example:

Slide width = 3 in.
Slide length = 15 in.

$$\begin{aligned} \text{B.I.} &= (3 \times 15)(0.05) \\ &= 2.25 \\ \text{scfm} &= (3 \times 15)/800 \\ &= .056 \end{aligned}$$

Recommended: One 3-B.I. or .06-scfm reclassifier.

Chains

The lubrication rating for simple drive chains comprised of a drive sprocket and driven sprocket can be calculated by using Equation 8-24 or 8-25:

$$\text{B.I.} = \frac{\text{PDR}\sqrt{(S/100)^3}}{8} \qquad (8\text{-}24)$$

$$\text{scfm} = \frac{\text{PDR}\sqrt{(S/100)^3}}{320} \qquad (8\text{-}25)$$

where P = chain pitch, in. (Figure 8-23)
D = diameter in either sprocket, in. (Figure 8-24)
R = chain rows for multiple strand roller chains
S = speed in rpm of the sprocket used for "D" (if speed is less than 200 rpm, use 200 rpm in calculations)

$$\text{B.I.} = \frac{WD\sqrt{(S/100)^3}}{15} \tag{8-26}$$

$$\text{scfm} = \frac{WD\sqrt{(S/100)^3}}{600} \tag{8-27}$$

where W = chain width, in.
D = diameter of either sprocket, in.
S = speed of the same sprocket in rpm (if speed is less than 200 rpm, use 200 rpm in calculations)

If the chain is completely enclosed, only one half of the bearing-inch rating as calculated need be used.

For each sprocket beyond two, the total reclassifier rating should be increased by 10%.

At surface speeds up to 2,000 fpm, standard reclassifiers can be used. From 2,000 to 3,000 fpm, pressure-jet reclassifiers can be helpful. For speeds greater than 3,000 fpm, consult the oil-mist vendor.

On single roller chains, the lubrication rating as determined from Equations 8-24 through 8-27 should be divided so that one reclassifier points at each row of side plates. Oil-mist application to a double-row roller chain is illustrated in the lower right-hand sketch of Figure 8-24.

On double-row and wider chains the center rows of side plates should get twice as much lubrication as each outside row. For instance, a triple-row chain requiring 24 bearing-inches should have 4 bearing-inches on each outside row. Thus, the reclassifiers across the chain width would read 4-8-8-4.

Silent chains should have equally rated reclassifiers every 1/2 in. of width, starting 1/4 in. in from the outside edges.

On all chains, the reclassifiers should point slightly against the chain motion and should be within 1 in. of the chain. The preferred point of application is inside the chain as it leaves the drive sprocket, since here the chain is slack and the oil can penetrate (see Figure 8-25). By applying oil on the inside surface, centrifugal force around the next sprocket will tend to pass the oil through the chain.

Before running a new chain, it should be washed free of grease and then soaked in oil.

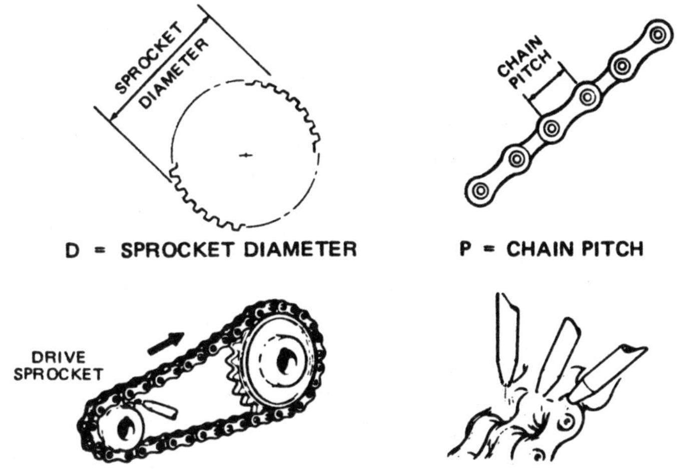

Figure 8-24. Oil-mist application details for typical drive chains. *(Source: C.A. Norgren Company.)*

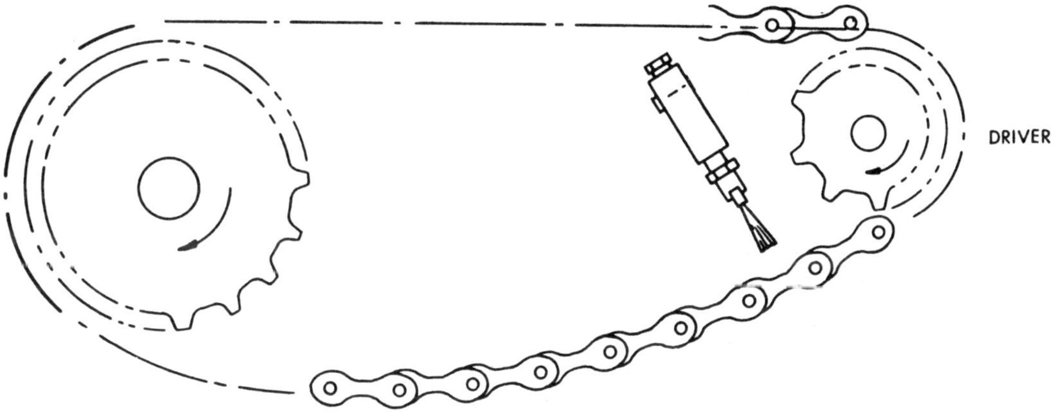

Figure 8-25. Preferred point of oil-mist application is inside the chain as it leaves the drive sprocket. *(Source: C.A. Norgren Company.)*

Chapter 9

Electric Motor Lubrication

By the mid-1970s, oil mist had demonstrated its superior suitability for lubricating and preserving electric motor bearings [14]. By that time, petrochemical plants in the U.S. Gulf Coast area, the Caribbean, and South America had converted in excess of one thousand electric motors to dry-sump oil-mist lubrication. In 1986, there were more than 4,000 electric motors on oil-mist lube in the U.S. Gulf Coast area alone.

However, universal acceptance did not come overnight. On the one hand, it seemed logical to extend oil-mist lines from centrifugal pump bearing housings to be adjacent electric motor bearings. On the other hand, concern was voiced that lube oil would enter the motor and cause damage to winding insulation or cause overheating until winding failure occurred. Initial efforts were, therefore, directed towards developing lip seals or other barriers confining oil mist to only the bearing areas.

When occasional seal failures were experienced on operating motors, oil mist entered the stator and coated the windings with lube oil. The potential explosion hazard was again investigated on this occasion and confirmation obtained that the oil/air mixture is substantially below the sustainable burning point. Experiments had shown the concentration of oil mist in the main manifold ranging from .005 to as little as .001 of the concentration generally considered flammable. The fire or explosion hazard of oil-mist lubricated motors is thus no different from that of NEMA-II motors. No signs of overheating were found, and winding resistance readings conformed fully to the initial, as-installed values.

Today's epoxy motor winding materials will not deteriorate in an oil-mist atmosphere. This has been conclusively proven in tests by several manufacturers. Windings coated with epoxy varnish were placed in beakers filled with various types of mineral oils and synthetic lubricants. Next, they were oven-aged at 170°C (338°F) for several weeks, and then cooled and inspected.

Final proof was obtained during inadvertent periods of severe lube oil intrusion. In one such case, a conventional oil-lubricated, 3,000 hp, (~2,200 kW), 13.8 kV motor ran well even after oil was literally drained from its interior. The incident caused some increase in dirt collection, but did not adversely affect winding quality.

But, experimentation with motor winding and cable terminations has shown that Teflon® wrap should be used for best results. Other materials, including silicone tape, exhibit a tendency to swell or become gummy when exposed to oil mist.

It was also found advantageous to provide sealant between the motor frame and conduit box to reduce the harmless, but nevertheless unsightly, mist emissions at the conduit enclosure. Mist supply and condensed oil drain ports should be made accessible without the need for removing fan covers and guards. A simple pipe nipple or similar extension was found to do just fine. Figures 9-1 and 9-2 show clean oil-mist installations that meet these requirements.

Finally, large-scale users of oil-mist lubricated electric motors discovered that space heaters with low sheath temperatures (less than 200°C, or 392°F) would not cause excessive smoking or coking of oil mist coming in contact with the sheath. Specifications for new motors should take this into account, but if motors are being converted to oil mist, it is not necessary to be overly concerned with this item.

Converting Electric Motors from Grease Lube to Oil-mist Lube

Converting to dry-sump oil-mist lubrication does not necessarily require that the motor be removed and sent to the shop. Motors with regreasable bearings are easiest to convert because they generally incorporate neither oil rings nor bearing shields. Figure 9-3 shows a typical oil-lubricated bearing that can be modified for dry-sump lubrication by adding only the piped oil-mist inlet, vent, and overflow drain passages. Oil rings must be removed because there is, of course, no longer an oil sump from which oil is to be fed to the bearing. Figure 9-4 shows the bearing shields removed in order to establish unim-

Figure 9-1. Pipe nipples or similar extensions supply pure oil mist to inaccessible electric motor bearings.

Figure 9-2. Vertical electric motor showing pure oil-mist supply and condensed oil drain lines.

peded passage from the oil-mist inlet pipe through the bearing rotating elements and finally the vent pipe to atmosphere. However, recent experience shows that the inboard bearing shield need not be removed to ensure a successful installation. It has been demonstrated in tests that oil mist did not have to be forced through a bearing, even a double-row thrust bearing. Oil mist completely fills a bearing housing and is condensed into larger particles by contact with the rapidly rotating bearings [70]. Nevertheless, it is recommended that oil mist indeed be forced through the rotating bearing, i.e., the supply and vent openings should be located on opposite sides of a bearing for more efficient deposition of oil in the bearing.

Figure 9-5 illustrates a petrochemical plant area with a series of vertical motors. One such motor, rated 125 hp, 3,560 rpm, experienced frequent thrust bearings failures with conventional oil lubrication. Installation of dry-sump oil mist apparently solved the chronic lubrication problem. Bearing housing temperatures were lowered from 160°F (71°C) to 110°F (43°C) after the conversion to dry-sump lubrication.

The simplicity of extending dry-sump oil-mist lubrication from general purpose pumps to their drivers or vice versa is evident from Figure 9-6. Sloped stainless steel inlet tubing is used from the distribution block to the additional bearings which must be served by the oil mist.

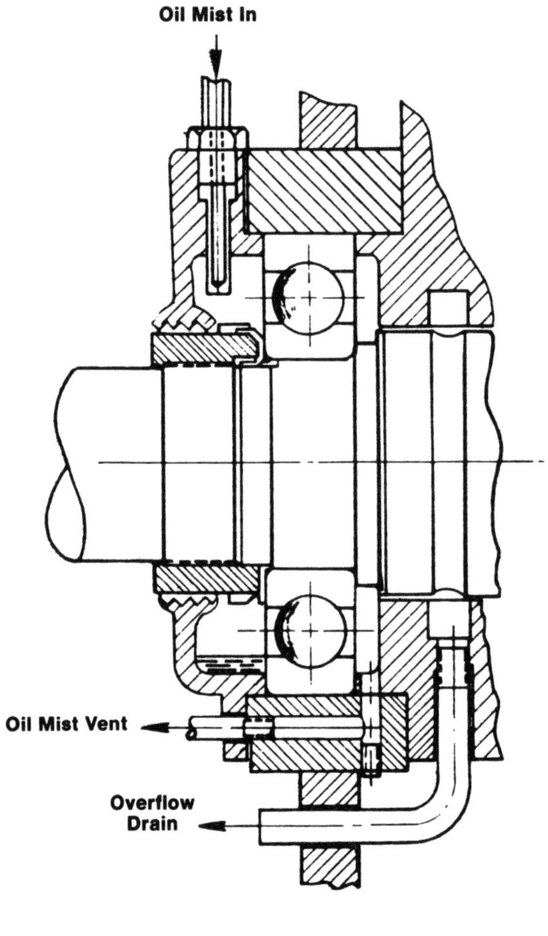

Figure 9-3. Large electric motor bearing after conversion from liquid oil lubrication to dry-sump oil-mist lubrication. *(Source: Reference 16)*

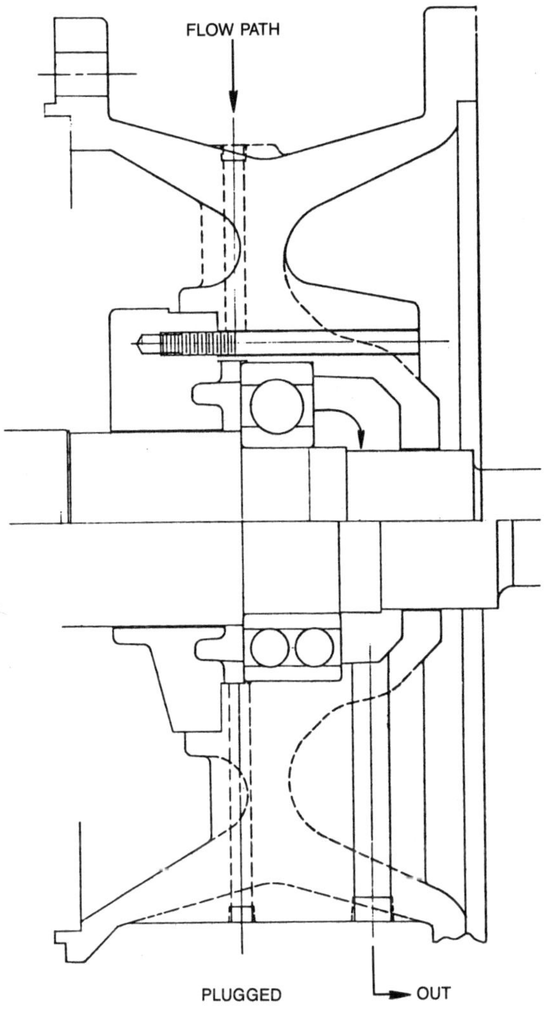

Figure 9-4. Electric motor bearings with both shields removed to promote unimpeded passage from the oil-mist inlet pipe through the rotating elements to vent pipe and atmosphere.

It may be anticipated that a properly installed and maintained oil-mist lubrication system will result in a high percentage reduction in bearing failure rates. It must be noted, however, that such bearing failure reductions will not be achieved if the basic bearing failure problem is not lubrication related. Oil mist cannot eliminate problems caused by defective bearings, incorrect bearing installation, excessive misalignment or incorrect mounting clearances.

We can summarize by noting that dry-sump oil-mist lubrication is extremely well-suited for rolling element bearings in electric motors. Tests have shown that bearing temperature rises can be reduced as much as 50% to 65% when a change from grease to oil-mist lubrication is made on a totally enclosed fan-cooled A-C electric motor.

The German Siemens Company considers this lubrication method advantageous for use in its electric motors with power ratings up to 3,000 kW (Figure 9-7). Again, oil mist excels as a preservative, preventing the ingress of atmospheric contaminants into standby

Figure 9-5. Petrochemical plant with vertical motors. Oil mist has been responsible for lowering thrust bearing temperatures on many of these motors.

Figure 9-6. Distribution blocks facilitate the extension of oil-mist systems from driven equipment to drivers and vice versa.

equipment. Bearing friction losses are kept low, and with through-flow oil-mist lubrication (Figure 9-8) electric motor bearings tend to run considerably cooler than with grease or oil ring lubrication.

Any rolling element bearing can be converted to dry-sump oil mist as long as three basic requirements are not overlooked:

1. On highly loaded bearings (thrust bearings), the venting arrangements of the bearing before being allowed to escape to the surroundings.

2. The application fitting must produce oil droplet size and quantity required by the bearing being serviced. Classifier locations must be selected so as to ensure that the oil reaches the bearing elements. This may require that lube oil drains or vents be piped away from regions exposed to fan windage [14].

3. Before considering conversion of *older* motors to oil-mist lubrication, it should be ascertained that mist or oil cannot reach any wiring or that insulations are indeed oil resistant.

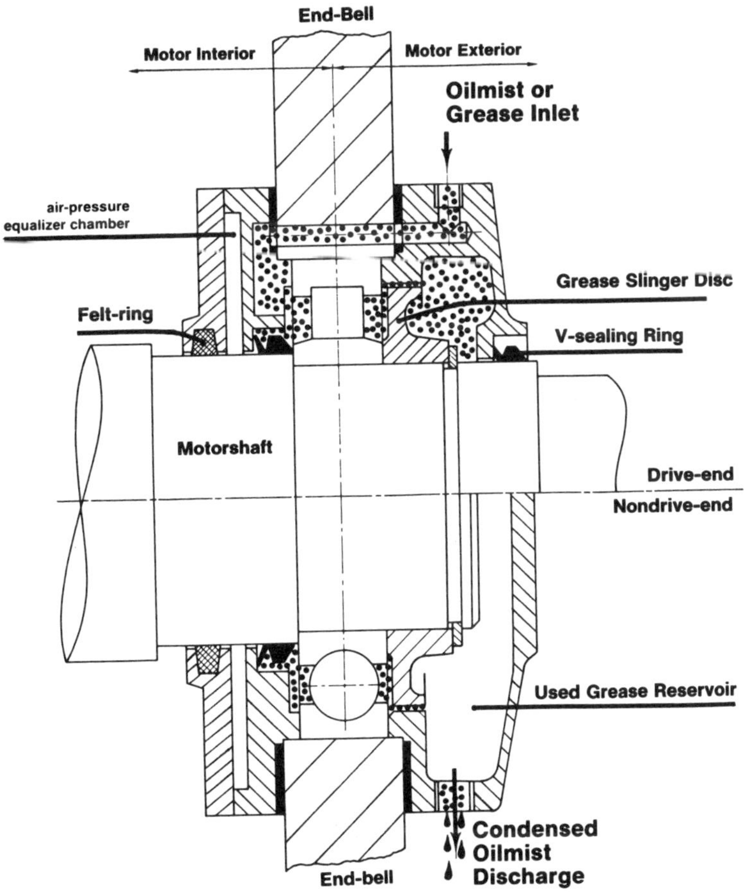

Figure 9-7. Oil-mist lubricated bearings furnished with Siemens TEFC motors from 18 to about 3,000 kW. *(Source: Siemens Technical Bulletin, also reprinted in Reference 16.)*

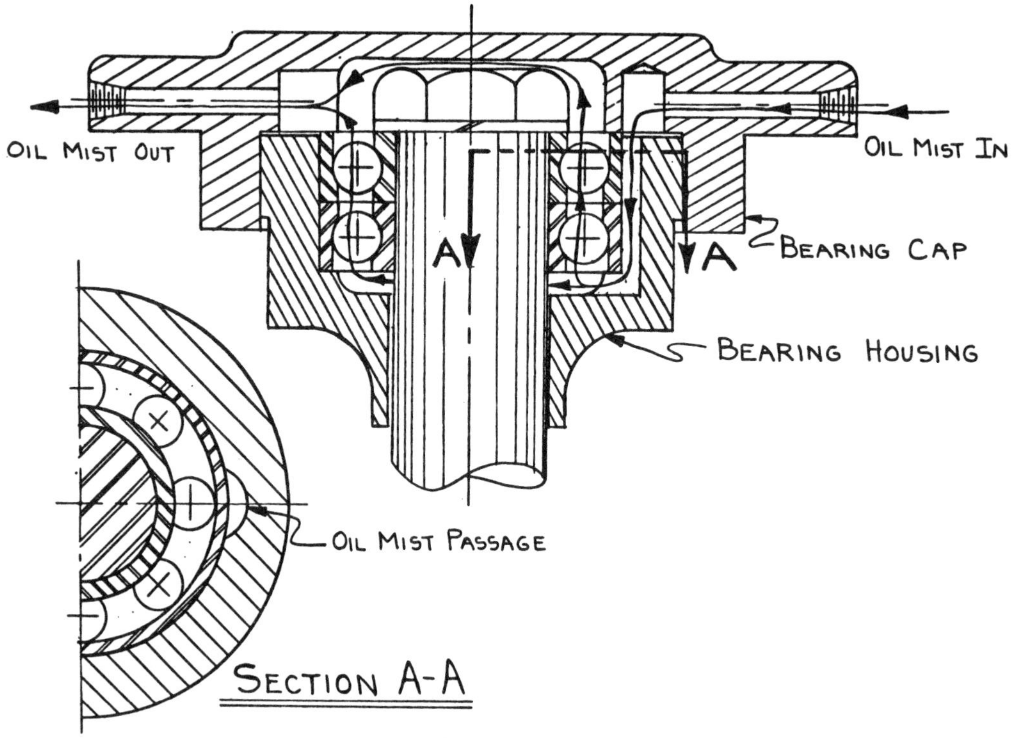

Figure 9-8. Electric motor thrust bearings arranged for through-flow of pure oil mist. *(Source: Reference 16.)*

Chapter 10

Closed-loop Oil-mist Installations

Oil-mist-lubricated equipment is usually executed for once-through or open-loop mist application. This is by choice, and certainly not by necessity. Closed-loop oil-mist application is feasible and is used whenever warranted by economic conditions or environmental considerations.

In the early 1980s, we demonstrated the feasibility of applying a slight vacuum at the drawoff port of the Inpro/Seal, Bearing Isolator shown earlier in Figures 6-6 and 6-7. This device comprises a noncontacting rotor-stator assembly, which acts as a bearing housing closure. It is mounted in such fashion as to allow the oil mist to first migrate through the rolling elements of an antifriction bearing. A major portion of the oil contained in the carrier air will thus have the opportunity to wet out on the bearing parts. The excess noncondensed oil mist can be evacuated through the drawoff port. From this drawoff port, the oil mist can be piped to an oil separator or electrostatic precipitator. The condensed oil can be filtered and can again be "misted." This closes the loop.

However, rotor-stator seals are not a prerequisite for drawing off the excess oil mist. Figures 6-8 and 6-9 illustrated how an elastomeric lip seal and magnetic face seal, respectively, can be used in conjunction with a drilled passage to accomplish the same purpose.

Oil separators generally incorporate a small high-speed fan, which is located above a coalescing pad or cartridge. The oil mist is induced to flow through the coalescer medium where most of the oil collects in droplet form and flows back into the separator housing. The remaining oil mist will now impinge on the fan blades and only air will be expelled into the environment. For oil removal, the separator may include a final charcoal filter.

Either an oil separator or electrostatic precipitator could be used in conjunction with the lube-oil reservoir shown in Figure 10-1. The large amount of oil mist escaping through its breather vent could thus be condensed or precipitated (Figure 10-2), and the entire system thereby made into a true closed-loop arrangement.

The lube oil thus collected can be reused if it is first processed through a lube-oil reclaimer or purifier unit. One such unit, operating on the vacuum dehydrating principle, is shown in Figure 10-3; References 17 through 20 give details.

Electrostatic precipitators represent an efficient and economical method of removing airborne smoke and mist. These precipitators are designed to be mounted directly on high-speed machinery or any convenient oil-mist discharge location. One such electrostatic precipitator is shown in Figure 10-2. Note again that mist-laden air enters this unit through the inlet and passes through an ionizing section where all the particles are given an electrical charge. The mist then passes through the collecting cell, where the charged particles are attracted to and collected on plates or opposite polarity. Cleaned air passes

Figure 10-1. Oil mist escaping from the breather vent of this large lube oil reservoir could be condensed by an oil separator or suitably connected electrostatic precipitator. This would convert this quasi-closed system into a truly closed-loop lubricating system.

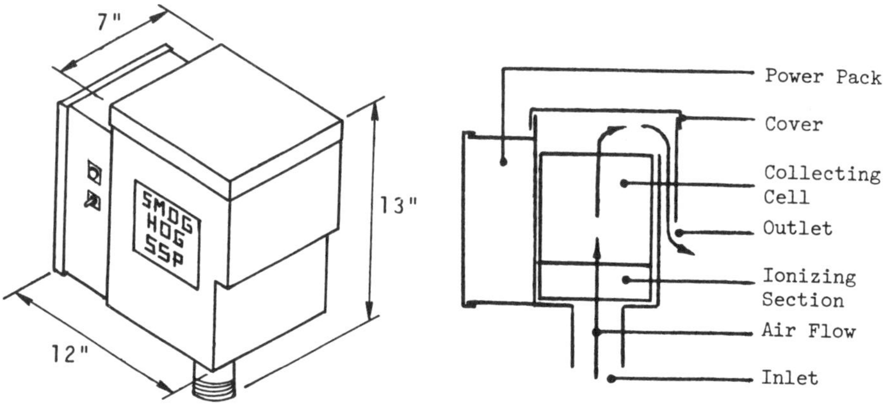

Figure 10-2. Electrostatic precipitators can effectively condense oil mist. *(Source: United Air Specialists, Inc., Cincinnati, OH.)*

through the outlet. The collected oil, now in liquid form, runs off the collecting cell plates and drains back down the inlet.

The range of effectiveness of a typical electrostatic precipitator is shown in Figure 10-4. This particular model ("Smog-Hog," by United Air Specialists, Inc., Cincinnati, Ohio) covers smokes and mists ranging in particle size from 0.01 to 100 μ. Note that the respirable fraction ranges from 0.01 to 10 μ, while oil mist typically ranges in size from 0.5 to 2.5 μ.

Figure 10-3. Condensed oil mist and other lubricating oils can be purified and reused after processing through suitable vacuum dehydrators. *(Source: Allen Filters, Inc., Springfield, MO.)*

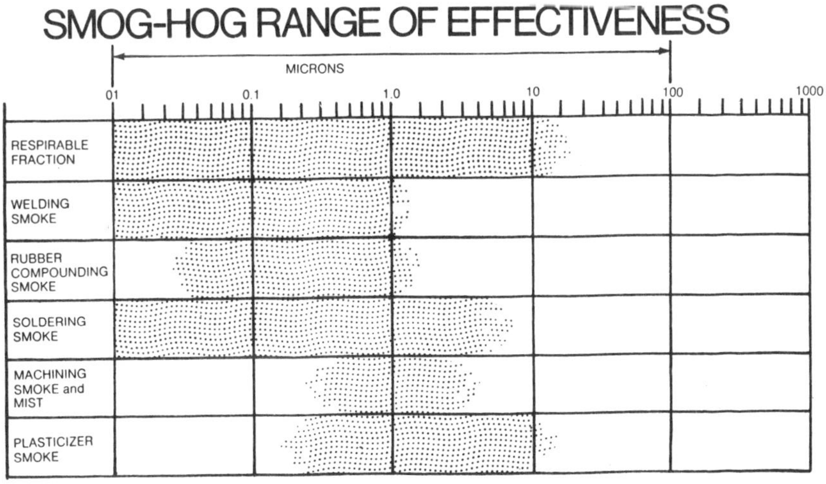

Figure 10-4. The effectiveness of electrostatic precipitators covers smokes and mists ranging in particle size from 0.01 to 100 microns. This range totally covers typical oil mists, which range in size from 0.5 to 2.5 microns. *(Source: United Air Specialists, Inc., Cincinnati, OH.)*

OIL-MIST SYSTEMS FOR TEXTILE MACHINERY

The best and most interesting operational example of a closed-loop oil-mist lubrication system is found in the textile industry. It is here where draw rolls are often lubricated with oil mist because this lubrication method is ideally suited for high-speed ball bearings that operate with metal temperatures of 280°F (138°C) for years. To prevent contamination of machine and processed synthetic fibers by escaping oil mist, closed systems are sometimes chosen.

All draw units fed by the oil-mist device are connected in parallel. Oil mist is fed to headers that run the length of a machine. Lateral pipes lead from the header to groups of draw rolls and individual tubing lines feed oil mist to each draw roll. Return lines are arranged as quasi mirror images of supply lines. Both are sloped (Figure 10-5) and interconnected by orificed bridgeover lines.

Figure 10-5 shows a closed-loop oil-mist system on a modern European-made textile machine. Oil mist is generated as described in earlier chapters. ① represents the mist

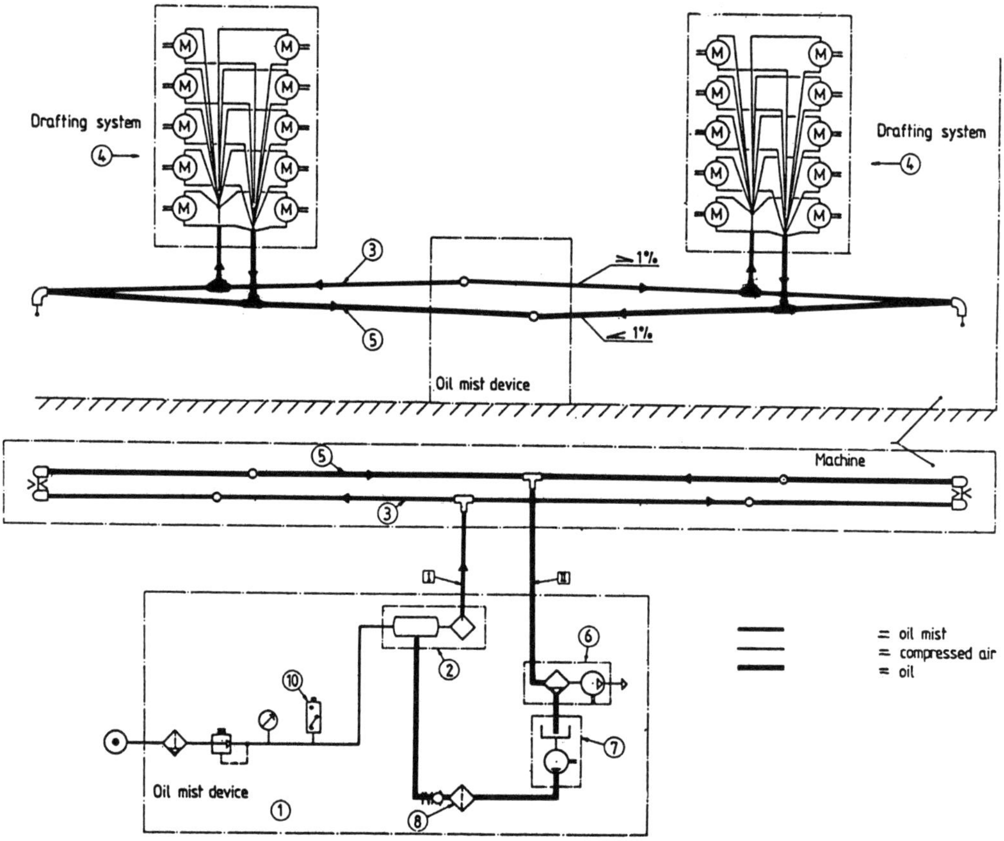

Figure 10-5. Closed-loop oil-mist system on a modern textile machine. A drafting system comprising 96 twin-bearing draw rolls is dependably lubricated by dry-sump oil mist.

console boundaries. Air and oil are combined in a mist head-reservoir assembly ②. The resulting oil-mist flows at a pressure of approximately 4 in. H_2O (1 kPa) in header pipes ③ and associated lateral branches to groups of draw rolls ④. Distribution tubing allows the oil mist to migrate towards each individual draw roll. The mist now flows through the bearings, to the return header ⑤, and into the oil separator ⑥. The oil separator operates at a slight vacuum which, of course, induces the mist not only through the bearings, but also prevents stray mist from escaping into the surroundings. A demisting vessel for the textile industry is shown in Figure 10-5a [69]. Although not shown on the diagram (Figure 10-5), the return-line vacuum can be readily created by using the same motive air, which is subsequently combined with oil in the mist head assembly. From the oil separator, the oil is piped into a small storage tank ⑦. This storage tank contains a pump which, upon being actuated by a level switch, forces returned oil through the filter ⑧ and into the mist head assembly reservoir ②.

One large textile machine equipped with 192 high-speed bearings requires only about 4 gal (15 l) of lube oil per month. The oil filter is screw-on automotive type and requires changeout only every 2 years. Typical oil-mist supply pressures are in the vicinity of 4 to 6 in. H_2O, while return lines operate at a slight vacuum of 2.5 to 3.5 in. H_2O.

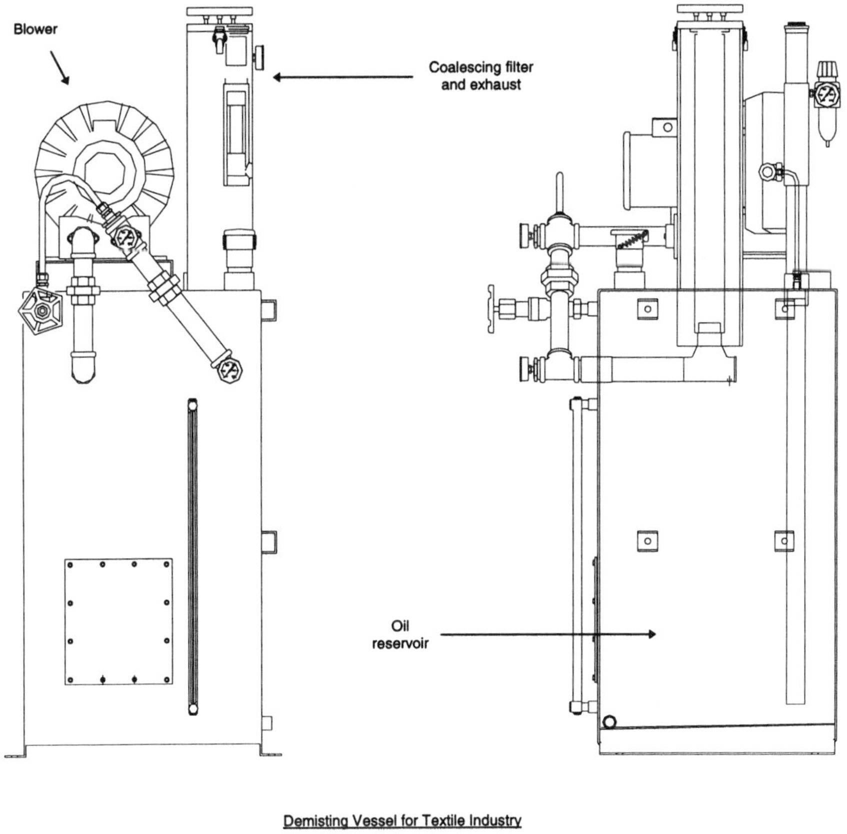

Figure 10-5a. A demisting vessel for the textile industry

Of course, any good textile machine lube system will be adequately protected by low oil level and low oil-mist header pressure instrumentation. We have omitted these only so as not to clutter the sketch.

Machine Tool Lubrication

Modern machine tolls sometimes employ a centralized "mixed lube" lubrication system that is derived from conventional "measured-shot-at-timed-intervals" systems with some of the advantages of oil-mist systems added.

The principal advantages of mixed-lube systems are the virtual freedom from stray mist emissions and the ability to process high-viscosity lubricants without heaters. Mixed-lube systems require more elaborate and expensive components than conventional continuous systems, partly due to the timing and metering instrumentation. Moreover, Reference 21 states that the air consumption of mixed-lube systems exceeds that of conventional continuous-mist systems.

One such mixed-lube system, shown in Figure 10-6, is manufactured by the Lubriquip Division of Houdaille Industries. Similar systems find extensive use in single- and multi-spindle machining heads operating at spindle peripheral speeds as high as 10,000 fpm (50.8 m • s^{-1}. The reader will recognize some similarity with the pressure-jet reclassifier illustrated in Figure 7-2. Lubriquip calls its system Spindl-Gard® and considers its nozzle-tube assembly the key component. This assembly, shown in Figure 10-7, uses principles based on fluid dynamics and the friction forces of air. When the system is in action, the dynamic forces of air moving through the nozzle-tube slowly distribute an oil film around the inner wall of the tube and move the oil toward the bearing surface. With this method it is claimed that as little as 0.005 $in.^3$ (0.082 cm^3) of oil injected into the

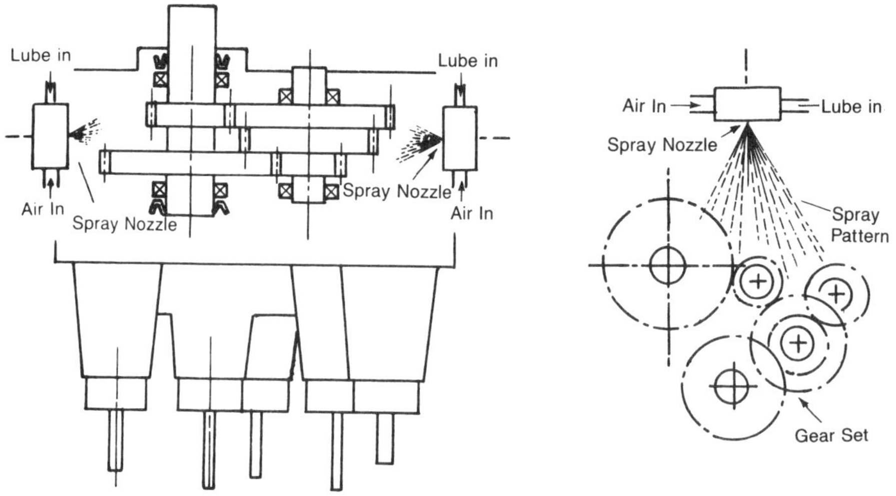

Figure 10-6. Machine tool lubrication using nozzle-tube assemblies and principles based on fluid dynamics and the friction forces of air. *(Source: Lubriquip Houdaille, Cleveland, OH.)*

mixing tee can be accurately applied to the bearing surface at the rate of 8.5×10^{-8} in.^{3} per minute (13.9×10^{-4} cm^3 • min^{-1}).

Figure 10-8 will serve as a guide to understand how such an oil mist derivative system functions and how its various components interact. The pump (1), reservoir (2), and controller (3) are mounted on a stationary part of the machine. An external air supply is required to operate the pump (275-1,030 kPa or roughly 40 to 150 psi) and to provide a supply of clean, dry air (approximately - 40° dew point, 5 μ filtration, at 200 kPa, or 30 psi) to pass through the Spindl-Gard® Monitor (5) for eventual delivery to the spray nozzle (6).

The divider valve (4) with cycle indicator pin (4A) and cycle switch (4B), as well as the valve-mounted Spindl-Gard® Monitor (5) are mounted on the movable machining head. Flexible lines bring the oil to the feeder valve inlet and the air to the inlet port at one end of the monitor.

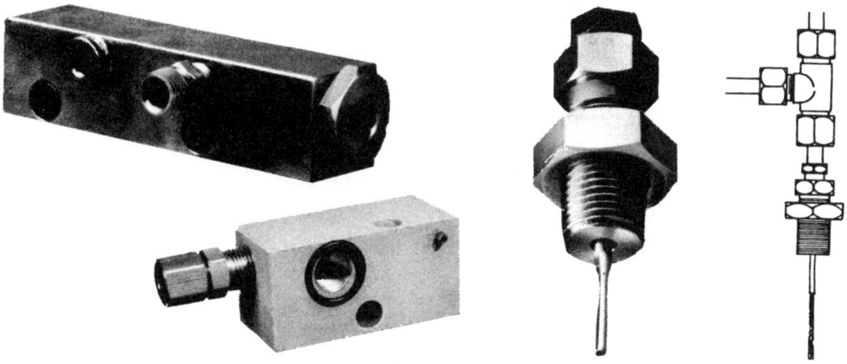

Figure 10-7. The nozzle-tube assembly is the key to Lubriquip Houdaille's machine tool lubrication system.

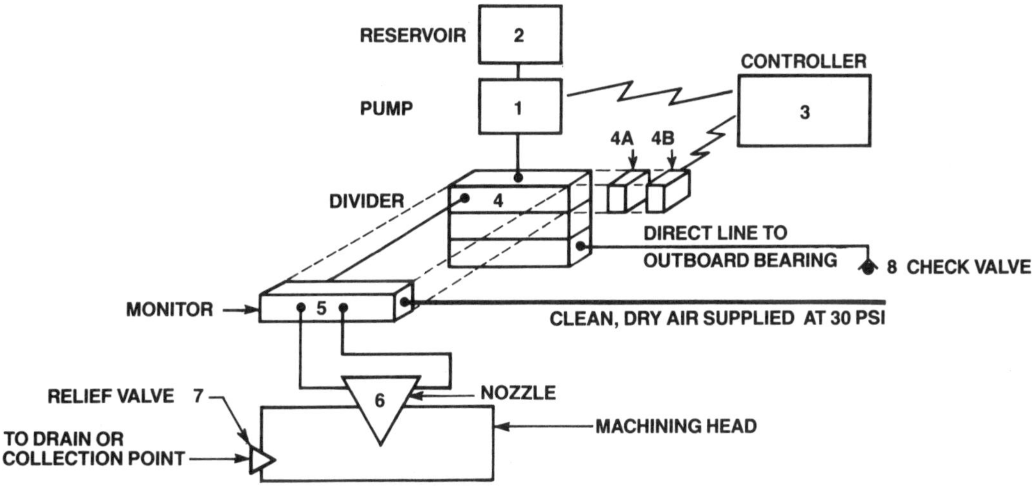

Figure 10-8. Principal components of an oil/air derivative system used for machine tool lubrication. *(Source: Lubriquip Houdaille, Cleveland, OH.)*

Oil destined for the spray nozzle is precisely metered in the top segment of the divider valve. Second and/or third segments in the valve that will cycle in sequence may be used to deliver oil directly to the outboard bearing, tool holder, etc. Any blockage in the valve itself or in any of the lube lines coming out of it will cause the cycle indicator pin to register that fault and cause the cycle switch to generate an electrical fault signal.

The main air supply is fed into the end of the monitor. It then passes through a sensing device before being channeled out the front of the monitor and into the line that serves the spray nozzle. A blocked or broken air line on either side of the monitor will cause a pressure drop or buildup that unbalances a sensing piston. This, in turn, stops the cycle indicator pin and triggers the cycle switch to signal a fault.

The lube line coming out of the monitor is connected to one of two inlet ports in the nozzle assembly, which is inserted in the machining head. The air line coming out of the monitor is connected to the other inlet port in the nozzle assembly.

The force of the air coming into the nozzle assembly at 10 to 15 psi picks up and transports minute particles of oil through the nozzle apertures in a round spray pattern to lubricate the components inside the machining head.

Figure 10-9. Multiple oil-mist units supply dry-sump oil mist to rolling mill trains. *(Source: DeLimon Fluhme & Company, Dusseldorf/Germany.)*

A relief valve (7) inserted at or near the bottom of the machining head keeps air pressure within the head from exceeding 2 psi (13.8 kPa). It also provides a simple means to drain off any oil that would otherwise collect in the sump.

Soft-seat check valves (8) should be installed in any lines coming out of the valve that will be delivering oil directly to an outboard bearing or similar component. These checks prevent contaminants and coolant from being forced back into the head or system if a backpressure condition should develop.

Rolling Mill Bearing Lubrication

Many work roll and backup roll bearings in ferrous and nonferrous rolling mills are lubricated by oil-mist systems. Some of these are closed-loop systems. Figures 10-9 and 10-10 show the manifolding that is required if the condensed oil is to be collected for re-use. Figure 10-11 depicts how oil mist is applied to massive four-row tapered roller bearings in the steel industry.

Figure 10-10. Oil-mist feed lines and manifolding at the cooling bed of a central rolling mill train in a German steel plant. *(Source: DeLimon Fluhme & Company, Dusseldorf/Germany.)*

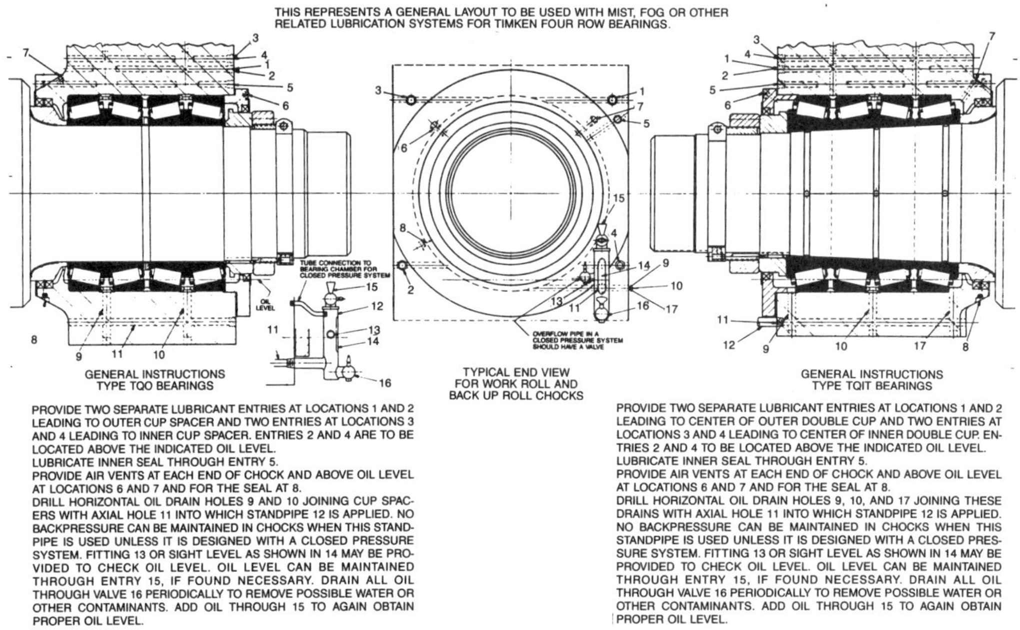

Figure 10-11. Massive four-row tapered roller bearings using oil mist in conjunction with a small wet sump in major steel plants. *(Source: Timkin Roller Bearing Company, Canton, OH.)*

Chapter 11

Sparing and Redundancy Considerations

A system specification should give the user's requirements or project philosophy on spare, or backup, units. There are four ways to have the desired redundancy:

1. The user may specify a spare console to be installed and piped adjacent and parallel to his main console. This is illustrated in Figures 11-1 and 11-2, representing a large petrochemical plant in the U.S. Gulf Coast region. The spare unit differs very little from the main console. Only the annunciator lights have been deleted from the spare unit. In all other respects, the spare and main consoles are the same.

2. Two oil-mist consoles can be installed next to each other at the common border of plot plan areas A and B (Figure 11-3). Although each of the two consoles would normally serve only its designated area, they both can be oversized by a factor of 2 to 2.5 and provided with bridgeover piping, which would allow console A alone to

Figure 11-1. Totally redundant, spared oil-mist consoles installed at a large petrochemical plant. *(Source: Lubrication Systems Company.)*

Figure 11-2. Totally redundant, spared oil mist consoles differ only by the lack of annunciator lights in one of the two units. *(Source: Lubrication Systems Company.)*

serve areas A and B, and console B alone to serve areas A and B. Figure 11-3 illustrates this option on a plot plan.

3. The oil mist console could be provided with dual, parallel-mounted, and parallel-piped mist generating heads. This so-called "piggyback arrangement" is depicted in Figures 11-4 and 11-5.

4. A portable console (Figure 11-6) can be brought in and temporarily hooked up.

Obviously, there are advantages and disadvantages to each of the four alternatives. Option 1 was used in the 1970s when a very large grassroots plant was planned and executed with hundreds of dry-sump oil-mist-lubricated pumps and electric motors. The installation was to have total redundancy and enough money had been allocated for this purpose.

Option 2 is attractive in many installations. It does, however, require favorable plot layout. Equipment to be served by either of the two consoles must be located within the maximum accepted operating range (150 m, approximately 500 ft) from the consoles. Moreover, the mist generating heads must be sized to accommodate and adjust for standard flow at perhaps 40% of emergency capacity. Regardless of flow adjustment, the proper oil-to-air ratio must be maintained. We hasten to add that compliance with these requirements is not a problem for quality mist-generating equipment from competent manufacturers such as Alemite, Lubrication Systems, Norgren, and others.

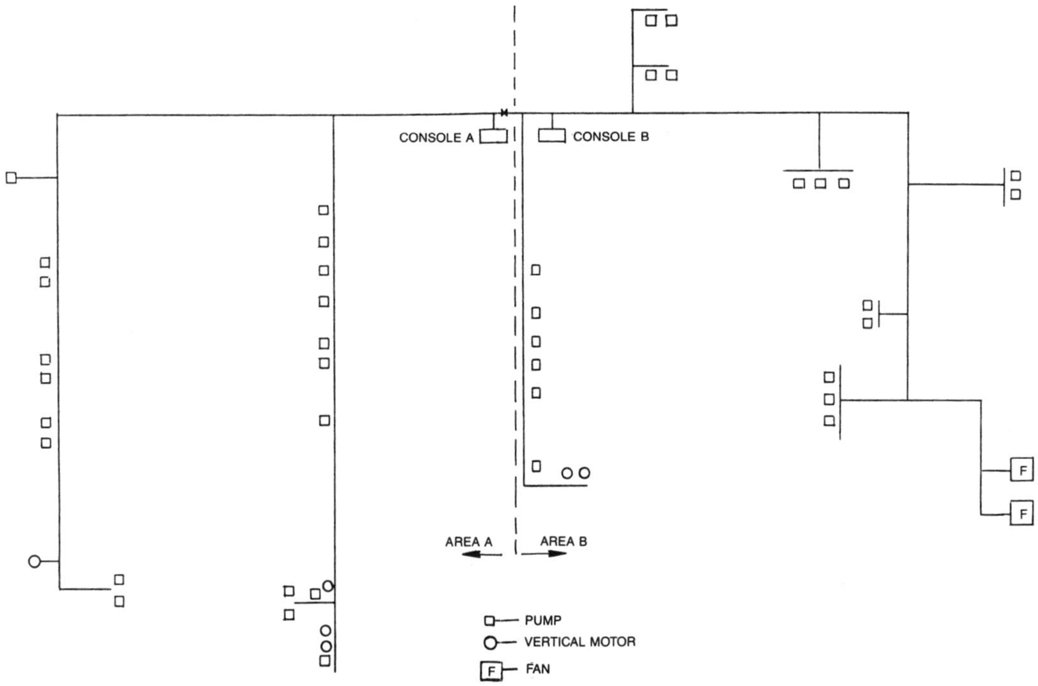

Figure 11-3. Consoles A & B sized for total coverage in case of emergency.

Figure 11-4. Oil mist console provided with dual, parallel-mounted ("piggyback") and parallel-piped oil-mist generator unit. *(Source: Lubrication Systems Company.)*

The "piggyback" execution explained as Option 3 has the advantage of lower cost over Option 1, and fewer location-related restrictions than Option 2. With oil-mist lubrication technology having proven its dependability and technological maturity, we frequently recommend Option 3 over the other two possibilities. This judgment is influenced by failure statistics that rank oil-mist system reliability at the top of the equipment list in modern petrochemical plants. In 1983, a facility incorporating 17 Option 1 systems (i.e., 34 oil-mist consoles) reported one brief downtime event in three years of operation. The malfunction occurred in a float switch and resulted in low oil level in the small oil reservoir located inside the oil-mist console. As intended, the annunciator alarm went off and operations personnel commissioned the spare console long before oil starvation could damage any of the equipment bearings associated with the system.

Option 4 may be economical if a user plant owns a well-engineered portable oil-mist console for purposes of preserving machinery in a storage yard. In an emergency, the portable console can be positioned near a defective unit and be temporarily connected to the header system.

Experience in modern petrochemical plants indicates, however, that well-engineered oil-mist consoles are extremely reliable. Moreover, failures or systems interruptions of up to 8 hours duration have repeatedly proven the "survivability" of rolling element bearings in petrochemical plants.

Few, if any, spare parts are needed by user plants in the United States that preinvest in the kind of redundancy previously outlined. The principal oil-mist systems manufacturers carry stock and can respond quickly to unforeseen emergencies. Overseas installations may have to carry a small spare parts inventory to make up for their remoteness from the source of supply.

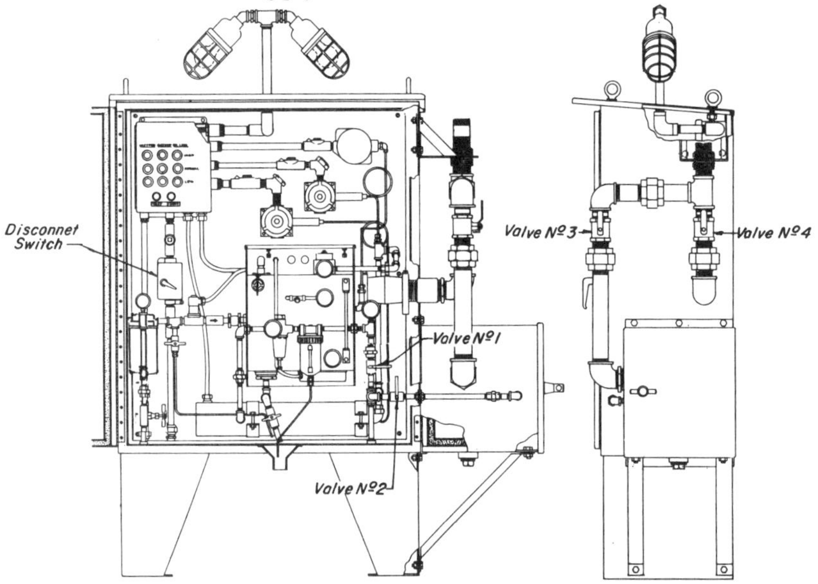

Figure 11-5. Redundancy of oil-mist generator assembly has been provided by sideways-mounting of spare unit. *(Source: Alemite Division of Stewart-Warner Corporation.)*

Figure 11-6. Portable oil-mist console can serve as a spare unit or as a supply console for oil-mist preservation of field-stored machinery. *(Source: Lubrication Systems Company.)*

Chapter 12

Specifications for Oil-mist Systems

Oil-mist systems projects, both retrofit and grass-roots type, are generally executed by first developing an inquiry specification. One such specification is shown in Appendix A.

Inquiry specifications for oil-mist systems can represent a standard format or can be tailored to meet specific, unique requirements. A standard format has the advantage of uniformity and saves time if similar projects are to be executed in the future. However, even the standard format specification must be supplemented by listings of certain specific requirements. Some of these are indicated by asterisks in the sample form represented in Appendix A.

Users and oil-mist system vendors usually cooperate in developing a listing of the equipment to be oil-mist lubricated. One such listing is represented in Figure 12-1. Also, the equipment is often shown on a plot plan (Figure 12-2). The user or contractor may also opt to include an abbreviated sketch or isometric diagram of the entire system, including the mist generator console, distribution piping, oil-mist application details, etc. A typical isometric sketch is shown in Figure 12-3.

If the user is experienced in oil-mist application, he may wish to specify reclassifier locations and sizes. Alternatively, he may request that the bidder or vendor provide sketches or other information on reclassifier location and sizes.

Backup, or spare units should be specified in harmony with the user's project philosophy. Most users will be best served by the "piggyback option," as specified by item 5.1 in Appendix A and described in Chapter 11. It is not customary to equip spare units with instrumentation fully duplicating the main unit, because it can be assumed that upon failure of the main unit, a maintenance technician or process operator will be engaged in troubleshooting efforts at the console site. This person would manually adjust and supervise the operating parameters of the spare unit, making it unnecessary to have a full complement of redundant instrumentation and controls.

Oil and air heaters may not be required in mild climates or installations that use oils with suitably low viscosities. Typical guidelines for the use of oil and/or air heaters are given in Figure 12-4.

Depending on plant size and project philosophy, oil-mist lubricant can be stored in off-site tanks with capacities exceeding a full year's requirements (Figure 3-5) or in skid-mounted tanks with a capacity of typically 275 gal (approximately 1,000 l). One such portable tank is shown in Figure 12-5.

A good specification also addresses the fabrication and installation of oil-mist piping. The user or purchaser of the oil-mist system may opt to specify the slope of pipe

Manufactured By
Lubrication Systems Co.
P. O. Box 19294
Houston, Texas 77024
713-464-6266

MIST BEARING LIST

Sheet No. 1 of 7
Estimate No. QD-526
Job No. 83-42-L

Firm PROTOCARB SYNTHETIC CO
Machine or Equipment MFG UNIT 17, CORAL CITY, TX
Builder
Minimum Ambient Temperature
Operating Temperature Range

Date Oct. 7, 1985
Reference Dwg. A-20101
Oil Type SYNESSTIC-100
Oil Viscosity ISO-100

REF. No.	No. LUBE POINTS	DESCRIPTION OF LUBE POINT: IDENTIFICATION — TYPE OF ELEMENT (NOTE 1)	SIZE (NOTE 1)	SPEED (RPM)	SERVICE (NOTE 2)	SCFM PER LUBE POINT	RECLASSIFIER FITTINGS PER LUBE POINT: TYPE	RATING: BI	RATING: SCFM	QTY.	TOTAL SCFM
1	2	S.O. PUMP TUR.		3600	PURGE		501	3	.09	2	.18
2	2	L.O. PUMP TUR.		3600	PURGE		501	3	.09	2	.18
3	2	MAIN L&S PU.		3600	PURGE		501	3	.09	2	.18
4	1	OIL TANK		—	PURGE		502	6	.18	1	.18
5	2	PUMP		1800	PURE		503	10	.30	2	.60
6	2	TURBINE		1800	PURGE		502	6	.18	2	.36
7	1	GOVERNOR		—	PURGE		501	3	.09	1	.09
8	2	PUMP		3600	PURE		503	10	.30	2	.60
9	2	TURBINE		3600	PURGE		502	6	.18	2	.36
10	1	GOVERNOR		—	PURGE		501	3	.09	1	.09
11	1	PUMP		1800	PURE		504	15	.45	1	.45
									TOTAL SYSTEM SCFM		

NOTE 1: SIZE REQUIREMENTS—SUGGESTED ABBREVIATIONS: ANTI-FRICTION BEARINGS (AF)—List bore diameter and number of rows; PLAIN BEARINGS (PL)—List bore diameter, length and diametral clearance; SLIDES—List slide length and width; ROLLER CHAIN (ROL CH)—List chain pitch and number of rows plus small sprocket PD and RPM; SILENT CHAIN (SIL CH)—List chain width plus small sprocket PD and RPM; CONVEYOR CHAIN (CNVR CH)—List chain width and length plus drive sprocket diameter.

NOTE 2: TYPES OF SERVICE: Light (L), Moderate (M) and Heavy (H).

Figure 12-1. Typical tabulation of oil-mist lubricated equipment in a process plant.

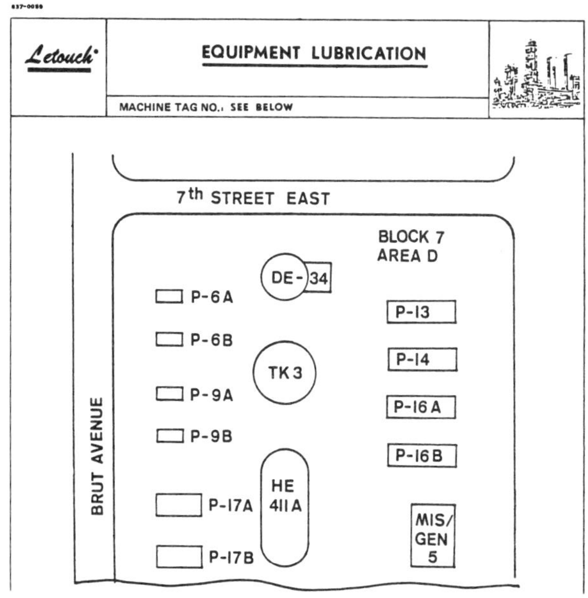

Figure 12-2. Equipment to be oil-mist lubricated can be conveniently shown on a simplified plot plan. *(Source: Reference 22.)*

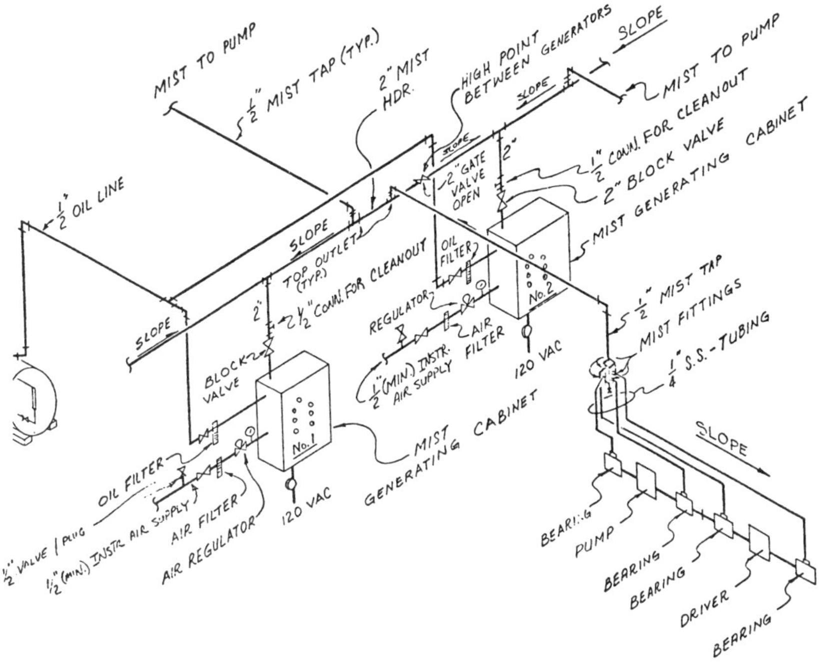

Figure 12-3. The user or contractor may opt to include in the specification package an isometric diagram. Either a part or an entire system can be illustrated in this manner.

headers and branch lines (Table 4-1). Alternatively, it may be left to an experienced system vendor or turnkey contractor to determine appropriate slopes. The vendor's proposed slope values can then be compared to accepted industry values (Table 4-1 and Figure 4-6) and significant deviations resolved between parties.

Some specifications include equipment tabulations, lubrication summaries, piping details, isometric sketches, and other data. Clearly, the degree of detail depends on the owner's expertise as well as contractor experience and reputation. There are no set rules that fit all circumstances, and the reader is advised to consider the following chapters before deciding on the degree of detail desired.

An effective owner-contractor interaction may favor the exchange of the owner's design basis memorandum for the oil-mist contractor's detailed proposal. As the term implies, the design basis memorandum would briefly state the overall requirements perceived by the user. The contractor's response would describe in detail his proposed field implementation strategy, provide equipment tabulations, lubrication summaries, pipe sizes, material selection, etc. We call this question-and-answer process an acceptability review, or completeness audit. This process allows user/purchaser and vendor/supplier to gauge their respective understanding of scope of supply, rationale for specification clauses, job execution philosophies, cleanliness and inspection requirements, and a host of other details. An acceptability review meeting would precede issuance of a final specification by the purchaser. Next, the vendor would adjust his final pricing—generally anticipated as shown in Figure 12-6—and the signing of contract documents would seal the agreement.

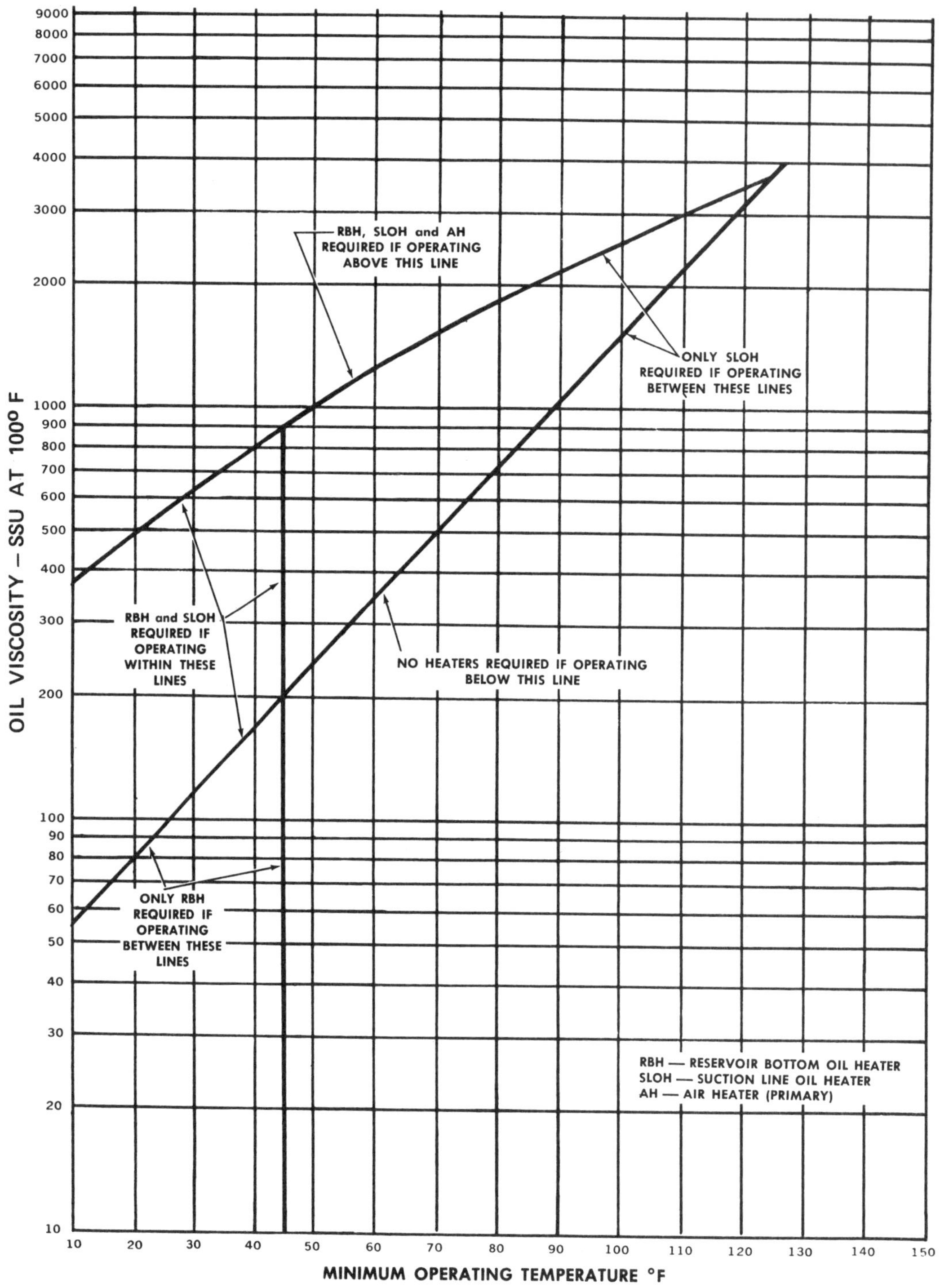

Figure 12-4. Definition of air and oil heater requirements. *(Source: Lubrication Systems Company.)*

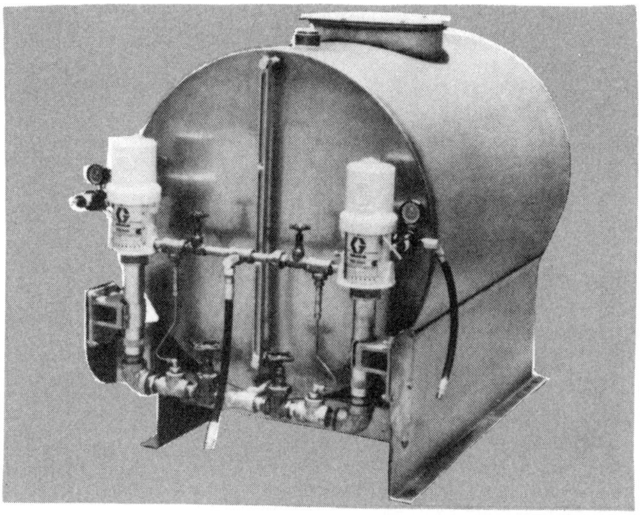

Figure 12-5. Skid-mounted ("portable") oil-mist tank with an approximate capacity of 275 gallons or about 1,000 liters. *(Source: Lubrication Systems Company.)*

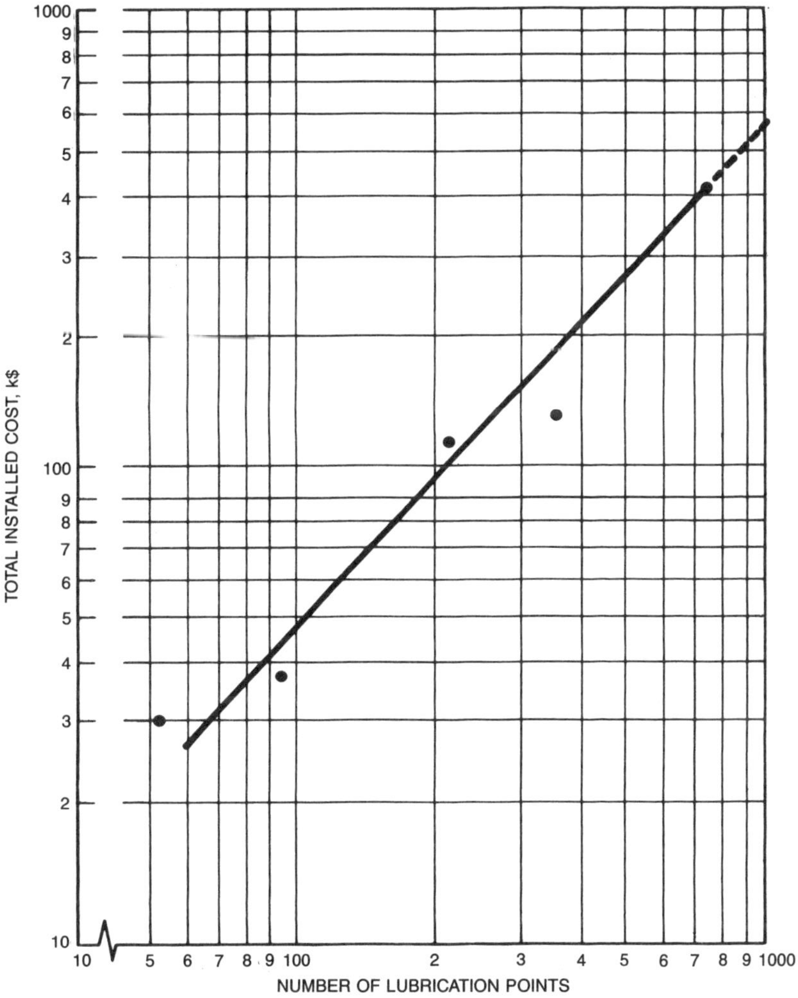

Figure 12-6. Total installed cost (1982-$) versus number of lubrication points for plant-wide oil mist systems in the USA. Anticipated 1998 cost would be twice that shown here.

Chapter 13

Field Implementation

Oil-mist lubrication projects are executed very similarly to other piping-intensive projects. The equipment to be lubricated is first identified on any convenient plot plan, as are major piping runs and oil-mist console, oil supply tank, source of motive air, etc.

Piping isometric sketches as shown in Figures 13-1 and 13-2 are prepared next. Piping isometrics define pipe diameters, lengths, slope values, support locations and whatever else might be of interest to field implementation personnel.

Both Figures 13-1 and 13-2 depict a small fraction of a large-scale refinery installation that was commissioned in 1972. It differs very little from installations implemented in 1998. More specifically, an up-to-date plant would delete the block valves shown at the

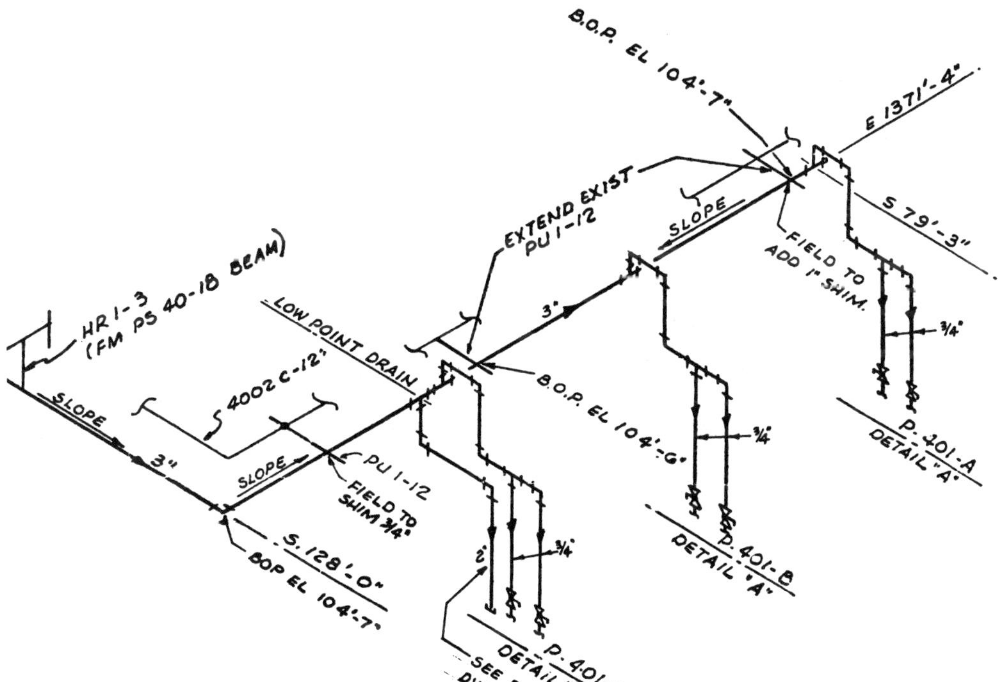

Figure 13-1. Well-planned piping isometric dating back to 1971. Newer systems commissioned since 1976 have deleted the ball valves from the end points of each 3/4-in. pipe drop. However, as of 1998 there have been no other changes.

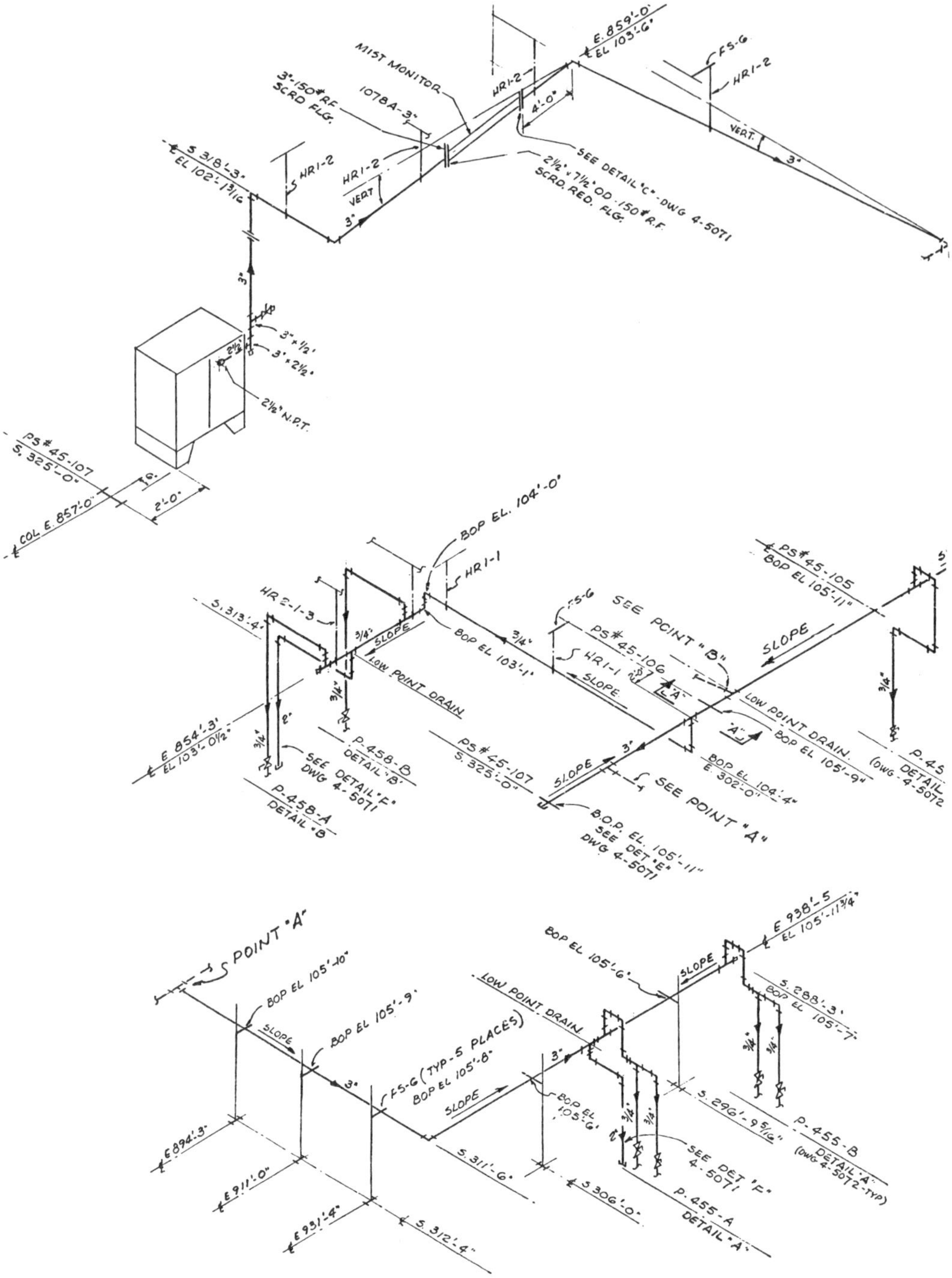

Figure 13-2. Minor portion of a large-scale refinery oil-mist lubrication project commissioned in South America in 1972.

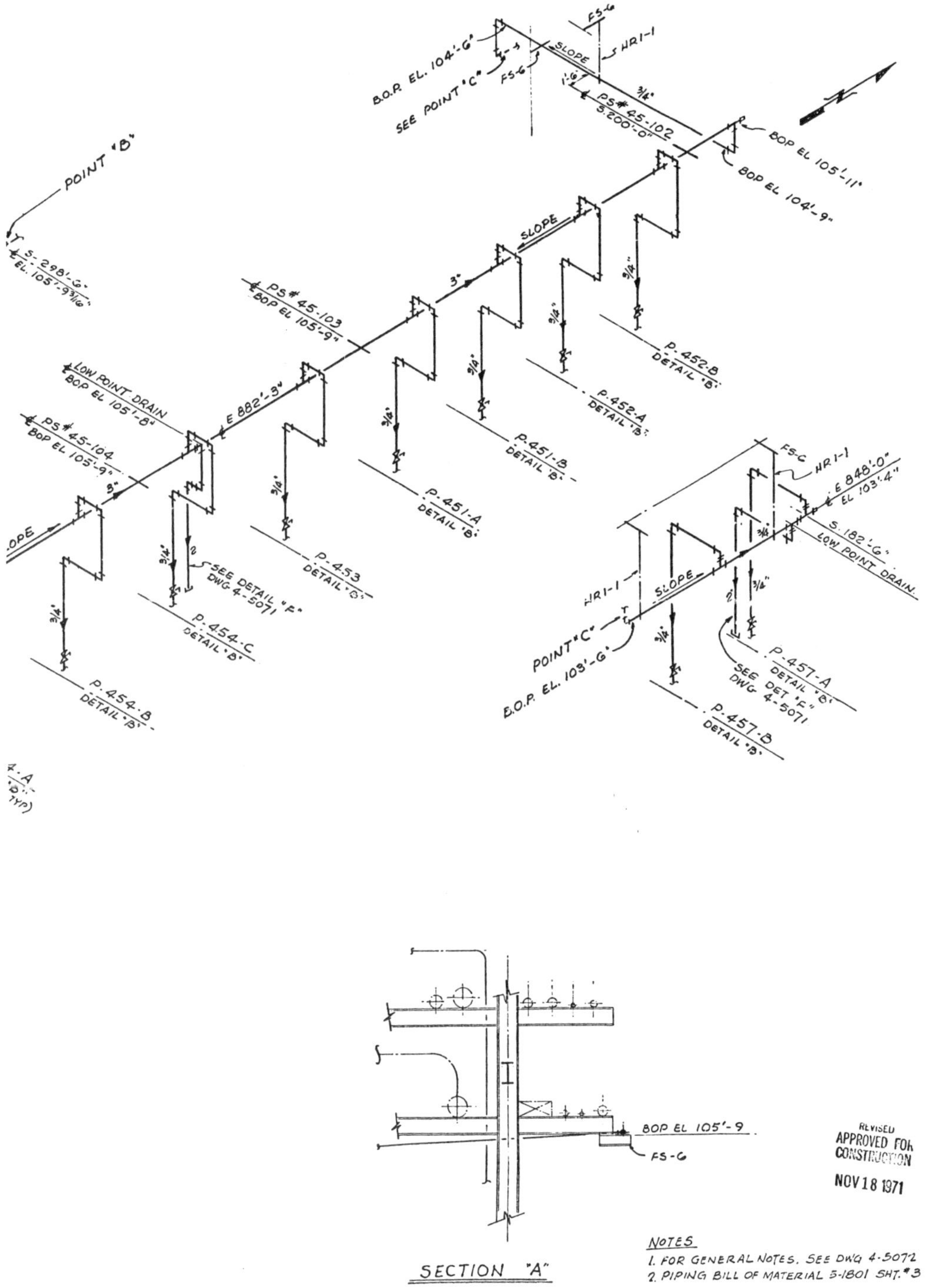

Figure 13-2. (*Continued*)

termination point of each vertical pipe run—sometimes called a "drop"—but would essentially duplicate all other details.

A typical bill of materials for a specialties unit in a chemical plant is shown in Figure 13-3. This bill of materials describes a turnkey system comprising about 40 pumps. Using the cost data given in Figure 12-6, its 1982 total cost, including labor and materials but excluding any potential change requirements on the equipment to be lubricated, would have been about $36,000, and perhaps $75,000 in 1998.

Changes to existing equipment may be necessary if conventionally lubricated equipment is to be connected to a newly installed oil-mist lubrication system. Such conversions are well within the capability of many oil-mist systems contractors. Their conversion activities should, however, be predefined by sketches or similar specification documents. Figure 13-4 and 13-5 illustrate typical guideline sketches.

Total Req.	Qty. for One	Part Number	Description
1		EXP-CBH-DABA-E	LubriMist Console in Aluminum
			Enclosure w/VBU-C-DE in 77-500-801
			Enclosure Assembly, mounted and pre-
			piped
1			120# Lubster Tank w/Pump
400'			2" Std. Galv. Pipe T&C
2			2" Galv. Tees
8			2" Galv. 90° Ells
3			2" x 3/4" Red. Bushing
38			2" x 2" x 3/4" Red. Tees
800'			3/4" Std. Galv. Pipe T&C
40			3/4" x 6" Galv. Nipples
80			3/4" Galv. 90° Ells
40		77-500-215	4 Outlet Manifold Block Assembly
30		77-800-502	Reclassifiers
10		77-800-503	Reclassifiers
2		77-800-504	Reclassifiers
42		77-500-752	1/4"P x 1/4"T S.S. Fittings
231'			1/4" x .035 T-304 S.S. Welded Tubing
118		U-119-AC	1/8" Plugs
120'			2" x 2" x 1/4" Galv. Angle Iron
45			3/8" x 2" U-Bolts
84			3/8" x 3/4" U-Bolts
1		F-628-2	Air Combo F.R.L.
1		203-266	
1		205-418	
1			1/2" x 2" Galv. Nipple
1			1/2" Galv. Tee
2'			1/4" S.S. Tubing
1		SS-5-TA-1-8	Adapter
1		SS-4CA-150	Check Valve
1		77-500-990	Cuno Oil Filter Assembly
1		77-500-992	Cuno Air Filter Assembly
60'			1/2" Std. Galv. Pipe T&C
2			1/2" Galv. Tees
6			1/2" Galv. 90° Ells
2		77-500-490	1/2" Ball Valve
10			3/8" x 1/2" U-Bolts
			Pipe and Pipe Fittings

Type of Lubricant

Oil — Heavy Light

Grease — NLGI #

Name:

Figure 13-3. Typical bill of materials for oil-mist systems. *(Source: Lubrication Systems Company.)*

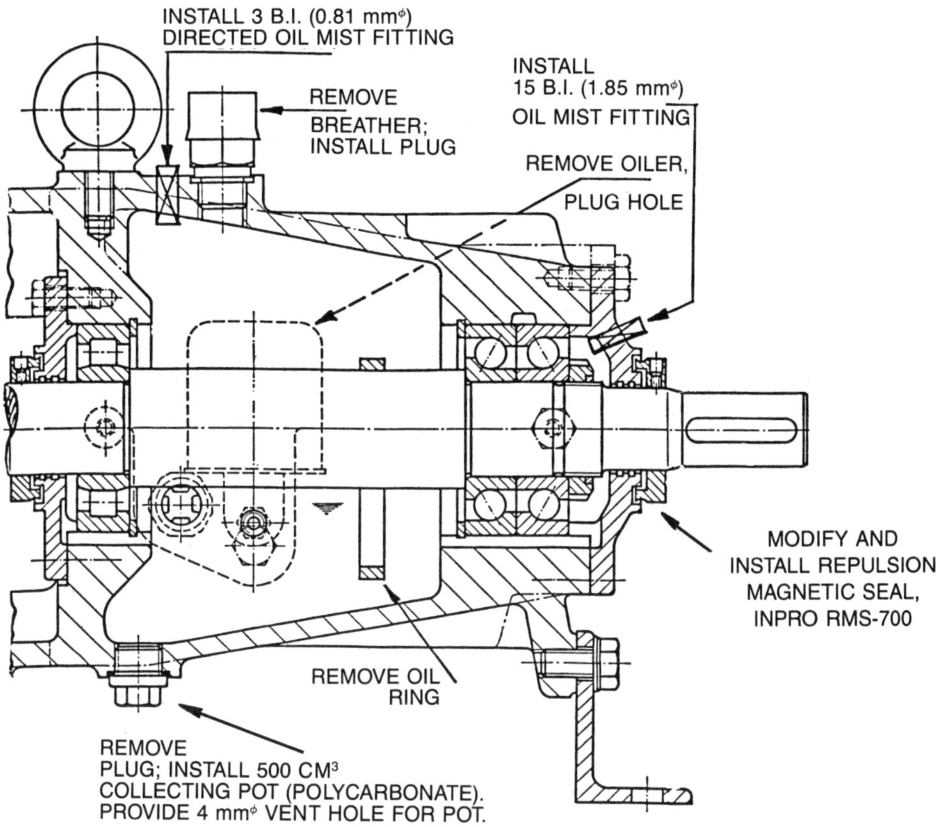

Figure 13-4. Conversion instructions for heavy-duty centrifugal pump. *(Source: Sulzer-Weise, Bruchsal, Germany, Adapted)*

EQUIPMENT TABULATIONS AND LUBRICATION SUMMARIES

Equipment tabulations and lubrication summaries are the most important ingredients of the systems planning activity. There are four essential steps that lead to good equipment tabulations and lubrication summaries:

1. Determine mist requirements in B.I., L.U., or scfm units for each bearing or machine element to be lubricated.

2. Select the size and type of reclassifier fitting for each bearing—whether mist, spray, or condensing.

3. Determine vent size and method of venting. Alternatively, verify that existing vent path is acceptable.

4. Determine total mist requirement.

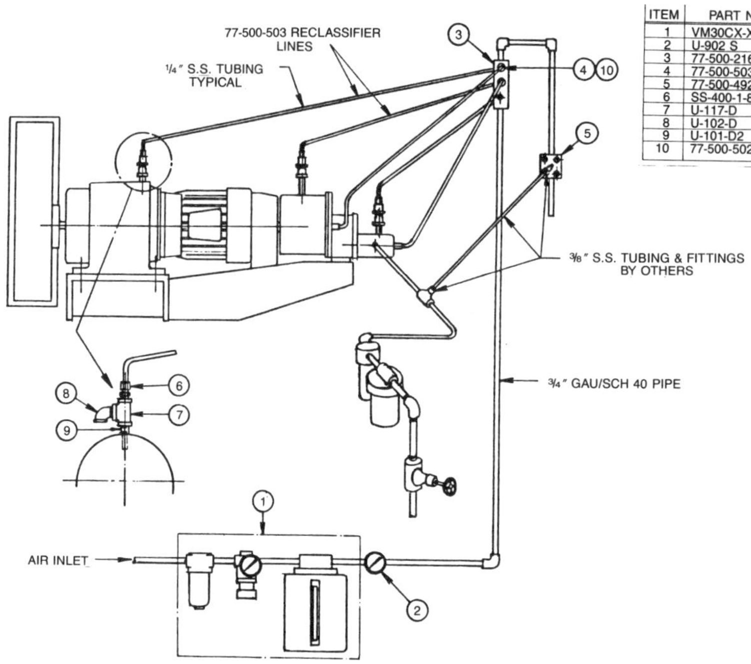

ITEM	PART NO.	DESCRIPTION
1	VM30CX-X-BXA	MIST GENERATOR
2	U-902 S	AIR GAUGE
3	77-500-216	MIST DISTRIBUTION MANIFOLD
4	77-500-503	RECLASSIFIER–2 REQ'D
5	77-500-492	AIR OPERATED VALVE 2-WAY NORM CL
6	SS-400-1-8BT	¼"T × ½"P S.S. MALE CONN.
7	U-117-D	½" STD GALV TEE
8	U-102-D	½" GALV STR ELL
9	U-101-D2	½" × CL GALV PIPE NIPPLE
10	77-500-502	RECLASSIFIER–3 REQ'D

Figure 13-5. The turnkey oil-mist contractor should identify equipment conversions on sketches or other specification documents. This was done for this air preheater drive, which is wet-sump lubricated. Note wet-sump vent execution, items 7 and 8. *(Source: Lubrication Systems Company.)*

In addition to Figure 12-1, Figures 13-6 through 13-8 represent suitable formats for these tabulations and summaries. Having read the preceding chapters, the reader will be well equipped to proceed with the tabulation task or to determine the acceptability and accuracy of tabulations submitted by others.

Once the total mist requirement has been firmed up, the systems planner can select an appropriately sized mist lubricator unit from vendor curves of the type shown in Figures 13-9 and 13-10. When selecting from curves such as shown in Figure 13-10, pick a unit that can produce the desired output with the air bypass valve in the closed position (see heavy black curve.) Note air pressure supply pressure in psi. This is also the regulator gauge pressure. Next, determine plant air supply pressure required by adding the pressure loss through air accessories to the regulator gauge pressure. Let us assume, for example, that a machine requires 3.5 scfm of mist for lubrication. A smaller unit could be used but a 300-B.I. unit provides more margin. Plant pressure should be at least 28 psi (18 psi for regulator setting plus 10 psi for losses in air accessories). Air bypass curves should be referred to when large quantities of air are needed for cooling purposes.

Figure 13-6. Oil-mist system worksheet with special column for calculations. *(Source: C.A. Norgren Company.)*

WORKING SHEET
For Use With Norgren Application Design Manual NT-1

FIRM ______________________ DATE ______________________

ADDRESS ______________________ REFERENCE DWG ______________________

MACHINE ______________________ LUBRICANT USED ______________________

SYSTEM REQUIRED ______________________ SKETCH ATTACHED YES ________ NO ________

AMBIENT TEMP. ________ °F ________

FIG REF. NO. *	IDENTIFICATION OF LUBRICATION POINT	QUAN.	SIZE (IN.) †	TYPE & MANF	SPEED (RPM)	CALCULATIONS	RECLASSIFIER		BEARING INCHES	RECLASSIFIER TYPE ‡
							RATING ×	QUANTITY =		
1	Ball Bearing (no preload)	4	2.187 Shaft D.	SKF	2-1850 2-1160	$(2.187 \times 1) = 2.187$ B.I.	4	4	16	Four 18-009-008
2	Gear Pair	1	2″ Face Width 6″ P.D. & 9″ P.D.	SPUR	1750	$\frac{(6 + 9) \times 2}{4} = 7.5$ B. I.	8	1	8	One 18-009-034
3	Roller Chain	1	.75″ Pitch Single Row 8″ Sprocket Diam.		1160	$\frac{.75 \times 8 \times 1 \sqrt{\left(\frac{1160}{100}\right)^3}}{8} = 29.6$	16	2	32	Two 18-009-032
4	Plain Bearings Load = 102 lbs./in.²	2	4″ Shaft D. 4″ Len.		350	$\frac{4 \times 4 \times 2}{8} = 4$	4	2	8	Two 18-009-001
5	Horizontal Slide	1	5″ Length 3″ Projected Width			$\frac{5 \times 3}{20} = 0.75$	1	2	2	Two 18-009-033
6	Plain Bearing Load as for #4	1	1.5″ Shaft D. 1″ Length			$\frac{1.5 \times 1 \times 2}{8} = 0.38$	1	1	1	One 18-009-033
7	Plain Bearing Load as for #4	1	2″ Shaft 1.5″ Length			$\frac{2 \times 1.5 \times 2}{8} = 0.75$	1	1	1	One 18-009-033
							TOTAL BEARING INCHES		68	

* Reference number should be used on sketch or drawing for clarification of points being lubricated.
† Dimensions Required:

PUMP OIL MIST LUBRICATION SUMMARY

OIL MIST CONSOLE DESIGNATION: MIS/GEN 5-101 LOCATION: PLASTICS UNIT

NOTE: USE DIRECTED OIL MIST IF BEARING BORE VELOCITY EXCEEDS 2000 FPM

ITEM	MFR. AND MODEL	RPM	HP	V>2000FPM (SEE NOTE)	ROLLING ELEMENT BRG CHARACTERISTICS: INBOARD ID	INBOARD LOAD FACTOR	INBOARD ROWS	INBOARD BRG INCHES	OUTBOARD ID	OUTBOARD LOAD FACTOR	OUTBOARD ROWS	OUTBOARD BRG INCHES	APPLICATION FITTINGS: INSTALLATION LOCATION AND DIAMETER, INCHES: IN-BOARD	OUT-BOARD	COMMON TO INB & OUTB	OIL MIST FLOW TOTAL TO ITEM (SCFM)
P-6A	GOULDS (OH)	3560	55		1.7	1	1	2	1.7	1	2	4			.060	1.52
PM-6A	A-C-365 TS	3560	75		1.9	1	1	2	1.9	1	1	2	.032	.032		1.62
P-6B	GOULDS (OH)	3560	55		1.7	1	1	2	1.7	1	2	4			.060	1.52
PM-6B	A-C-365TS	3560	75		1.9	1	1	2	1.9	1	1	2	.032	.032		1.62
P-9A	BINGHAM (OH)	1760	15		1.8	1	1	2	1.8	1	2	4			.047	1.19
PM-9A	A-C-284T	1760	25		1.7	1	1	2	1.7	1	1	2	.032	.032		1.62
P-9B	BINGHAM (OH)	1760	15		1.8	1	1	2	1.8	1	2	4			.060	1.52
PTU-9B	ELLIOTT YSD	1760	25		2.9	2	1	6	2.3	1	1	3	.047	.036		2.10
P-17A	PACIFIC (BB)	3570	2600		3.3	2	2	14	3.3	1	1	4	.076	.040		3.05

Figure 13-7. Oil-mist lube tabulation used by experienced petrochemical company.

Pipe Sizing and Configurations

Lines can be made from black pipe, plain steel tubing, or galvanized pipe. The inside of black pipe should be free of scale and oiled to prevent corrosion during storage.

Lines should typically conform to Table 13-1. If lines with smaller bores or greater lengths are used, there will be some risk of uneven lubrication. Lube points near the mist lubricator will be over-lubricated while those farther away will be under-lubricated. Also, the greater pressures needed to force the mist through the lines could increase turbulence and result in oil condensation. It should be noted, however, that the guidelines given in Table 13-1 reflect only the recommendations of one manufacturer. Others apply different criteria, which generally results in the same sizes being used.

The distances given in Table 13-1 are intended to be from the oil-mist generator to the farthest lube point and include both the main supply header plus the branch line to the farthest lube point. For example, if a 1-in. pipe has been selected as the main supply line, the combined run of main supply line and branch line should be limited to 140 ft for a system with a pressure of 20 in. H_2O. However, these criteria are known to be quite conservative. Properly installed oil-mist systems in the U.S. Gulf Coast region often oper-

OIL-MIST SYSTEM WORK SHEET								
ITEM NO.	QUAN.	TYPE OF ELEMENT	SIZE & SPEED	CFM PER ELEMENT	APPL. FITTING TYPE & SIZE	FITTING CFM @ DMP	METHOD OF VENTING	TOTAL CFM
1	6	2 ROW SPHERICAL ROLLER BEARING #22222C	4-7/8" SHAFT DIA. 430 RPM	0.242	(1) -8 SPRAY	0.300	OUTBOARD LABYRINTH	1.800
2	4	2 ROW TAPERED ROLLER BEARING	3-1/16" SHAFT DIA. 660 RPM	0.153	(1) -5 SPRAY	0.159	VENT FITTING	0.636
3	6	1 ROW CYLINDRICAL ROLLER BEARING #A1316TS	3.1496 SHAFT DIA. 660 RPM	0.079	(1) -3 SPRAY	0.078	DIRECT SPRAY NO SEALS	0.468
4	1	SPUR GEARS	4x7 P.D. PINION 1130 RPM 4x12 P.D. DRIVEN GEAR	0.475	(3) -5 SPRAY	0.159	SEMI-ENCLOSED	0.477
5	2	PLAIN BEARING	5 DIA. X 7 LONG 400 RPM	0.350	(2) -4 COND.	0.206	GROOVE BRG. SLEEVE	0.824
6	2	1 ROW DRIVE ROLLER CHAIN	3/4" PITCH 8" DRIVE SPROCKET 660 RPM	0.317	(2) -5 SPRAY	0.159	SEMI-ENCLOSED	0.636
							TOTAL SYSTEM CFM:	4.851

Figure 13-8. Oil-mist system worksheet. *(Source: Alemite Division of Stewart-Warner Corporation)*

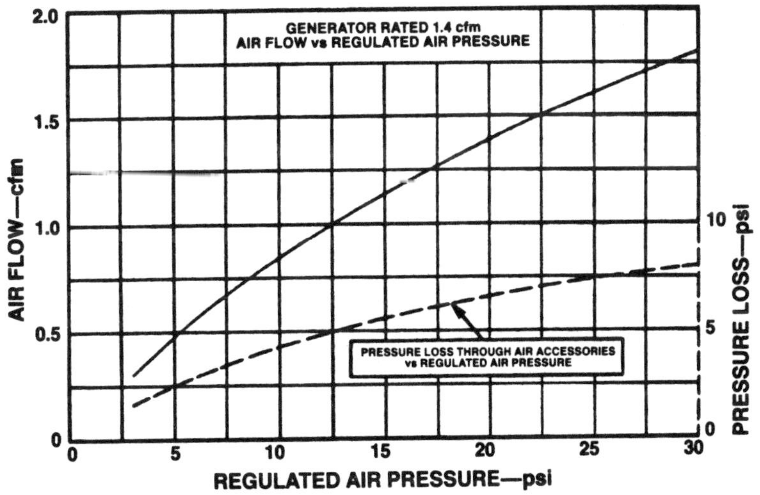

Figure 13-9. Performance curve (air flow versus regulated air pressure) for small oil-mist generator. *(Source: Alemite Division of Stewart-Warner Corporation.)*

ate with up to 500 ft. (>150 m) separating oil-mist console and final point of application.

The oil-mist distribution lines or branches connecting to main headers must be installed so that condensation of oil in the lines is minimized. This can be achieved by keeping the inside of lines free of projections and by avoiding sharp changes in direction. Figures 13-11 and 13-12 show good versus bad executions.

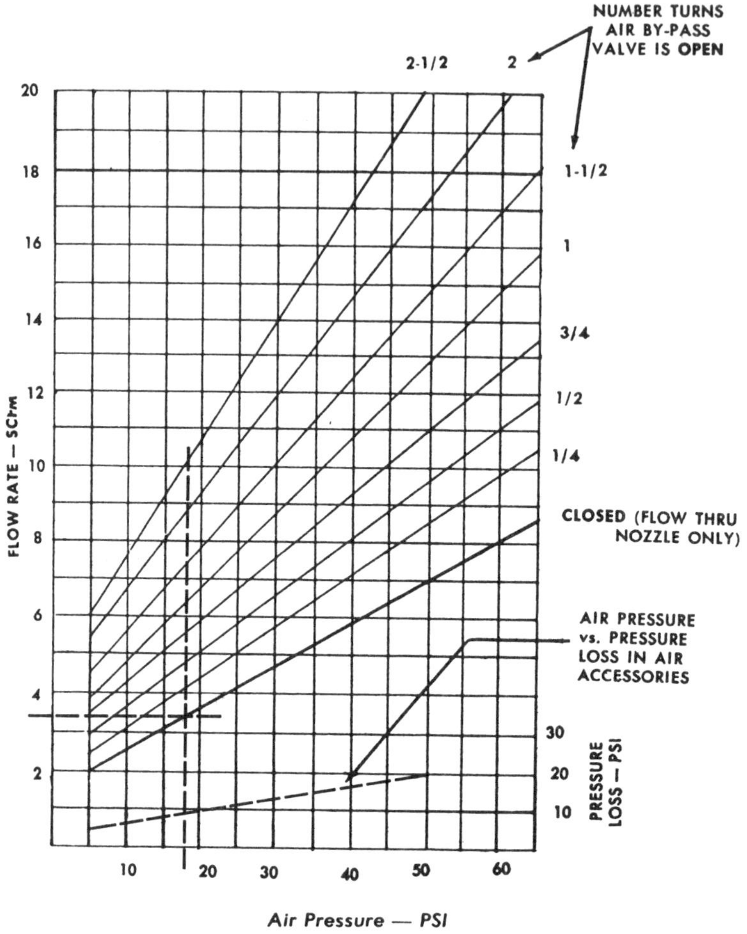

Figure 13-10. Flow rate versus regulated air pressure for a 300-B.I. oil-mist generator. *(Source: Lubrication Systems Company.)*

Sloping guidelines were described earlier in Table 4-1 and Figure 4-6. Another way of restating these guidelines would be to slope the main line toward the mist control unit for a distance equal to 300 times line I.D. Most oil condensation occurs in this length. The sloping criteria must be observed because mist flows in the direction opposite of the oil and slows the oil's movement. Other lengths, in which oil mist and condensed oil move in the same direction, are not critical in this regard. On horizontal runs without slope it is necessary to install drains at intervals of about 300 times line I.D. The drain can lead to a bearing or container. Avoid low spots or traps in which oil can collect, increasing line pressure drop and thus decreasing mist flow. Even shallow traps that do not appreciably affect pressure can be troublesome. They can cause locally increased flow velocity and the turbulence that can result causes increased wetting-out of oil, reducing the downstream oil/air ratio.

Branch or feeder lines should connect to the top of headers or main feeder lines to

Table 13-1. Pipe Sizing, Mist Flow and Distance Criteria for Oil-Mist Systems

Nominal Size (in.)	Bore Diameter (in.)	Bore Area (in.^{2})	Max. Flow Rate-scfm	Max. Distance to Farthest Lube Point (feet) 20-in. Manifold Pressure	35-in. Manifold Pressure
Tube					
3/16	.1235	.012	.12	20	30
1/4	.186	0.27	.27	20	30
3/8	.311	.076	.76	40	50
1/2	.402	.127	1.26	45	65
5/8	.465	.170	1.70	50	75
Pipe					
1/8	.269	.049	.5	27	45
1/4	.364	.104	1.0	36	60
3/8	.493	.191	1.9	60	80
1/2	.622	.304	3.0	80	100
3/4	.824	.534	5.3	105	135
1	1.049	.866	8.7	140	170
1-1/4	1.380	1.497	15.0	160	225
1-1/2	1.610	2.036	20.0	205	265
2	2.067	3.365	34.0	245	300
2-1/2	2.469	4.792	48.0	270	300

reduce the chances of transporting solid contaminants, scale, etc. to application fittings where they might plug small orifices. If branch lines slope toward bearings, add a drain to the end of the line. Pipe tees with reducing fittings can be used to connect feeder lines.

Piping installations should disallow the use of thread sealing compound or Teflon® tape. These sealing means are not necessary for oil-mist systems operating at low pressures and could easily lead to contamination of distribution lines and application fittings if improperly applied.

A 0-to-100-in. water-column gauge should be installed at the end of at least one branch line to check mist pressure.

It should be noted that pressure deviations of ±10% from the manufacturer's standard recommended header pressure are generally allowed. Supervisory instrumentation should activate an alarm only when header pressures are approximately 20% away from the desired standard. Calibration accuracy should typically be verified in one-year intervals.

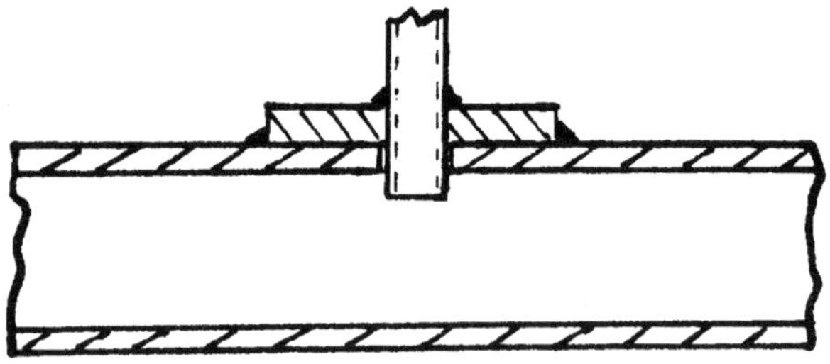

Figure 13-11. A protruding branch line connection does not permit optimum oil-mist performance.

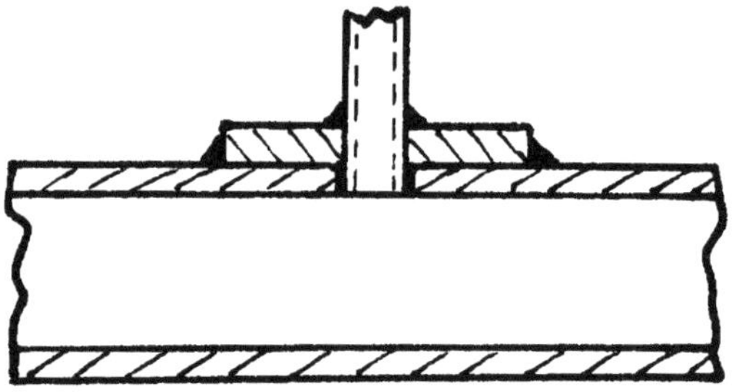

Figure 13-12. Smooth, nonprotruding branch take-offs promote satisfactory oil-mist flow.

Finally, the system must be flushed clean before connecting feeder lines to reclassifiers, which because of their small bore size (0.016 to 0.084 in. diameter) could easily become plugged with dirt. Pipe may be flushed with solvent or else blown with steam and dried with nitrogen or instrument air. See also Appendix A for additional cleaning recommendations.

Chapter 14

Shipping and Storing Oil-mist-lubricated Equipment

Machinery equipped with rolling element bearings and later expected to operated on dry-sump oil-mist lubrication must be shipped and stored with bearings properly preserved. Although it has sometimes been suggested to simply envelop the entire bearing with grease, this has not proved to be as risk-free a procedure as one might assume. Light greases do not provide long-term acceptable corrosion protection in wet and tropical environments, and heavy greases would have to be carefully removed from the bearing components prior to operation on oil mist.

One of the best means of preparing pump, fan, and electric motor bearings for shipping and short-term storage is to clean the bearings in a chlorinated solvent before assembling them on the shaft. A petroleum-base preservative such as "Product C" (see Table 14-1 for more details) should then be hand-sprayed into the bearings as they are being rotated. After the preservative has dried, a premium quality electric motor bearing grease is injected into only the space between two adjacent balls and the shaft rotated so as to distribute the grease charge more evenly.

Although this amount of grease is only about 10% of the volume of grease normally used in a typical bearing, it has proved perfectly adequate to allow three five-hour heat runs of electric motors without overheating or damage.

If oil mist is connected to equipment that has been preserved and lubricated as just indicated, the time period between initial application of oil mist and wetting out of a sufficient coating on the rolling elements will be bridged by the small amount of grease. Using only a small amount of grease will ensure that the mist actually flows through the bearing. Excessive amounts of grease risk plugging and are certainly detrimental to long-term satisfactory operation of equipment.

Coating with "Product C" allows extended storage of bearing systems in wet climates. This was verified by testing in humidity chambers and extended field experimentation in the U.S. Gulf Coast area. In each test or experiment it was shown that bearing temperature rise and shock pulse vibration (incipient defect) readings were entirely as expected for motor operation under ideal conditions.

Preserving Equipment with Oil Mist

Inadequate machinery preservation during pre-erection storage or long-term deactivation (mothballing) is very likely to increase machinery "mortality" at start-up. Many

Table 14-1. Characteristics of Conventional Storage Preservatives

	Storage Preservation			
Storage condition and/or severity	Outdoor storage, general exposure to elements	Indoor storage under severe conditions, or outdoor storage (partial shelter) under moderate conditions, or outdoor storage with exposure to elements for short term only	Indoor storage under moderate conditions	Outdoor storage with exposure to elements under the severe conditions
	A	**B**	**C**	**D**
Product and typical characteristics	Firm coating, resistant to abrasion	Soft coating (self-healing)	Thin oily film	Asphaltic film, needs removal before part is used
Density				
kg/m³ at 15.6°C	868.5	923.7	876.9	922.5
lb/gal at 60°F	7.25	7.71	7.32	7.70
Viscosity,				
cSt at 40°C	—	—	14	149
cSt at 100°C	24.8	33.1	3.3	—
SSU at 100°F	—	—	79	800
SSU at 210°F	123	162	37.4	—
Flash Point, °C	279	260	166	38 } Volatiles
°F	535	500	330	100 } Volatiles
Melting or pour point, °C	73	66	-4	—
°F	164	151	+25	—
Unworked penetration at 25°C (77°F)	75	245	—	—
Film thickness, mil	1.6	1.6	0.9	3.0
Approximate coverage				
m²/liter	26	26	44	11
sq ft/gal	1000	1000	1800	450
Non-volatiles, %	99	99	—	55
Methods of application/ temperature, °C	dip/85, brush, swab/60-71	dip/77 swab/18-27	roller coat, brush, mist	spray, dip or brush/ambient
Maximum time until inspection and possible reapplication under condition				
Mild	Extended	Extended	6-12 Months	Extended
Moderate	1-3 Years	1-3 Years	1-6 Months	1-3 Years
Severe	6-12 Months	6-12 Months	Not recommended	6-12 Months

times, machinery arrives at the plant site long before it is ready to be installed at its permanent location. Unless the equipment is properly preserved, scheduled start-up dates may be jeopardized, or the risk of failure is increased.

Long-term storage preservation by nitrogen or oil mist purging is often applied by industry. Generally, this method of excluding moisture is used indoors for small components, such as hydraulic governors, but also for large components, such as gears (Figure 14-1), or even turbomachinery rotors kept in metal containers as illustrated in Figure 14-2. The purge medium consumption rate is governed by its rate of outward leakage and may be kept at a low, highly economical rate if the container is tightly sealed. The container needs to be pressurized to only about 10 in. H_2O (2.5 kPa), although a more typical rate is approximately 1 to perhaps 5 psig (7 kPa to 35 kPa). It may be fitted with a safety relief valve to prevent overpressures. Alternatively, the container could be furnished with an orificed vent to promote through-flow of nitrogen or oil mist at very low pressure. This is called preservative gas sweep.

Even more often we may be interested in economically preserving field-stored and field-installed inactive machinery. Here, the application of an oil-mist purge as shown in Figures 14-1 through 14-14 is particularly useful [15]. As was brought out earlier, this highly advantageous preservation method consists of a centralized system which utilizes the energy of compressed air to supply a continuous feed of atomized rust preventive or lubricating oil to multiple points through a low-pressure distribution system at approximately 5 kPa or 20 in. H_2O. The volumetric ratio of air to lube oil is roughly 200,000:1. After leaving the header system, oil mist passes through a small diameter nozzle before

Figure 14-1. Gear units at this storehouse are preserved with oil mist. *(Source: Alemite Division of Stewart-Warner Corporation.)*

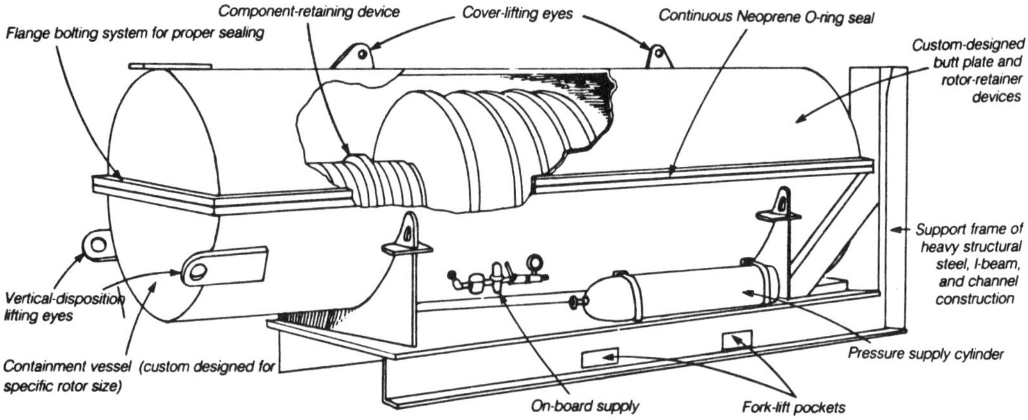

Figure 14-2. Details of rotor storage container. Rust protection and long-term preservation by low-pressure oil mist is safe and economical. *(Source: Hickham Industries, La Porte, TX.)*

Figure 14-3. Outdoor yard for temporary storage of machinery in a wet climate. *(Source: Reference 15 and Lubrication Systems Company.)*

entering the cavity to be preserved. This nozzle, or application fitting, meters the oil-mist stream so that the cavity or housing is pressurized to a level just above atmospheric pressure. Figures 2-1 through 2-3 earlier illustrated the oil-mist generation principle in schematic form.

When used as a storage preservation medium, oil mist can be made to enter the cavity (e.g. bearing housing, seal housing, trip valve mechanism, coupling enclosure, machinery casing, etc.) at any convenient location other than the bottom drain. The mist will typically exist at a pressure of approximately 0.5 to 1 in. H_2O (125-250 Pa) and will mi-

Figure 14-4. If there is a cavity, it is being protected by oil mist. *(Source: Reference 15 and Lubrication Systems Company.)*

Figure 14-5. Plastic tubing slopes downward from the distribution block in this temporary outdoor storage facility. *(Source: Reference 15 and Lubrication Systems Company.)*

Figure 14-6. A temporary shelter and hundreds of small bore plastic tubing lines provide oil-mist preservation and protection to mixers, pumps, motors, turbines, valves, governors, etc. *(Source: Reference 15 and Lubrication Systems Company.)*

Figure 14-7. Large oil-mist-preserved outdoor storage yard. *(Source: Reference 15.)*

grate toward the surrounding atmosphere. In essence, it performs two equally important functions: it prevents the ingress of atmospheric air that might contain moisture and airborne contaminants, and it coats the machinery components with a corrosion-inhibiting premium lubricant.

Storage preservation using oil-mist methods is used outdoors and under protective shelters. Figures 14-3 through 14-8 show machinery storage yards in petrochemical plants in the U.S. Gulf Coast area.

Figure 14-8. Major machinery preserved by oil mist. Note cabinet and lube oil storage tank in foreground. *(Source: Reference 15.)*

Figure 14-9. Combined reservoir-control module for oil-mist preservation at a mothballed ethylene plant. *(Source: Phillips Petroleum Company, Sweeny, TX and Lubrication Systems Company.)*

Figure 14-10. Combined modules such as this can be used to preserve a single, large machine. *(Source: Alemite Division of Stewart-Warner Corporation.)*

Oil-mist storage preservation systems are bare-bones oil-mist lubrication systems. Virtually no electric controls or supervisory instrumentation are needed since a temporary outage would be of no serious consequence. Typical combined reservoir-control modules are illustrated in Figures 14-9 and 14-10. Although both of these combination modules have generally been used in outdoor storage yards, they have frequently been used to preserve or mothball major turbomachinery trains that have been decommissioned indefinitely. If major machinery is to be dependably preserved and yet kept ready to be returned to service without undue loss of time, oil-mist preservation merits serious consideration.

Figures 14-11 through 14-13 illustrate major machinery in an ethylene plant that is being preserved in this manner; oil-mist headers run the length of the platform. Lateral pipes branch out from the header and distribution blocks form the pipe terminus. Stainless steel or plastic instrument tubing connects the distribution block with the small oil-mist application fitting at the point to be preserved.

While application fittings for operating equipment must be sized to provide sufficient lube oil to satisfy a given bearing size or configuration, the sizing of application fittings for preservation systems can safely be left to cursory estimate. As a matter of practical experience, even sizable bearing housings, governors, valve mechanisms, etc., are usually served by fittings with a bore diameter of 0.047 in. (1.2 mm), and only on the largest casings of steam turbines, gears, etc., would larger fittings with a bore diameter of 0.060 in. (1.5 mm) be used. At a header pressure of 20 in. (approximately 5 kPa), the smaller application fitting delivers 0.18 cfm (0.3 m^3/hr) and the larger one 0.30 cfm (0.5 m^3/hr) of oil mist to the cavity to be preserved.

Figure 14-11. Machinery being preserved on a compressor platform in an ethylene plant. *(Source: Phillips Petroleum Company, Sweeny, TX.)*

Figure 14-12. Vertically oriented header, distribution block and stainless steel tubing for long-term mothballing of major turbomachinery. *(Source: Phillips Petroleum Company, Sweeny, TX.)*

Figure 14-13. Bearings and couplings on compression equipment "mothballed" and preserved by oil mist. *(Source: Phillips Petroleum Company, Sweeny, TX.)*

The properties of oil-mist preservatives need be no different from those of oil-mist lubricants. However, if premium-grade oil-mist lubricants are not available, a naphthenic-base premium-grade turbine lubricant will prove entirely satisfactory as a preservative oil. Since naphtha-basestock reduces the probability of wax plugging of small application fittings, it is important that these lubricants be specified instead of oils containing a paraffinic base.

Also, diester-base synthetic lubricants can be used in oil-mist preservation systems. These lubricants are especially suited for low-temperature storage conditions because they do not form wax and have extremely low pour points.

Determining Oil and Air Consumption

Figure 14-14 can be used to conservatively determine the rate of oil and air consumption. A system comprising 100 large or 167 small application fittings would consume 30 cfm (50 m^3/hr) of oil mist. And, since a suitable volume ratio of air vs. oil would be 200,000:1, this particular installation would consume approximately 250 cm^3 or 8.5 fluid ounces of lube oil per hour. In one year, the system would use about 580 gal (2,200 liters) of lube oil.

Based on 1998 cost data in the United States, we would expect to pay approximately $3 for one gal (3.8 l) of lube oil and $0.40 per 1,000 ft^3 ($0.014 per m^3) of air. Motive air consumed would thus be estimated at $6,600 per year. This figure would include equipment maintenance, power, depreciation, etc. Just for comparison: The average cost for repairing a single pump damaged by corrosion has been estimated at $6,000.

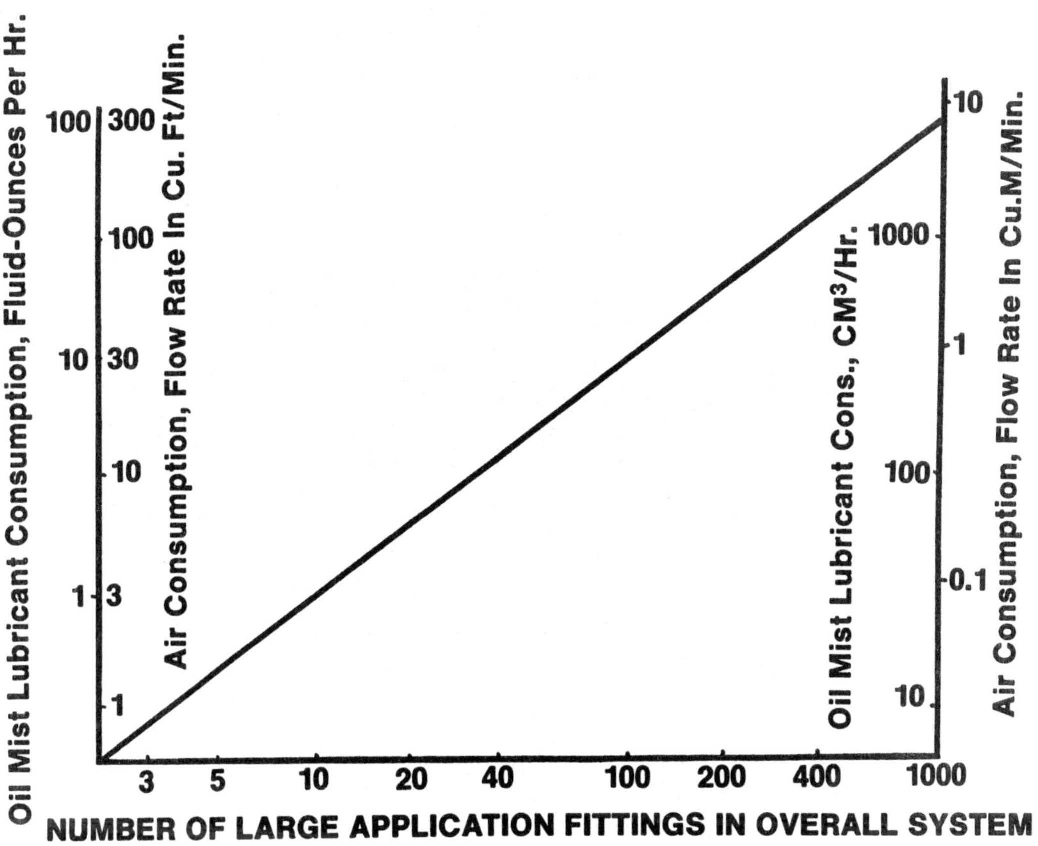

Figure 14-14. Oil-mist lubricant and air consumption of oil-mist preservation systems. *(Source: Referenced 15.)*

Cost of Oil-mist Preservation

The cost of implementing and operating an oil-mist preservation system must be weighed against the probable cost of having to repair or recondition unprotected machinery. Here are some figures.

- A 24-point oil-mist preservation system incorporating the generator-reservoir module shown in Figure 14-9 was installed on an integral gas engine-reciprocating compressor for $3,500. Application points included six power cylinders, six compressor cylinders, the crankcase, and the distance piece housings.

- The storage yard preservation system shown in Figure 14-6 included approximately 400 ft (125 m) of header pipe and plastic tubing runs to over 400 points of application. It was installed and commissioned for not quite $10,000, exclusive of the air compressor.

- A special skid package (Figure 11-6) was assembled for a South American refining complex. This totally self-contained system cost $21,000 and was initially used for storage protection. It will also serve as an emergency backup unit for plant-wide oil-mist lubrication systems or could be used for long term preservation of installed but temporarily deactivated machinery trains.

We can certainly conclude that oil-mist preservation of machinery is an essential task for oil-mist systems. With the escalating cost of downtime and machinery repairs, proper storage preservation and machinery mothballing techniques are becoming more important. In the vast majority of cases, a conscientiously executed preservation program will pay for itself very rapidly.

Chapter 15

Economic Justification for Dry-sump Oil-mist Lubrication

Economic comparisons of competing lubrication methods rely primarily on failure statistics. Only to a lesser degree are the comparisons influenced by differences in lubricant consumption, utility needs, and manpower requirements.

Bearing performance data have been carefully collected at Shell Oil Company over a three-year-period [6]. Results are shown in Table 15-1 for units that were new when started on oil mist and for units that were converted to oil mist after first running on conventional lubrication. Percent bearing failures per year is a relative value and is defined as the average number of bearings failed per year of operation divided by the total number of pumps. Bearing failure results for wet sump applications in new units at Shell were inconclusive because of the limited number of pumps that use wet sump. However, when results from the conversion units are compared with results from the new units, it can be seen that considerably more bearing failures occur with wet sump than with dry sump. These limited results also show that a 90% reduction in bearing failures is not unusual when converting from conventional lubrication to dry-sump oil-mist lubrication. Other observers have reported similar results for petrochemical plants in the United Stated [12]. The plot in Figure 15-1a shows what happened to pump repair costs attributable to bearing failures after oil mist was installed on a US refinery crude unit in early 1990. After the mist system was fully commissioned and brought on-line, these repair costs dropped by over 90%. The plot in Figure 15-1b shows annual pump bearing repair costs for three process units in a US refinery. The data represents the average annual repair costs for the two-year periods immediately before and after oil mist was installed on these units. An overall 65% reduction in costs was measured [69].

Lubricant consumption in plants with dry-sump oil-mist lubrication has been estimated as much as 40% below that for equipment having only liquid oil-sump lubrication. A very conservative estimate would assume the consumption figures to be about equal. However, the cost of premium-grade oil-mist lubricants and especially the cost of superior dibasic ester synthetics will exceed that of conventional oils and must be considered in any cost comparison.

Some cost comparisons are factoring in the value of compressed air. However, we have found that the incremental compressor power input requirements are close to the power saved due to reduced friction of oil mist lubricated vs. conventionally lubricated antifriction bearings. Therefore, the differences may be nearly offset.

Dry-sump oil-mist lubrication clearly requires less manpower than conventional lubrication for routine servicing and surveillance. Although dry-sump oil-mist lubrication may free maintenance workers for other tasks, it can often be justified for general purpose machinery in the petrochemical industry on maintenance credits alone. Plants that subject pump bearings to periodic preventative maintenance replacement may indeed justify oil-mist lubrication on the basis of discontinuing preventive maintenance altogether.

Table 15-1. Performance of Oil-Mist Systems

New Units					
Unit	Pumps on Oil Mist	% on Dry Sump	Years Experience	% Bearing Failures per Year: Dry Sump	% Bearing Failures per Year: Wet Sump
A	85	94	1-1/2	2.5	0
B	45	87	2-1/2	7.2	13.3
C	31	87	3	8.6	33.3
D	17	65	2-1/2	7.3	20.0
E	13	85	2	4.5	0
			Weighted Average =	5.3	16.8
Conversion Units					
Unit	Pumps on Oil Mist	% on Dry Sump	Bearing Failures per Year: Before	Bearing Failures per Year: After	Percent Reduction
F	200	98	31	1	97
G	70	30	7	<1	>86
H	58	0	No Data	>5	—

In petrochemical plants in the United States Gulf Coast area, the main incentive for dry-sump oil-mist lubrication of rolling element bearings is the reduction in equipment failures. This failure reduction and the accompanying repair cost avoidance can be attributed to the use of clean, fresh lubricants in a once-through fashion. Also, this lubricant is applied with considerably greater reliability and uniformity than could be expected from manual application by maintenance or operation personnel. Most importantly, equipment bearings are surrounded by a slightly pressurized preservative at all times. This virtually rules out the ingress of airborne contaminants such as dust and water vapor, thus reducing oxidation and contamination risks by orders of magnitude. Conservative assumptions would expect reductions in pump bearing failures of 80%, electric motor outage events of 90%, failure reductions on blowers and fans of 30%, and about 75% fewer failures of cooling tower fan gears to which an oil-mist purge has been added.

Here are some typical examples from a small petrochemical process unit:

EXAMPLE 1 - SMALL UNIT

Basis:

35 pumps oil misted
1 turbine oil mist purged
22 motors on these pumps oil misted

$9,000/repair on pumps (cost data for early 1998).

$2,400/repair on motors (based on typical costs for bearing replacement only. Five motors averaging 100 hp).

Pump Conversion Incentives (35 pumps)

Mean time between shop repairs: .... 3.0 operating years/pump

Number of pumps failing/year: .... 35/3 = 11.7

Shop failures due to bearing failures
(conventional lubrication); failures/year: .... 35% = 4.1

Reduction of failures due to
conversion to dry-sump oil-mist: .... 65% = 2.7/year

Maintenance cost credits: .... ($9,000)(2.7/year) = $24,300/year

Estimated production loss reduction credits: .... None, spared equipment

Estimated Fire Loss Reduction Credits:

Low Flash Point Pumping Services

48 incidents over 15 year period out of total pump population of 6,720. Total dollar losses—$16,838,000. Loss per pump per year: (16,838,000)/(15)(6720) = $167.00.

High Flash Point Pumping Services

26 incidents over 15 year period out of total pump population of 6,380. Total dollar losses—$1,872,000. Loss per pump per year: (1,872,000)/(15)(6380) = $19.60

Average fire loss per pump per year:
(167.00 + 19.60)/2 = $93.30

Outages due to bearing failures—35%
(.35)($93.30) = $32.66

Oil mist will reduce this cost by reducing bearing failures 80%, hence (0.80)($32.66/pump year) = $26.2/pump year

Basis 1998 dollars (35 pumping units)($26.2/pump year) = $917.00/yr.

[Assume appropriate inflation to the years beyond 1998: $(917)(x)^n$, where x = (1 + inflation rate), and n = number of years beyond 1998]

Motor Conversion Incentives (22 motors)

Motor outages per year due to bearing failures with conventional lubrication (10% failures per year due to bearing failures):	2.2
Reduction of motor outages due to conversion to dry sump oil mist:	80% = 1.76/year
Cost credits:	($2,400)(1.76) = $4,224/year

Small Steam Turbines Conversion Incentives (1 turbine)

Outages before conversion to wet sump purge (28%)	0.28/year
Outages after conversion to wet sump purge (10%)	0.1/year
Cost credits:	(0.18)($11,600) = $2088

Reduced Manpower Incentives

Approximately 10% of one man's time required on all three shifts to check lubrication and add oil to 35 pumping units as required without oil mist:

(0.10)($258,000/yr cost to man one job 24 hours)	$25,800

Total Oil-Mist Credits

Total credits (including production losses)	$58,192/year

Oil-Mist System Cost

Total cost	$101,000

Payout Period

$$\text{Payout} \frac{\$101,000}{\$58,192} = 1.73 \text{ years}$$

Another series of example calculations will illustrate how systems would be justified at a major plant that is to be constructed "grass-roots" [12].

Example 2 - Large Plant

Pump Conversion Incentives (800 pumps)

Mean time between shop repairs:	1.5 operating years
Number of pumps failing:	= 267/year
Shop repairs due to bearing failures: (conventional lubrication)	35% = 94/year
Credit due to fewer shop repairs after conversion to dry-sump oil-mist:	(.8)(94) = 75/year
Maintenance cost credits:	($8,000)(75) = $600,000/year
Estimated production loss reduction credits	= $180,000/year
Estimated fire loss reduction credits*	= $230,000/year
	= $1,010,000/year

Motor Conversion Incentives (760 motors)

Motor outages due to bearing failures with conventional lubrication	= 152/year
Credit due to fewer motor outages after conversion to dry-sump oil-mist:	(.9)(152) = 137/year
Cost credits:	($1800)(137) = $246,600/year

* Statistics at this plant showed one pump-induced fire event per 1,000 pump failures

Miscellaneous Blowers, Fans, Etc. (352 items)

Approximate number of repair incidents:	87/year
Incidents attributed to bearing failures:	32% of 87 = 28/year
Estimated number of bearing failures after conversion to appropriate oil-mist lube method:	7/year
Estimated cost credits:	(21)($20,150) = $423,150/year
Estimated production loss reduction credits:	= $200,000/year
Total	= $623,150/year

Cooling Tower Fan Gear Conversion Incentives (24 gears)

Gear outages before conversion to wet-sump *purge*:	4/year
Gear outages after conversion to wet-sump *purge*	1/year
Cost credits:	(3)($22,970) = $68,910/year
Motor outages before conversion to dry-sump oil mist:	5/year
Motor outages after conversion to dry-sump oil mist:	1/year
Cost credits:	(4)($2,000) = $8,000/year
Drive shaft repairs (pillow block) before conversion:	6/year
Drive shaft repairs needed after conversion to dry-sump oil mist:	1/year
Cost credits:	(5)($3,700) = $18,500/year
Total savings	= $95,410/year

Small Steam Turbine Conversion Incentives (40 turbines)

Outages before conversion to wet-sump *purge*:	11/year
Outages after conversion to wet-sump *purge*:	4/year
Cost credits:	(7)($11,230) = $78,610/year

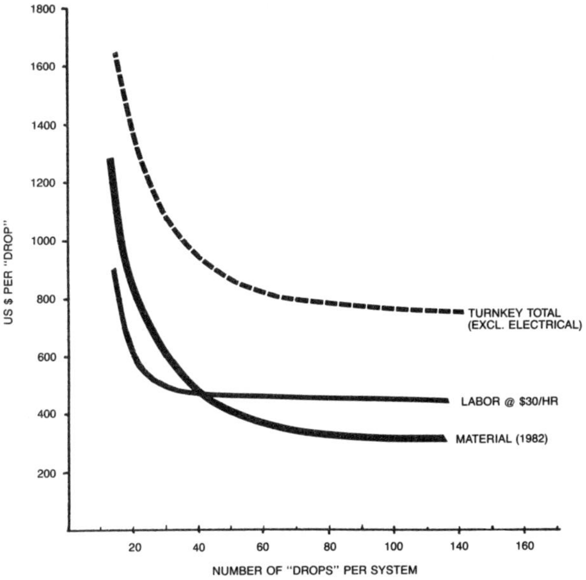

Figure 15-1. Budget costs for entire turnkey oil-mist systems can be estimated from this graph. (For a definition of "drop" refer to Figure 13-1 where each of the three pumps is served by two pipe drops.)

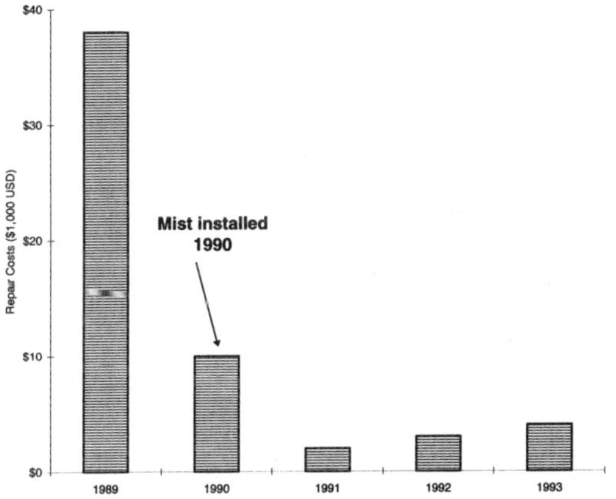

Figure 15-1a. Reduction of pump repair cost attributable to bearing failures for a US refinery crude oil unit.

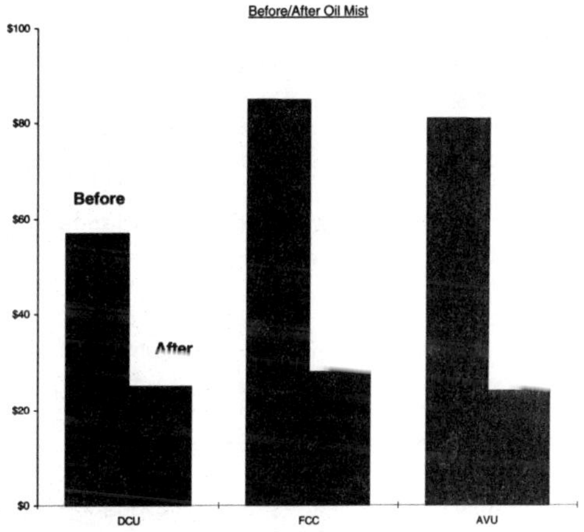

Figure 15-1b. Annual pump bearing repair cost for three process units in a US refinery.

ESTIMATED TURNKEY COST OF OIL-MIST LUBRICATION SYSTEMS

The estimated cost of applying approximately 4,000 oil-mist lubrication points to a plant with 2,000 items of rotating equipment is $3,300,000. This very rough estimate is based on an extrapolation of the curve represented in Figure 12-6, whose data points represent actual installations datiang back to the 1980's.

Based on these values, we anticipate the payout period for the demonstration plant just described to be:

$$\frac{\$3{,}300{,}000/\text{year}}{\$2{,}053{,}770} = 1.6 \text{ years}$$

A similar cost estimate could be made on a unit-by-unit basis if we knew or assumed the number of drops comprising a unit. Generally, one pipe drop will be found to serve 1.5 to 2.5 lubrication points. Figure 15-1 approximates the installed turnkey cost of oil-mist systems as a function of the number of pipe drops in 1980. The graphs provided in Figure 15-1 further divide total cost into labor and materials components.

The following worksheet may be used to estimate cost and savings with oil mist lubrication system (Source: Lubrication Systems Company).

PUMP CONVERSION INCENTIVE

• Annual Maintenance Cost Savings:				
— Pump population	1.	40	1.	
— Annual bearing failures due to lubrication (10% of line 1)	2.	4	2.	
— Reduction with LubriMist	3.	90%[1]	3.	
— Total cost of a pump/bearing repair	4.	$4,500[2]	4.	
— Maintenance cost savings Line 2 × Line 3 × Line 4	5.	$16,200	5.	$
— Number of pumps in hot service	6.	5	6.	
— Cooling water cost reduction ($300 per pump per year [3]) Line 6 × $300	7.	$1,500	7.	$
— Eliminate cooling water line maintenance ($200 per line per year [3]) Line 6 × $200	8.	$1,000	8.	$

MOTOR CONVERSION INCENTIVE

• Annual Maintenance Cost Savings:				
— Motor population	9.	45	9.	
— Annual bearing failures due to lubrication (10% of line 9)	10.	4.5	10.	
— Reduction with LubriMist	11.	90%[1]	11.	
— Total cost of motor/bearing repair	12.	$1,200[2]	12.	
— Maintenance cost savings Line 10 × Line 11 × Line 12	13.	$4,860	13.	$

STEAM TURBINE CONVERSION INCENTIVE

• Annual Maintenance Cost Savings:				
— Turbine population	14.	1	14.	
— Outtages per year due to lubrication (once every 4 years)	15.	0.25	15.	
— Reduction with LubriMist	16.	60%[4]	16.	
— Total cost of a turbine/bearing repair	17.	$5,800[4]	17.	
— Maintenance cost savings Line 15 × Line 16 × Line 17	18.	$870	18.	$

	EXAMPLE	ACTUAL UNIT
MANPOWER CREDITS		
• Elimination of oil sump changes:		
— Changes per pump per year	19. 2	19.
— Man hours per change	20. 1	20.
— Number of sumps	21. 42	21.
— Annual manpower credits + wage rate at $18/hour + benefits at 30% ($18/hr plus 30%) × Line 19 × Line 20 × Line 21	22. $1,965	22. $
• Eliminate routine oil sump filling:		
— Minutes per shift	23. 15	23.
— Annual manpower credits Line 23 ÷ 60 × 3 shifts × 365 × ($18/hr plus 30%)	24. $6,405	24. $
OTHER COST INCENTIVE AREAS		
• Annual reduction in lube oil consumption:		
— Normal pump consumption of lube oil	25. 18.5 gal./yr.	25.
— Oil consumption reduction with LubriMist	26. 40%[5]	26.
— Lube oil savings Line 1 × Line 25 × Line 26 × $4/gal)	27. $1,184	27. $
• Annual fire loss reduction credits:		
— Pumps in low flash point service	28. 10	28.
— Fire loss reduction credit per pump [6]	29. $25	29.
— Fire loss credit Line 28 × Line 29	30. $250	30. $
• Annual energy savings:		
— Power loss in pump reduced by 0.34 kw[7]		
— Cost of electrical energy	31. 5¢/kwh	31.
— Per cent time pumps operate	32. 50%	32.
— Energy credits Line 1 × 0.34 × Line 31 × Line 32 × 24 × 365	33. $2,978	33. $
• Production outtage: — Add as appropriate	34. TBD	34. $
SYSTEM ANNUAL OPERATING COSTS		
• Utilities -air:		
— SCFM of air per drop	35. 0.36	35.
— Number of drops	36. 46	36.
— Cost per 1000 SCFM of air	37. 0.32 cents	37.
— Total cost of air Line 35 × Line 36 × Line 37 × 60 × 24 × 365 ÷ 1000	38. $28	38. $
• Utilities -electricity:		
— "EXP" console require 1.4 kw		
— Total cost of electricity Line 31 × 1.4 × 24 × 365	39. $613	39. $

	EXAMPLE	ACTUAL UNIT
• Routine maintenance at recommended six month intervals:		
— Parts: oil and air filters, reservoir gasket and bulbs		
Yearly cost	40. $50	40. $
— Labor: 4 man hours per inspection		
4 × ($18 + 30%) × 2	41. $187	41. $

NET ANNUAL CREDITS

	EXAMPLE	ACTUAL UNIT
• Total all red boxes	42. $37,212	42. $
• Total all blue boxes	43. $878	43. $
• Net annual credits		
Line 42 minus Line 43	44. $36,334	44. $

LUBRIMIST SYSTEM TURN KEY INSTALLATION COST

	EXAMPLE	ACTUAL UNIT
• Average cost per slab, up to six points per slab	45. $1,000	45.
• Number of separate slabs		
— pump/motor pairs	46. 40	46.
— separate motors	47. 5	47.
— turbines/other	48. 1	48.
• Installation cost		
Line 45× (Line 46 × 47 × 48)	49. $46,000	49. $

PROFITABILITY MEASURES

	EXAMPLE	ACTUAL UNIT
• Payback period:		
— Installation cost (Line 49) divided by net annual credits (Line 44)	50. 1.27 years	50. years

• Pre-tax rate of return:
- — Based on ten-year life
- — Tax effects not included
- — Use following table to estimate:

Payback period-years	0.50	0.54	0.62	0.77	0.83	1.00	1.10	1.23	1.40	1.67	1.96	2.41	3.09	3.57	4.19
Rate of return %	200	180	160	140	120	100	90	80	70	60	50	40	30	25	20

Part 2: Scientific Principles

Chapter 16

Overview of Scientific Principles

Although oil-mist lubrication has been widely used for decades, the underlying scientific principles were not studied until the late 1980's and early 1990's. Part 2 of this comprehensive text deals extensively with this new knowledge.

As was explained in Part 1 of our text, the typical oil-mist system is exceedingly simple, effective, and reliable. The oil-mist generator is an "air blast" atomizer operating on either a vortex or a venturi principle. As dry compressed air passes through the vortex or venturi at very high speed, oil is introduced into the flow from a reservoir by siphoning action, and the high-speed air blasts the oil into oil mist (approximately 1 part oil per 170,000 to 240,000 parts air by volume). The oil mist coming out of the generator contains oil droplets or particles averaging less than three microns in diameter. The oil mist produced by the mist generator can be conveyed up to 500 ft (about 150 m) through pipes. These pipes are also known as main-headers and oil mist moves in these pipes at 20 to 35 inches (508 to 889 mm) of water column pressure. As oil mist flows through the pipes, the flow turbulence causes the larger particles to be deposited on the pipe wall. The rate of particle deposition increases with increasing turbulence induced by higher velocities. As oil mist flows through pipes, oil droplets also settle on the pipe due to gravity. Small pipes come off either from the bottom or, preferably, from the top of the main header, and supply oil mist to the point of lubrication. These drop-point pipes are connected to the point of lubrication through application fittings known as reclassifiers. Reclassifiers meter the supply of oil mist and also control the quality of oil mist, in terms of droplet size, discharged to the lubrication point. Some low-speed mechanical systems (such as chains and sprockets, gears, sliders etc.) are incapable of producing enough turbulence to cause coalescence of oil droplets for efficient wetting of the interacting surfaces. In this type of application, reclassifiers capable of inducing turbulence in the flow are used.

There are two types of oil-mist lubrication: the wet sump or oil mist purged technique and the dry sump or pure mist technique. In the wet sump technique oil mist is conveyed into the space above the oil level in a typical bearing housing, which lubricates and replenishes oil in the sump. In the dry sump technique, the same oil mist is supplied into a bearing housing from which the liquid oil has been drained completely. Among the two techniques, the dry sump technique is more widely used because it provides lower operating temperatures, a continuous supply of fresh oil and cleaner operation [23].

Advantages and Disadvantages of Oil-mist Lubrication

The advantages of oil-mist lubrication [23,29,30&31] may be summarized as:

Reduced Maintenance Requirement

Oil-mist is a centralized lubrication system which provides for essentially single location maintenance. The system does not have any moving parts and there is no excess lubricant to leak out. There are no pumps, sumps or filters in the supply system. All these reduce the maintenance requirements.

Lower Operating Temperature:

Oil-mist lubrication does not require an oil sump in the bearing housing. Accordingly with oil-mist lubrication the rotating bearing does not churn oil, which results in lower friction and lower operating temperature. The cooler oil mist flowing through the narrow passages of the hot operating surfaces also lowers the operating temperature of bearings through forced convection.

Reduced Abrasive Wear of Surfaces

In the case of conventional oil-sump lubrication, as the bearings wear out wear particles start accumulating in the lube oil sump. These work-hardened wear particles are recirculated along with the lubricating oil on the operating surfaces of the bearing and cause abrasive wear. However, oil-mist lubrication is a continuous flow lubrication system. In oil-mist lubrication, fresh oil and air thus continuously flush the operating surfaces, thereby removing the wear particles. This reduces abrasive wear of the operating surfaces.

Exclusion of Corrosive Gases from Operating Surfaces

In the case of oil-mist lubrication, a slight positive pressure is maintained in the bearing housing. This positive pressure keeps water vapor and other corrosive gases away from the operating surfaces, which reduces corrosive degradation of the operating surfaces, even when the system is not in operation. This characteristic makes the oil-mist lubrication system extremely attractive for the protection of "standby," or non-running equipment. Oil mist is highly suitable for outdoor applications and in locations containing corrosive gases.

Savings in Energy and Resources: Since no sump is required for oil-mist lubrication the rotating bearings do not churn oil. This results in lower frictional energy loss. Also, oil consumption is about 40% lower in oil-mist lubrication compared to conventional sump lubrication. Bearing failure rates have been reduced by 80 to 90% when conventional sump lubrication was replaced by the oil-mist lubrication [29]. All these translate into savings in energy and resources.

The disadvantages of oil-mist lubrication may be summarized as:

Capital Investment

New investments have to be made to install oil-mist generation and supply systems. In the short run the cost of new investments may be a problem, however in the long run oil-mist lubrication provides substantial advantages. Limited studies have shown that the typical payback period for the capital investment in oil-mist lubrication systems in large petrochemical plants is about 2 years [23].

Stray Mist

As oil mist passes through the bearings, the larger particles are deposited easily on the bearing surfaces, however some of the smaller particles go out with air, which is called stray mist. As much as 40 to 60% of the oil may go out as stray mist. All industrial applications so far have easily met the 1990's OSHA requirement of less than 5 mg of oil per cubic meter of air in inhabited areas [23], since oil-mist lubrication consumes very little oil and the stray mist contains very small amount of oil. With oil-mist lubrication, instead of splashing oil all over the bearing housing, oil is deposited only on the operating surfaces resulting in low oil consumption. There are systems available which can completely recycle stray mist; however they increase the cost of the oil-mist lubrication system. These systems are often referred to as closed loop oil-mist lubrication systems and are highlighted in Part 1 of this text.

Cost of Compressed Air

Oil-mist lubrication requires dry and compressed air to generate oil mist. The cost of compressed air may be significant. In one study it has been found that the cost of compressed air is about twice as high as the cost of oil delivered to the bearing [24]. However, the energy required to compress air may be offset by the energy savings due to reduced friction in bearings lubricated by the oil-mist lubrication system.

The reader will note that Part 2 of this book is divided into five "scientific basis" chapters dealing with,

(1) Generation of oil mist
(2) Supply of oil mist to the point of application
(3) Reclassification of oil mist
(4) Deposition of oil at the point of application and Stray Mist
(5) Tribological performance of oil-mist lubrication

Part 2 thus deals extensively with the underlying technical and scientific principles that make oil-mist lubrication possible. Tests setups and tests results are documented for the edification of serious scholars and academics interested in the analytical and experimental methods employed by one of the coauthors in the pursuit of work leading to a doctorate in mechanical engineering.

Chapter 17

Generation of Oil Mist

OIL-MIST GENERATOR

Two types of oil-mist generators are currently commercially available; these are the vortex type oil-mist generator and the venturi type oil-mist generator. In the oil-mist generator dry, compressed air is allowed to flow through a vortex or a venturi chamber. As shown in Figure 17-1, in the vortex type oil-mist generator air is given a swirling motion by the geometry of the chamber [25]. The swirling air flows out through an orifice at the bottom of the vortex chamber. Just ahead of the exhaust orifice the high speed air induces low static pressure. This low pressure pulls lubricating oil into the air flow stream perpendicular to the direction of the air flow. The high speed air blasts the film of lubricating oil into aerosol or oil mist. In the venturi type oil-mist generator the dry compressed air is accelerated in a venturi chamber. As shown in Figure 17-1, the lubricating oil is introduced at the lowest static pressure location, hence the highest air speed point along the venturi tube. In other words, the lubricating oil is induced into the air stream by the low pressure. In the venturi type oil-mist generator lubricating oil is introduced in the direction of the air flow [26] or perpendicular to the direction of the air [27] flow (Figure 17-1). In aerosol mechanics if oil is introduced in the direction of the air flow then the atomizer is referred to as the co-flow type air blast atomizer, and if oil is introduced perpendicular to the direction of the air flow then the atomizer is referred to as the cross-flow type air blast atomizer.

After flowing out of the vortex or the venturi chamber the oil mist impinges on a baffle which is positioned perpendicular to the flow path. This baffle collects oil droplets too large to stay airborne in the long supply pipe. The deposition of oil droplets in the supply line is discussed in Chapter 18 of this text. The mechanism of oil droplet deposition on the baffle is known as the inertial impaction mechanism of aerosol. This mechanism of aerosol deposition has been extensively researched and will be discussed later. After impinging on the baffle the oil mist flows into the main supply line.

Extensive research work has been done on the characteristics of plain-jet air blast atomizers for co-flow and for cross-flow configurations by Nukiyama and Tanasawa [32], Rizk and Lefabvre [33) et al. All the models developed by the above mentioned researchers indicate that the liquid viscosity influences the size distribution of the generated aerosol in plain-jet aerosol generators.

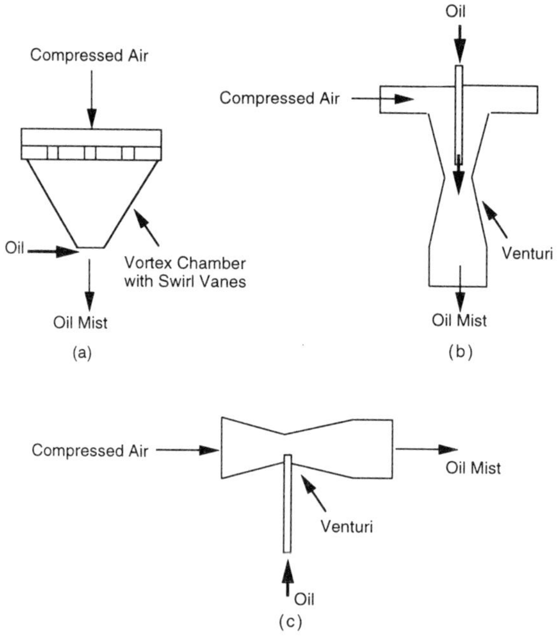

Figure 17-1. (a) Diagram of the basic principle of a vortex type oil-mist generator (Lubrication Systems Co., Houston, TX), (b) Diagram of the basic principle of a co-flow venturi type oil-mist generator (C.A. Norgren Co., Littleton, CO), and (c) Diagram of the basic principle of a cross-flow venturi type oil mist generator (Bill Spitzer & Associates, Houston, TX)

Characteristics of Generated Oil Mist

The characteristics of the generated oil mist may be described primarily by two parameters,

(1) Mass concentration of oil in oil mist and
(2) Particle mass or number size distribution of oil mist.

The variables which have been found to influence the characteristics of the generated oil mist are,

(i) Air flow rate through the generator,
(ii) Oil flow control valve position,
(iii) Oil level in the oil reservoir of the generator,
(iv) Oil viscosity and
(vi) Oil type and additive package.

For a particular geometry of the oil-mist generator the air flow rate determines the speed of air at the point where the oil is blasted into oil mist. If the air flow rate is reduced from the design point by reducing the supply air pressure, the oil-mist generation capability of the generator is reduced. For a particular geometry and air flow rate the amount of oil introduced in the air flow-path for oil-mist generation is determined by the low static pressure induced by the high speed air flow and the position of the oil flow control valve. The amount of oil introduced in the flow may be directly controlled by the oil flow control valve. For a particular geometry and air flow rate if the oil flow control valve is fully open, the amount of oil introduced in the air flow path will depend on the viscosity of the lubricating oil and the oil level in the reservoir.

Tests have been conducted to investigate the influence of oil viscosity and oil level in the reservoir on the characteristics of the generated oil mist [34]. The test bearing setup is shown in Figure 17-2. The generator used in these tests was a 40 BI (Bearing Inch) or 1.2 SCFM vortex type oil-mist generator [23] generating oil mist at 20 inch water column pressure (or at 5 kPa). In these tests the oil valve of the generator was kept completely open and the air bypass valve (which allows the supply air to bypass the vortex chamber) was kept completely closed. For the oil viscosity variation tests the oil level in the generator was kept constant. In these tests in order to measure the droplet size distribution of oil mist, an Andersen 1 ACFM (Actual Cubic Feet per Minute) nonviable cascade impactor was used. As aerosol flows through the stages of the cascade impactor the airborne particles are collected on various stages according to the Aerodynamic Equivalent Diameter (AED) sizes. The aerodynamic equivalent diameter is the diameter of the unit density sphere that has the same terminal settling velocity due to gravity as the particle (any shape, size or density) in quiescent air. The operating viscosity of oil was varied by varying the oil temperature in the reservoir of the generator. Five different oil temperatures 302.6, 310.9, 319.3, 327.6 and 335.9 K (85°, 100°, 115°, 130° and 145°F) were used in the experiments and the corresponding oil viscosities were 165, 100, 65, 45 and 32 centistokes respectively. For the oil level variation tests the reservoir oil temperature was kept constant at 303.2 K (86°F). The reservoir oil levels used in these tests were 1.3, 5.1, 8.9, 12.7 and 14.6 cm (1/2", 2," 3-1/2," 5" and 5-3/4") from the vortex chamber (from the top level). In the viscosity and the oil level tests synthetic diester base 100 viscosity grade (ISO VG 100) oil (without any special oil-mist additives) was used. The droplet mass-size distribution of the generated oil mist for a special blend of mineral 100 oil (with special additives for oil-mist lubrication systems) and a synthetic diester base 68 oil were also measured to compare the droplet mass-size distributions and mass concentration of oil in the generated oil mist. The special additives were added to the mineral oil to reduce stray mist. All the tests were repeated three times for each test setup.

Figure 17-3 shows the mass fractions of oil collected in the various stages of the cascade impactor for three commercially available lubricating oils. These oils are the special mineral 100 oil (with special additives for oil-mist lubrication), the synthetic diester base 68 oil (without any special additives for oil-mist lubrication) and the synthetic diester base 100 oil (without any special additives for oil-mist lubrication). The mass fraction is defined as,

$$\text{Mass fraction} = \frac{\text{Mass of oil collected in a stage}}{\text{Total mass of oil collected in the impactor}} \times 100$$

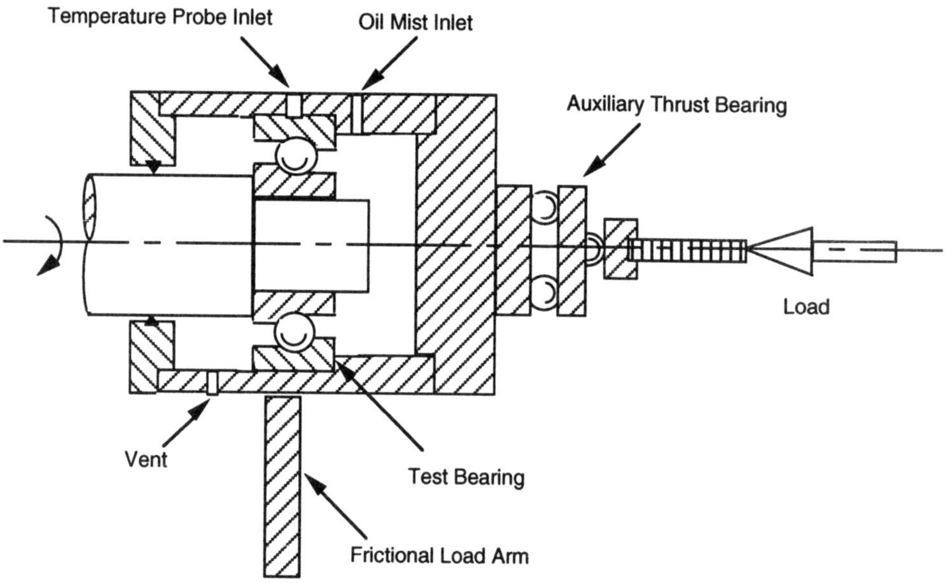

Figure 17-2. Simplified cross-sectional view of the test bearing arrangement.

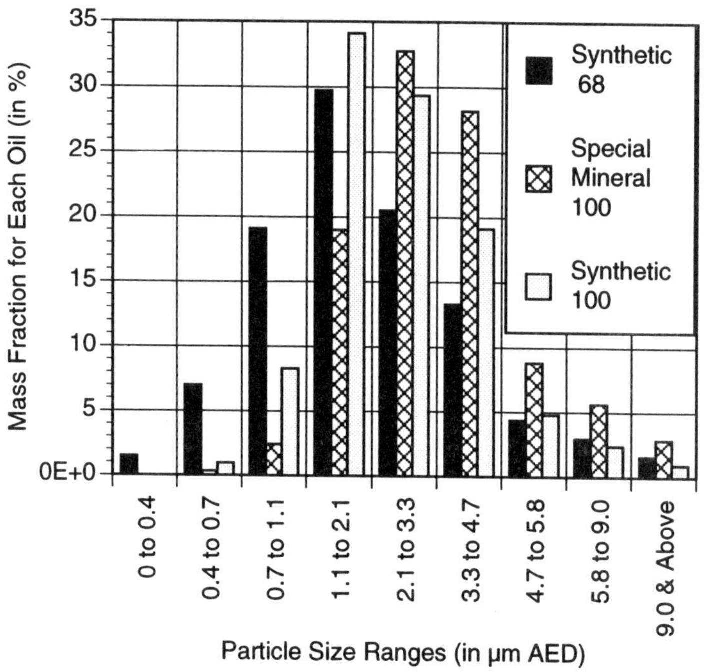

Figure 17-3. Mass-size distribution of the generated oil mist (vortex type generator) for three lubricating oils, where mass fraction is the fraction of the total mass of oil available in a size range for a particular oil. The Aerodynamic Equivalent Diameter (AED) is the diameter of a unit density sphere that has the same terminal settling velocity due to gravity as the particle (any shape, size or density) in quiescent air.

The mass concentrations of the generated oil mist for the three oils are shown in Figure 17-4. It seems that the special oil containing the additives to reduce stray mist produced the highest percentage of larger droplets in the generated oil mist; however, this oil mist also contained the least amount of oil. It is also evident from these plots that the oil mist generated by the mist generator has a single mode mass-size distribution. The influence of the variation of oil viscosity on the mass fraction of oil collected in the various size ranges of the cascade impactor is shown in Figure 17-5. It seems that with increasing oil viscosity the mass fraction of oil in the smaller size ranges decreases and in the larger size ranges increases. Figure 17-6 shows the variation of the relative mass concentration of oil in oil mist with variation of the viscosity of lubricating oil, where

$$\text{Relative mass concentration} = \frac{\text{Mass concentration}}{\text{Max. measured mass concentration in data set}}$$

Figure 17-6 clearly indicates that with increasing oil viscosity the mass concentration of oil in the generated oil mist decreases. The decrease in oil level in the generator reservoir decreases the mass concentration of oil in the generated oil mist which is evident in Figure 17-7. In essence, Figure 17-7 shows the variation of the relative mass concentration of oil in oil mist with variation of the reservoir oil level. The solid lines in Figures 17-5, 17-6 and 17-7 represent second order polynomial curve fit of the experimental data (R-square > 0.90 in all the plots).

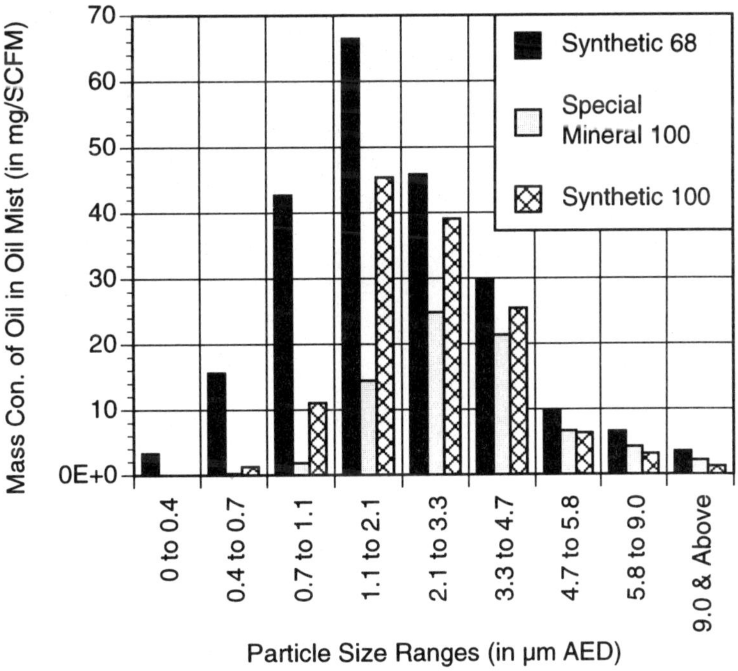

Figure 17-4. Mass concentration of oil in oil mist (vortex type generator) after generation for three lubricating oils.

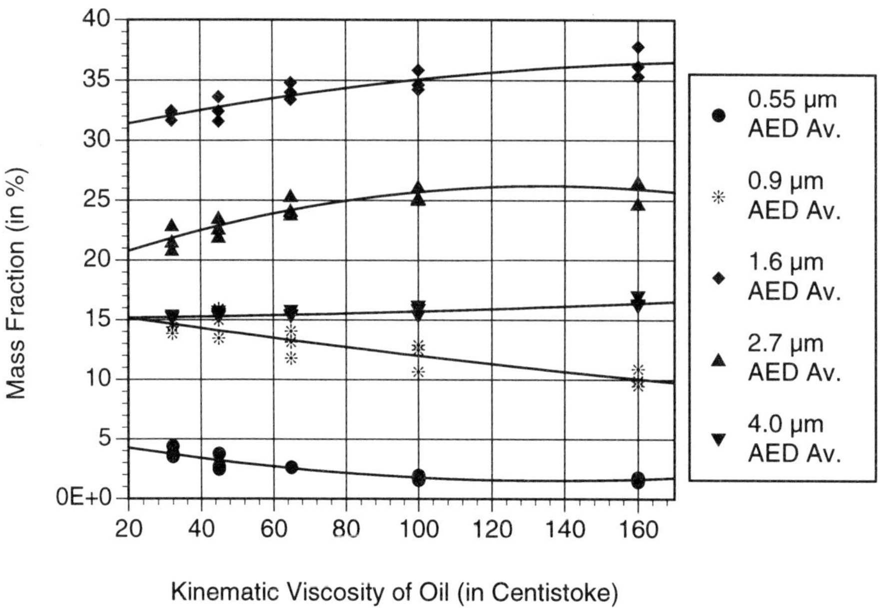

Figure 17-5. Variation of the mass fraction of the generated oil mist with variation of oil viscosity; where AED Av. is the average particle size for a size range (see Figure 17-3 or 17-4).

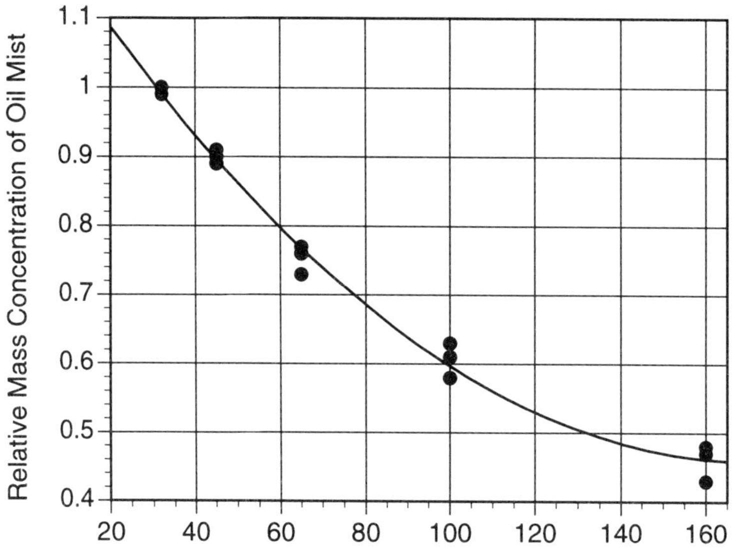

Figure 17-6. Variation of the relative mass concentration of the generated oil mist with variation of oil viscosity; where relative mass concentration is the ratio of mass concentration to the maximum measured mass concentration in the data set.

The type of oil and the additive package used in formulating the lubricating oil may strongly influence the characteristics of the generated oil mist [46]. Several development oils containing special additives have been studied. Coded names are used to identify the development oils. In the figures the oils are identified as Lube A, Lube B, Lube C, Lube D, Lube E, Lube F, Lube G and Lube H. The performance of the widely used Lube C was considered as the "bench mark." The other oil are variations of the Lube oil C. Following is a brief description of the development oils,

Lube A	Ester Lube (+ Polymer X)	Lube E	Ester Lube (+ Polymer 2Z)
Lube B	Ester Lube (+ Polymer Y)	Lube F	Ester Lube (+ Polymer X)
Lube C	Ester Lube (No Polymer)	Lube G	Ester Lube (+ Polymer 3X)
Lube D	Ester Lube (+ Polymer Z)	Lube H	Ester Lube (+ Polymer 5X)

All these oils are essentially ISO Viscosity Grade 68 centistokes at 40°C (ISO VG 68 oils).

Note: X, Y and Z refer to a certain polymer additive—3X means 3 times the X quantity, etc.

The droplet mass-size distributions of the oil mist after generation for the eight test lubricating oils are shown in Figure 17-8. All the oils shown in Figure 17-8 have the peaks of their particle mass-size distributions in the 1.1 to 2.1 μm AED size range. The fractions of the total amount of oil available in the sub-micrometer size (less than 1 μm) ranges are approximately the same for Lube C and Lube A. The Lube D and Lube E had the least mass fractions of oil in the sub-micrometer size ranges. These particles are useless in oil-mist lubrication and they mostly contribute to stray mist which is discussed in detail later. The overall mass concentration of oil in oil mist (as a percentage of the mass concentration for Lube C) after generation for all the test oils is shown in Figure 17-9. The mass

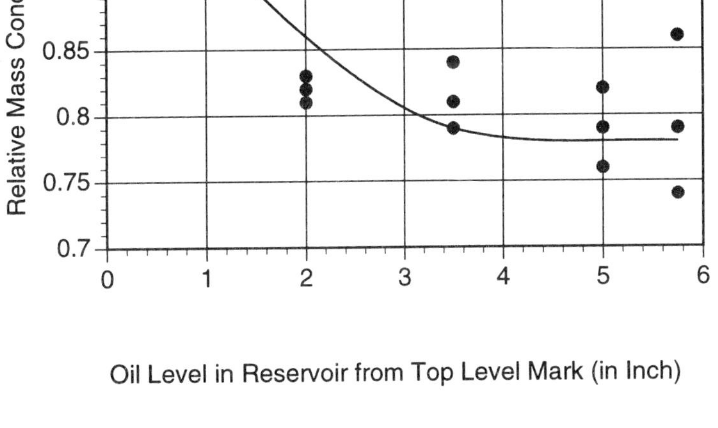

Figure 17-7. Variation of the mass concentration of the generated oil mist with oil level (from the level of the vortex chamber) in the mist generator reservoir.

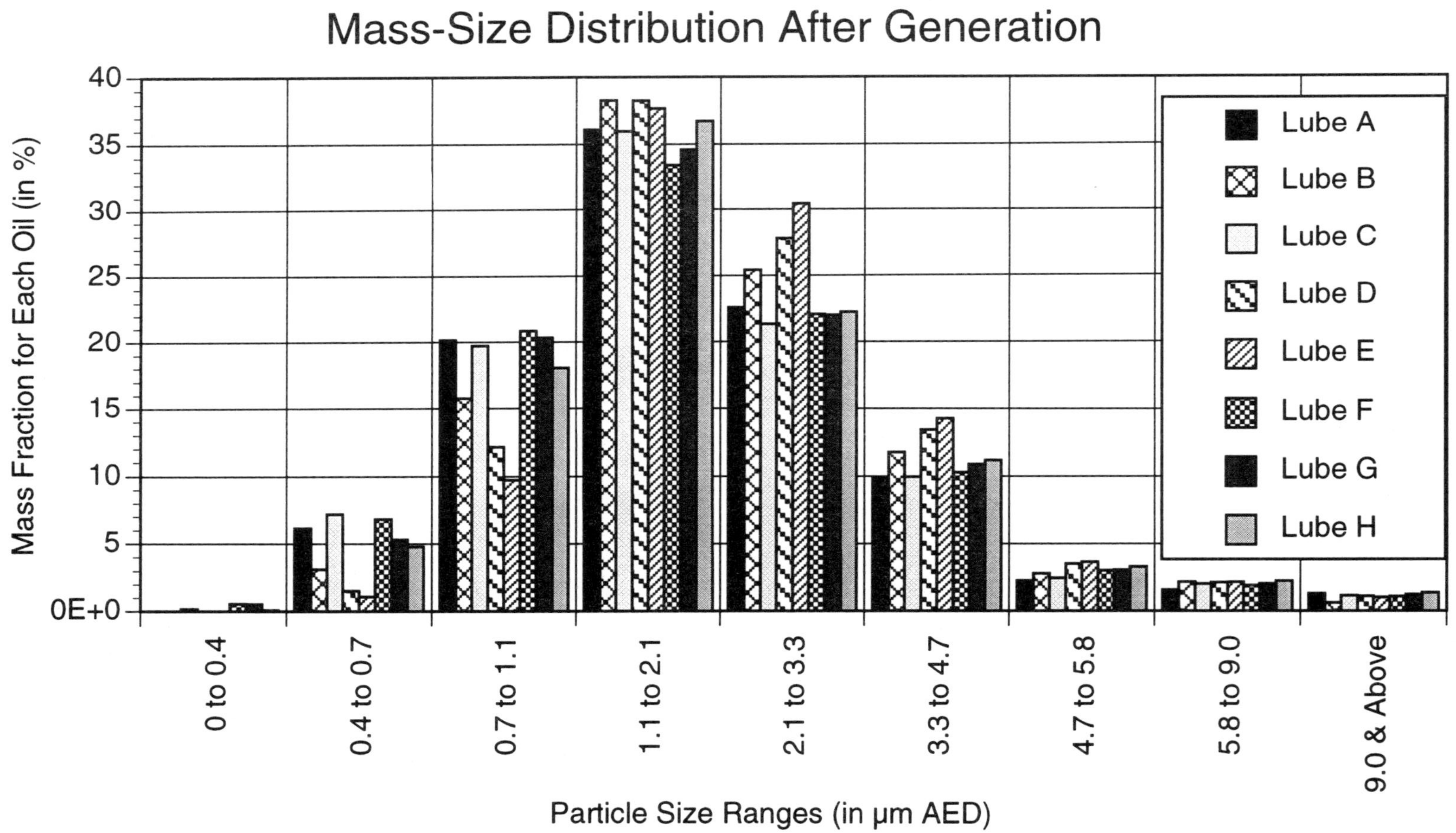

Figure 17-8. Mass-size distribution of the generated oil mist for several lubricating oils, where the mass fraction in the fraction of the total mass of oil found in a size range. The Aerodynamic Equivalent Diameter (AED) is the diameter of a unit density sphere that has the same terminal settling velocity due to gravity as the particle (any shape, size or density) in quiescent air.

concentration of oil in oil mist is the mass of oil available per unit volume of the oil mist at Standard Temperature and Pressure. Figure 17-9 shows that the mass concentration of oil in the generated oil mist for Lube C and Lube A are approximately the same. The mass concentration of oil in oil mist for all other oils was lower than for Lube C. Lube D and Lube E had about 74% and 62% of the mass concentration of Lube C, respectively. Figure 17-10 shows the mass concentration (in mg/SCFM) of oil in the generated oil mist in various size ranges. The small amount of oil in the sub-micrometer size ranges represents a very large number of small particles as shown in Figure 17-11. The number size distributions were calculated from the mass-size measurements by using the arithmetic average of a size range as the aerodynamic equivalent diameter of all the particles in that range. About 22% of the total number of particles in the oil mist for Lube C is in the 0 to 0.4 μm AED size range. Lube A does not have particles in the 0 to 0.4 μm AED. In the 0.4 to 0.7 μm AED size range, Lube C has about 42% and Lube A has about 50% of the particles. The presence of large numbers of smaller particles makes the oil mist appear dense. These tests clearly show that the characteristics of oil mist are strongly influenced by the additive package used in formulating the lubricating oil.

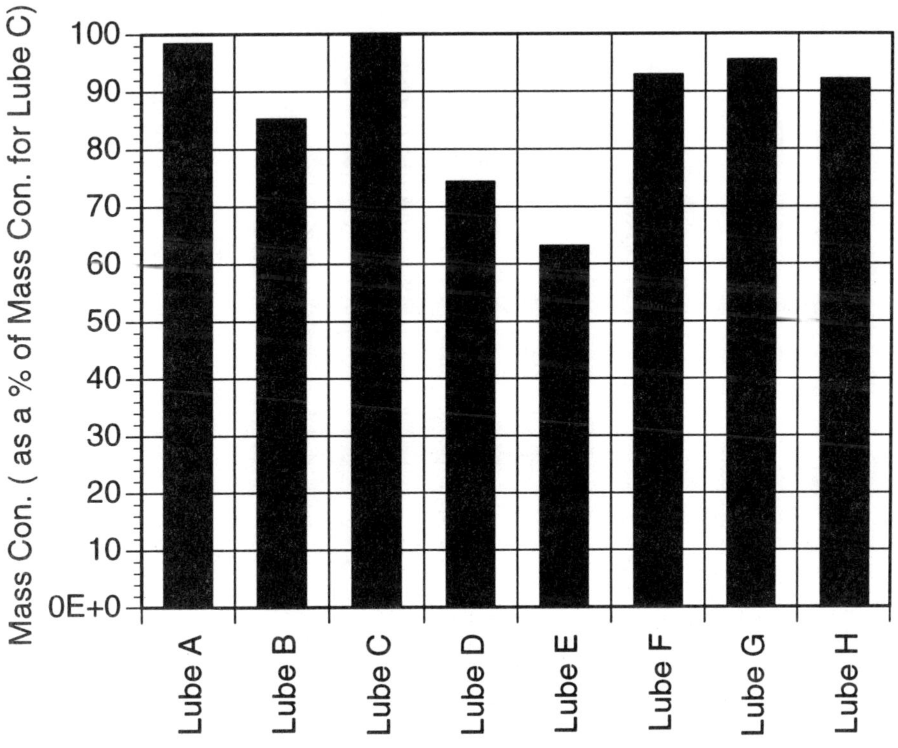

Figure 17-9. Overall mass concentration of oil in oil mist after generation for several lubricating oils. The Aerodynamic Equivalent Diameter (AED) is the diameter of a unit density sphere that has the same terminal settling velocity due to gravity as the particle (any shape, size or density) in quiescent air.

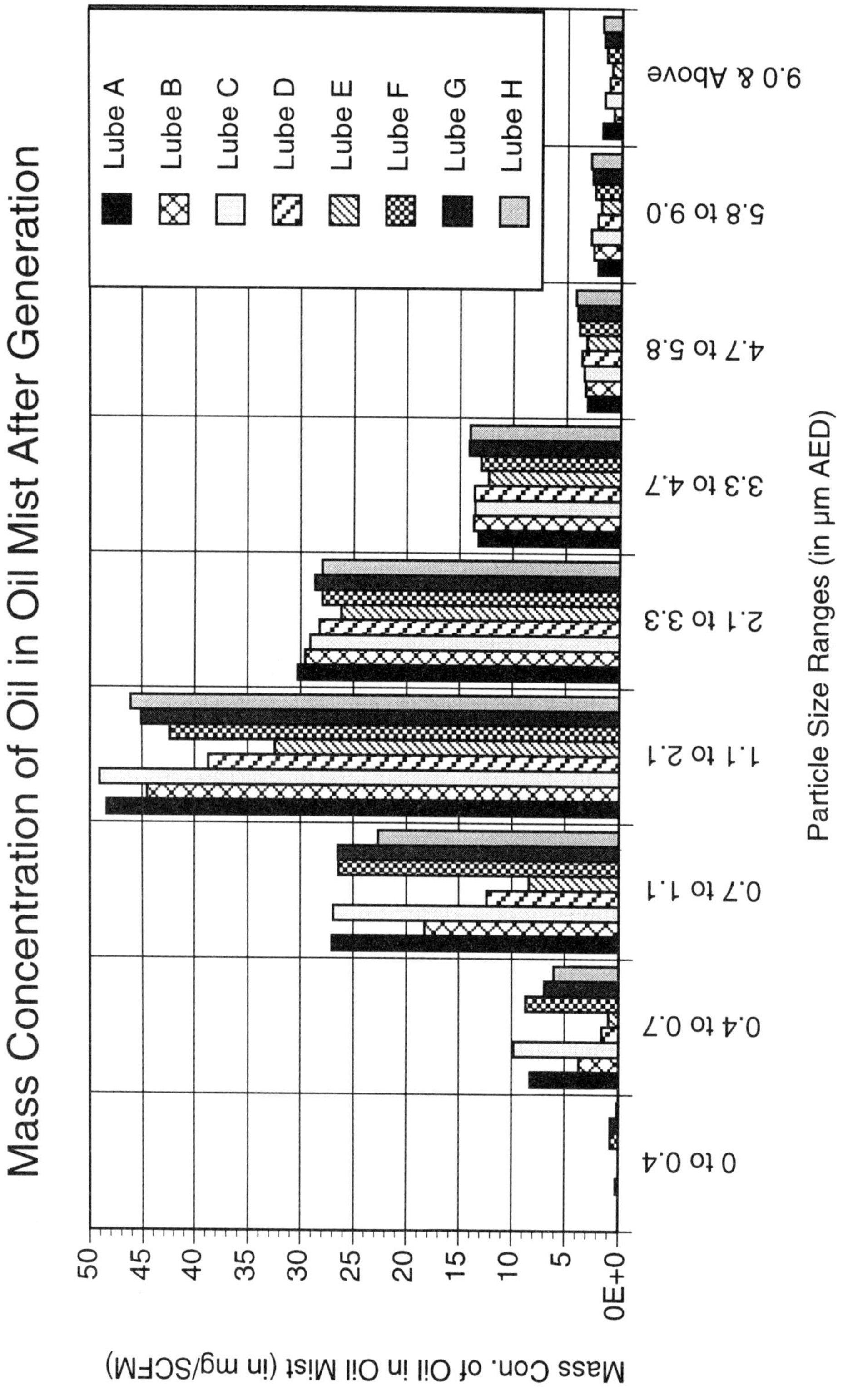

Figure 17-10. Mass concentration of oil in oil mist after generation for several lubricating oils.

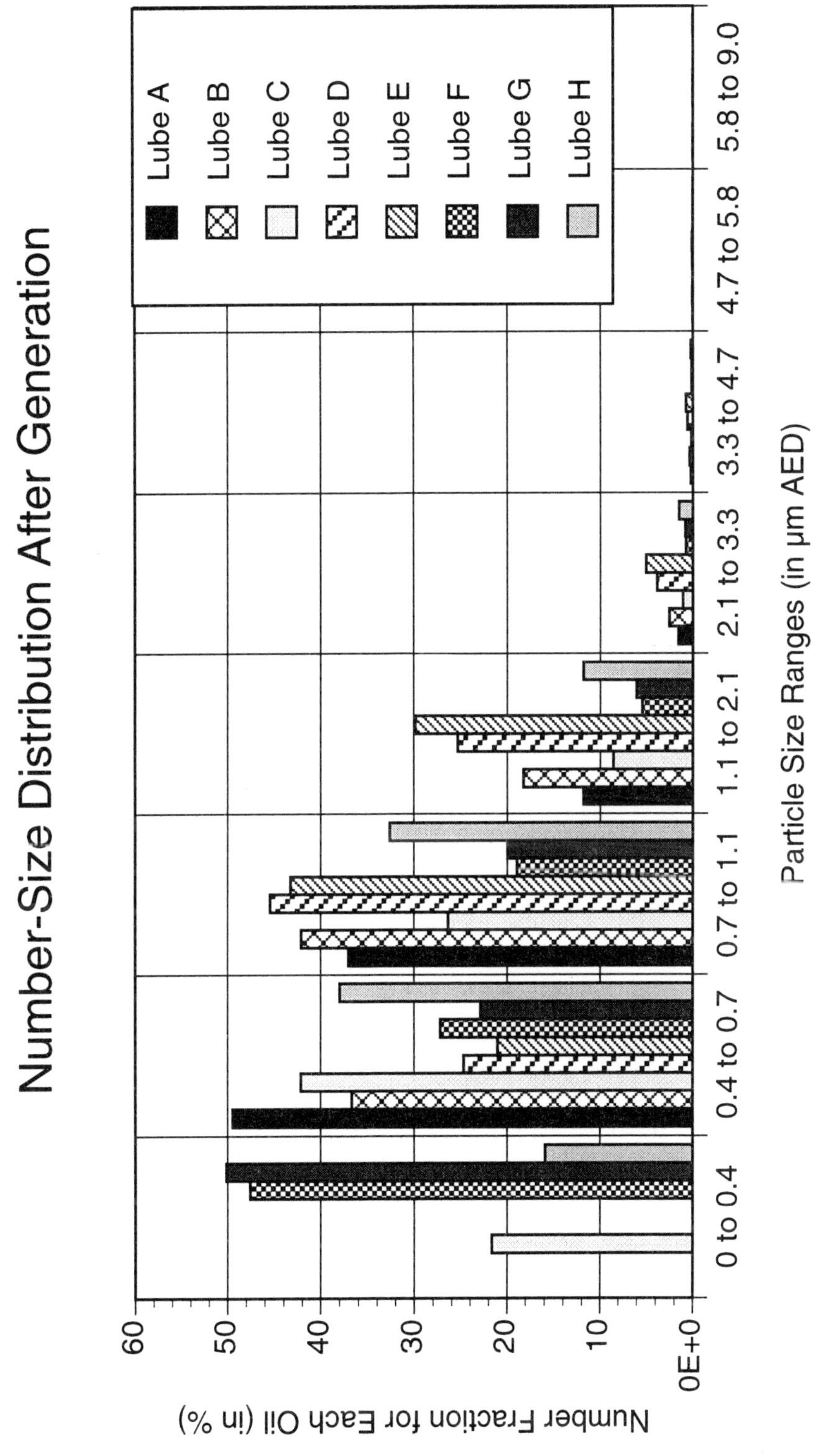

Figure 17-11. Number-size distribution of the generated oil mist for several lubricating oils, where number fraction is the fraction of the total number of particles found in a size range.

Chapter 18

Supply of Oil Mist to Application Points

PENETRATION OF OIL MIST IN THE SUPPLY LINE

After generation the oil mist is carried to various points of lubrication by pipes, also known as main headers. As oil mist flows through the main supply pipe the droplet mass or number size distribution and oil concentration in the oil mist change. Some of the important mechanisms which influence the penetration of aerosol in pipes are,

(1) turbulent deposition or turbulent eddy diffusion,
(2) gravitational settling,
(3) Brownian diffusion,
(4) kinematic coagulation and
(5) laminar gradient coagulation.

By far the most important mechanisms influencing the penetration of aerosol in pipe flows are (a) the gravitational settling and (b) the turbulent deposition [36]. The Brownian diffusion velocity of an aerosol particle of diameter Dp is [37]:

$$V_b = \left[\frac{3\,K_b\,T}{\pi^2\,D_p^3\,\rho_p}\right]^{1/2} \tag{18-1}$$

Where the Boltzmann coefficient $K_b = 1.38 \times 10^{-23}$ J/K, T represents the air temperature in degrees Kelvin and ρ_p represents the density of the particle. Here V_b is inversely proportional to the 1.5 power of the particle diameter. This diffusion velocity is responsible for the deposition of aerosol particles by the Brownian diffusion mechanism. It has been found that the influence of Brownian diffusion on penetration is negligible for particles greater than 1.0 μm AED [35]. The above equation is also known as Einstein's equation of diffusion. Kinematic or orthokinetic coagulation is coagulation that occurs as a result of relative particle motion caused by mechanisms other than Brownian motion. Particles of different size settle at different rates under the influence of gravity and thereby create relative motion between them which leads to collisions and coagulation. The kinematic coagulation caused by differential gravitational settling velocity of aerosol has been found to be very small in pipe flows [35]. The laminar gradient coagulation occurs due to flow velocity gradients. Particles flowing in a velocity gradient, even if they are the same size, will travel at different velocities because of their position in different (but close) streamlines. This relative motion leads to collisions as particles in a faster streamline overtake those in a nearby slower streamline. In the following three sections the most important mechanisms influencing the deposition of aerosol particles in pipe flows, namely, the gravitational settling, the turbulent deposition and a model which combines gravitational settling and turbulent deposition are discussed.

Gravitational Settling of Oil Mist in Pipes

According to Newton's law of resistance the drag force experienced by a spherical object of diameter Dp, traveling with a relative velocity V in a gaseous medium with density ρ_g is,

$$F_D = C_D \frac{\pi}{8} \rho_g D_p^2 V^2 \tag{18-2}$$

Where C_D is the drag coefficient. This equation is derived considering the rate of change of momentum of the gas displaced by the traveling spherical object. The drag coefficient (C_D) is constant for Reynolds numbers (based on sphere diameter) greater than 1000, i.e., when inertia forces dominate over viscous forces. However in aerosol mechanics almost always the Reynolds number for particle diameter is less than 1.0, i.e., for aerosols viscous forces dominate over inertia forces. This flow region is called Stokes flow regime. Stokes solved the Navier-Stokes equation by assuming, (1) inertia forces are negligible compared to the viscous forces, (2) fluid is incompressible, (3) there are no walls or other particles nearby, (4) motion is constant, (5) the particle is a rigid sphere and (6) fluid velocity at the particle's surface is zero. The Stokes equation for drag force acting on a spherical particle traveling in gas is,

$$F_D = 3 \pi \mu_g V D_p \tag{18-3}$$

Where μ_g is the viscosity of the carrier gas. Equation (18-3) is valid for Reynolds numbers (Re) less than 1.0 and Mach numbers (M) less than 0.3. The correction to Stokes law for nonrigid spheres such as oil droplets has been found to be insignificant [35]. By equating (18-2) and (18-3) one finds the value of the drag coefficient for Stokes flow,

$$C_D = 24 \frac{\mu_g}{\rho_g V D_p} = \frac{24}{Re_{Dp}} \tag{18-4}$$

When a particle is released in still air, it quickly reaches its constant terminal settling velocity. For oil-mist flow through a horizontal pipe, the velocity of the carrier air in the radial direction is zero; thus, when oil droplets settle they do not experience any air velocity in that direction. When the particles settling due to gravity reach terminal settling velocity, the drag force of the air on the particles become exactly equal and opposite to the gravitational force on the particles, i.e.,

$$F_D = F_g = m_p\, g \tag{18-5}$$

here F_g is the gravitational force, m_p is the mass of the particle and g is the gravitational acceleration. Assuming $\rho_p >> \rho_g$ (density of air) and substituting F_D in (18-5) from (18-6) one finds the terminal settling velocity of the particle,

$$Vg = \frac{\rho_p D_p^2 g}{18\,\mu g} \tag{18-6}$$

Equation (18-6) should be used for $D_p > 1.0\ \mu m$ and $Re_{Dp} < 1.0$. The particle relaxation time is defined as,

$$\tau = \frac{\rho_p D_p^2}{18\,\mu g} \tag{18-7}$$

so, $$V_g = g\,\tau \tag{18-8}$$

For very small particles ($D_p < 1.0\ \mu m$) the assumption of the Stokes law that the relative velocity of the gas at the surface of the sphere is zero is not satisfied; since for $D_p < 1.0\ \mu m$ the particle diameter approaches the mean free path of the carrier gas. This results in the "slip" of gas at the surface of the particle which causes the particle to settle faster. For $D_p < 1.0\ \mu m$ the relaxation time should be multiplied by the Cunningham slip correction factor. The Cunningham correction factor which corrects for the slip of flowing fluid on particle surfaces when particle radius approaches mean free path of the carrier gas molecules is given by,

$$= 1.0 + \frac{2\lambda}{D_p}\left[1.23 + 0.41 \exp\left(\frac{-0.55\,D_p}{\lambda}\right)\right], \text{ where}$$

Dp = Diameter of particle

λ = mean free path of gas molecule

Turbulent Deposition of Oil Mist in Pipes

If the aerosol flows downward through a pipe there can be no gravitational settling. However, if a particle comes within the stopping distance of the pipe wall the turbulent fluctuating component of the gas velocity normal to the pipe wall will cause the particle to be deposited on the wall [38]. Stopping distance is the distance a particle with a velocity will travel in a fluid medium before coming to a stop because of the drag force (in the absence of any other force). In order to calculate the deposition of aerosol due to turbulence in pipe flow, one has to calculate the deposition velocity (V_d). From experimental data Liu and Agarawal [39] gave relationships between the dimensionless deposition velocity V_+ and the dimensionless relaxation time τ_+. Here, $V_+ = V_d/U^*$ where U^* is the air friction velocity. Also, $\tau_+ = \tau\, U^{*2}/\nu_g$ where ν_g is the kinematic viscosity of air.

$$V_+ = 6.9 \times 10^{-4}\, \tau_+^{\,2} \qquad \text{for } \tau_+ \leq 15 \tag{18-9}$$

$$V_+ = 0.16/\tau_+^{\,0.086} \qquad \text{for } \tau_+ > 15 \tag{18-10}$$

$$\text{and } U^* = (f/2)^{1/2} U \tag{18-11}$$

here f is the friction factor and U is the average flow velocity of the aerosol in the pipe. For turbulent flow in a smooth pipe (Blasius equation),

$$f = \frac{0.316}{4\, Re_{D_t}^{\;1/4}} \tag{18-12}$$

$$\text{And} \qquad V_d = V_+ U^* \tag{18-13}$$

here Re_{Dt} is the pipe Reynolds number.

Combined Gravitational Settling and Turbulent Diffusion Model

Anand and McFarland [36] developed a model to predict deposition of aerosol particles in pipe flows. The model assumes that aerosol particles are deposited in inclined tubes because of turbulent diffusion and gravitational settling and it further assumes that losses caused by electrostatic effects and Brownian diffusion are inconsequential. The tubes are assumed to be hydraulically smooth. The flow is assumed to be turbulent and fully developed. The thermo-physical properties of the air and the aerosol particles are assumed to be constant. These assumptions are valid for oil-mist flow in the main supply line. It should be pointed out that the turbulent diffusion model has been developed based on the data from tests which used monodispersed aerosol (i.e., aerosol containing only one size particle). However, as discussed earlier the tests which measured the particle mass size distributions have clearly shown that oil mist is a highly polydispersed aerosol (i.e., aerosol containing particles of various sizes). The mass balance over a differential element (dx) of a pipe carrying the aerosol inclined at an angle (ø) with the vertical direction, with the concentration of aerosol entering the pipe section C_{no} and the concentration of aerosol C_n at a distance x from the starting point of the pipe section, may be written as,

$$\frac{dC_n}{dx} + \frac{4\,C_n V_e}{D_t U} = 0 \tag{18-14}$$

here, at $x = 0$, $C_n = C_{no}$

here V_e is the equivalent particle velocity at the wall, D_t is the pipe diameter and U is the flow velocity of aerosol. The equation (18-14) can be integrated using the initial condition to give the penetration (P),

$$P = \frac{C_n}{C_{no}} = \exp\left[\frac{\pi D_t V_e x}{Q}\right] \tag{18-15}$$

here Q is the volumetric flow rate of aerosol through the pipe. The effective deposition velocity (V_e) is the vector sum of the particle deposition velocity caused by the turbulent diffusion (V_d) and the terminal gravitational settling velocity (V_g). The mean effective deposition velocity at a cross-section [36] is,

$$V_e = \frac{1}{2\pi}\int_0^{2\pi}\left(V_d - V_g \sin\theta\right) d\theta \tag{18-16}$$

for $(V_d - V_g \sin\theta) > 0$

V_e is set equal to zero when $(V_d - V_g \sin\theta) \le 0$. This constraint is necessary; otherwise it would be implied that the mass could be transported through the upper half of the pipe from the outside environment. For inclined tubes,

$$V_g = g\,\tau \sin\phi \tag{18-17}$$

For horizontal tubes $V_g = g\,\tau$. Equation (18-7) and (18-8) can be used to calculate τ and V_g respectively. Equation (18-9) to (18-13) can be used to calculate V_d. In calculating the relaxation time (τ) Cunningham's slip correction factor was neglected since the additional refinement would not significantly enhance the quality of the model [36].

A computer program has been written using the equations given in the previous section to predict penetration of oil mist in the supply pipe [40]. The program calculates penetration of the aerosol through a pipe with inner diameter D_t when the flow rate is q for the various stages of the Andersen 1 CFM nonviable impactor. The Andersen 1 CFM nonviable impactor is an atmospheric aerosol particle mass-size measuring device marketed by Andersen Instruments Inc. of Atlanta, Georgia. This device operates on the principle of aerosol impaction mechanism which will be discussed later. The program uses the model for combined gravitational settling and turbulent diffusion; hence all the assumptions described in the previous section for the combined gravitational settling and turbulent diffusion model are applicable to this program. The program uses Simpson's one third rule to calculate the mean effective deposition velocity of a particle using equation (18-16). In the computer program the particle size range for an impactor stage (stage of the Inertial Cascade Impactor discussed earlier) was divided into ten sizes and then the penetration for each of the individual particle sizes was calculated. The arithmetic average of these penetration values was taken as the penetration for that particular impactor stage. In calculating the penetration of oil mist in the main supply line of the experimental setup (shown in Figure 18-1), the effect of the 90° pipe elbows used to connect the 1/2 inch PVC pipes was neglected for the following reasons: (1) As the aerosol flows through the bends the particle Stokes number (which is the ratio of the centrifugal to the fluid drag force on a particle) determines whether or not the particle will be deposited on the bend wall (due to dominant centrifugal force, i.e., high Stokes number). The Stokes number is given by,

$$S_{tk} = \frac{C \rho_p D_p^2 U_o}{18 \mu g R_c} \tag{18-18}$$

here, D_p = diameter of particle
C = Cunningham's correction factor
ρ_p = density of particle
U_o = velocity of aerosol jet
μ_g = viscosity of gas
R_c = radius of curvature

For the numerical modeling of oil mist penetration in the supply line the highest value of the Stokes number at pipe bends (for effective stage cutoff diameter) was found to be about 0.07, which is very low for any significant deposition in the pipe bend. (2) There have been several analytical studies on the deposition of aerosol in bends of circular cross section [41]. Most of these models including the most recent study by Cheng and Wang [42] indicate that for a Stokes number below 0.1 the penetration of aerosol through a 90° pipe bend is 100%. All the particle Stokes numbers for this numerical model were well below 0.1. The most recent model (based on experimental data) for penetration of aerosol in 90° pipe bends has been developed by Pui et al. [41]. This model is most suitable for particle Stokes number ranging from 0.3 to 1.4. In developing this model only two data points were taken below Stokes number 0.2 and they did not fit well to the model. The behavior of these data points was more like the Cheng and Wang model (which predicts zero deposition for Stokes number < 0.1). Considering the above mentioned reasons it was decided that the inclusion of the effect of the pipe elbows would not significantly enhance the numerical model of penetration of oil mist in the supply line.

Tests have been conducted to measure the penetration of oil mist through a 1/2 inch supply pipe system for five different flow rates (Figure 18-1). These flow rates were 495.3, 566.0, 660.3, 849.0 and 1132.0 cm^3/s at STP (1.05, 1.2, 1.4, 1.8 and 2.4 SCFM) which corresponded to pipe Reynolds number 2629, 3005, 3505, 4507 and 6009, respectively. For each

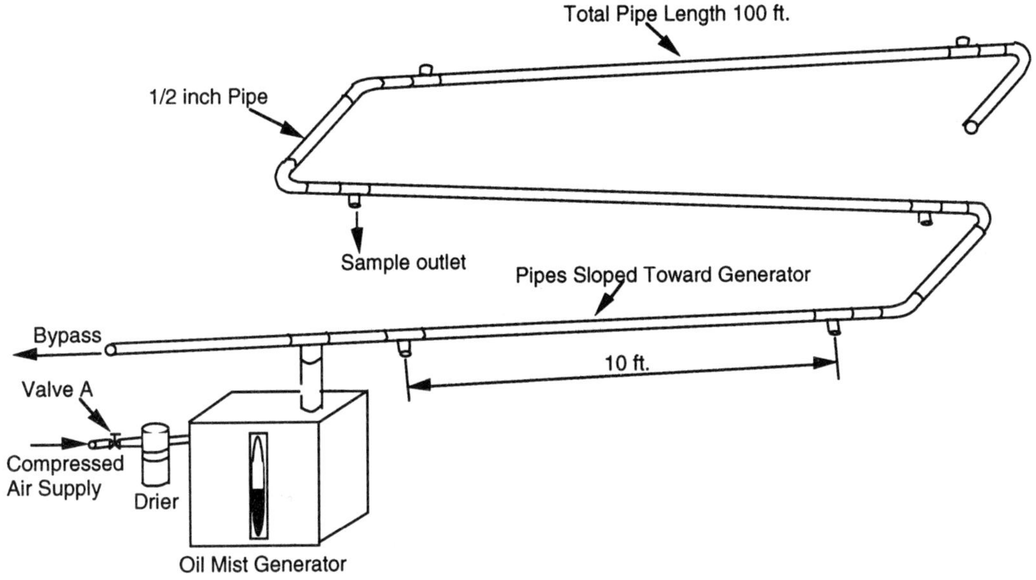

Figure 18-1. Diagram of the experimental setup for aerosol tests.

flow rate the oil droplet mass size distributions were measured at five different distances along the supply pipe. The distances were 0 m (beginning of the supply pipe), 3.05 m (10 ft), 9.13 m (30 ft), 18.3 m (60 ft), and 30.5 m (100 ft). The oil-mist pressure in the generator reservoir was maintained at 4.98 kPa (20 in. of water column) pressure. The oil temperature (303.2 K) and the oil level in the generator reservoir were kept constant. Synthetic 100 viscosity (ISO VG 100) lubricating oil was used in the tests. The oil valve of the generator was kept completely open and the air bypass valve was kept completely closed.

The variations of the masses of oil collected in various stages of the cascade impactor for three distances along the 1/2 inch supply pipe are shown in Figure 18-2 for 495.3 cm^3/s at STP (1.05 SCFM) flow rate. In this bar chart the amounts of oil collected in various stages have been expressed as fractions of the oil collected in the stage with 1.1 to 2.1 μm AED size range. The maximum mass of oil has been found in the 1.1 to 2.1 μm AED stage. The relative mass fraction is defined as,

$$\text{Relative mass fraction} = \frac{\text{Mass of oil collected in a stage}}{\text{Maximum mass of oil collected in a stage}}$$

The relative mass fraction plots for the other flow rates were found to be similar. In general, the mass of oil available in a size range reduces as the distance along the pipe increases. The penetrations of oil mist in the supply line versus the distance along the supply line for several impactor stages are shown in Figures 18-3, 18-4 and 18-5, where the oil-mist penetration to a distance (x) along the supply line for a particular impactor stage particle size range is,

$$\text{Penetration} = \frac{\text{Mass of oil collected in a stage at a distance x}}{\text{Mass of oil collected in the same stage at 0 dist}} \times 100$$

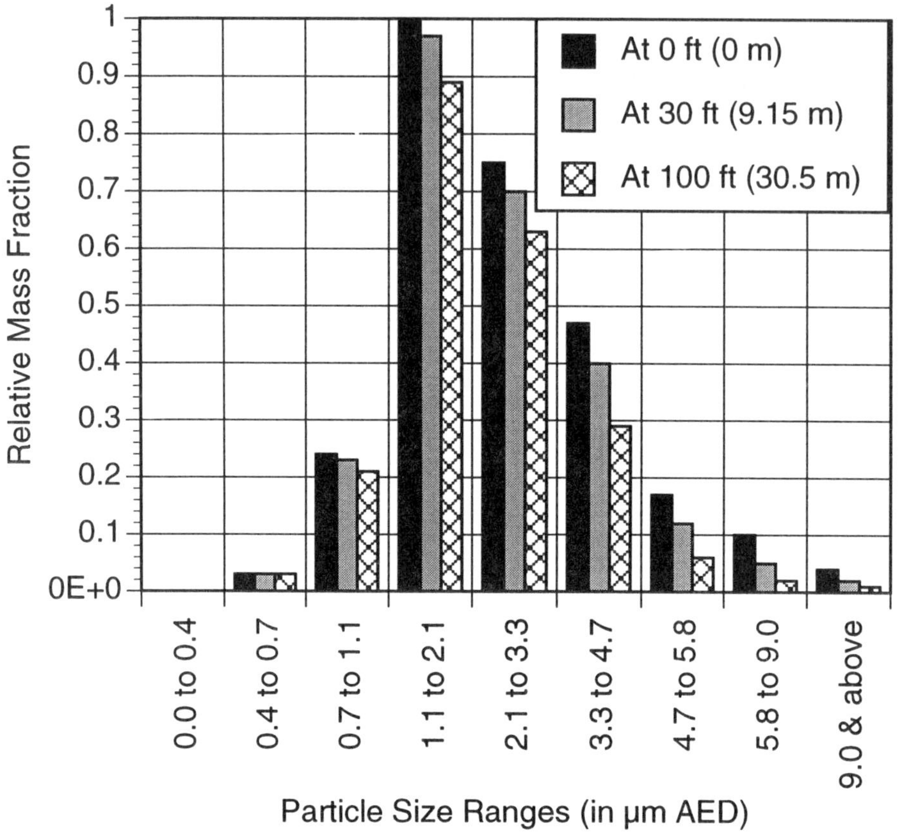

Figure 18-2. Variation of the relative mass fraction of oil in oil mist in various size ranges shown for three distances along a 1/2-inch pipe. For 1.05 CFM (496.3 cm^3/s) flow rate at Standard Temperature and Pressure (STP).

Figures 18-3, 18-4 and 18-5 have been plotted for 495.3, 566.0 and 660.3 cm^3/s at STP (1.05, 1.2 and 1.4 SCFM) flow rates through the supply pipe, respectively. The predicted penetrations calculated by the computer program for gravitational settling and turbulent diffusion of oil mist in the supply line are shown as solid lines in Figures 18-3, 18-4 and 18-5 and the data from the experiments are shown as symbols. Agreement between predicted and experimental values is best for 495.3 cm^3/s at STP flow rate (Reynolds number 2629). The difference between the predicted values and experimental data increases with increasing flow rate. The predicted penetration deviated significantly from the experimental values for 849.0 and 1132.0 cm^3/s at STP (1.8 and 2.4 SCFM) flow rates which correspond to Reynolds numbers 4507 and 6009. It has been reported by Anand and McFarland [36] that the model performs best at low Reynolds numbers (1620 to 2830). In the penetration tests all the other flow rates except 495.3 cm^3/s at STP were beyond Reynolds number 2830.

The penetration of oil mist in the supply pipe is plotted as a function of flow rate with impactor stage as the parameter in Figures 18-6 and 18-7. The solid lines represent the numerical model. The experimental values are shown as symbols. Figures 18-6 and

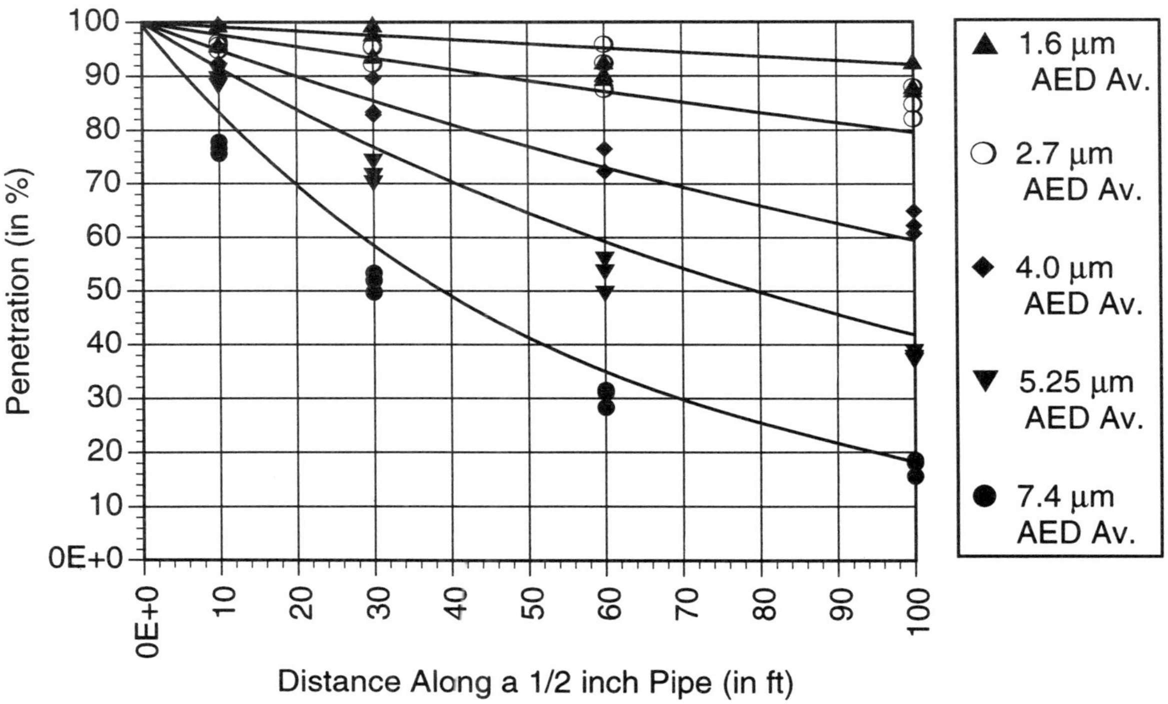

Figure 18-3. Penetration vs. distance with AED Av. as parameter. Here penetration is defined as the fraction of the total mass of oil traveling a distance along a pipe without deposition and AED Av. is the average particle size for a size range shown in Figure 18-2. The symbols represent experimental data and the solid lines represent combined gravitational settling and turbulent diffusion model. (For Reynolds No. 2629.)

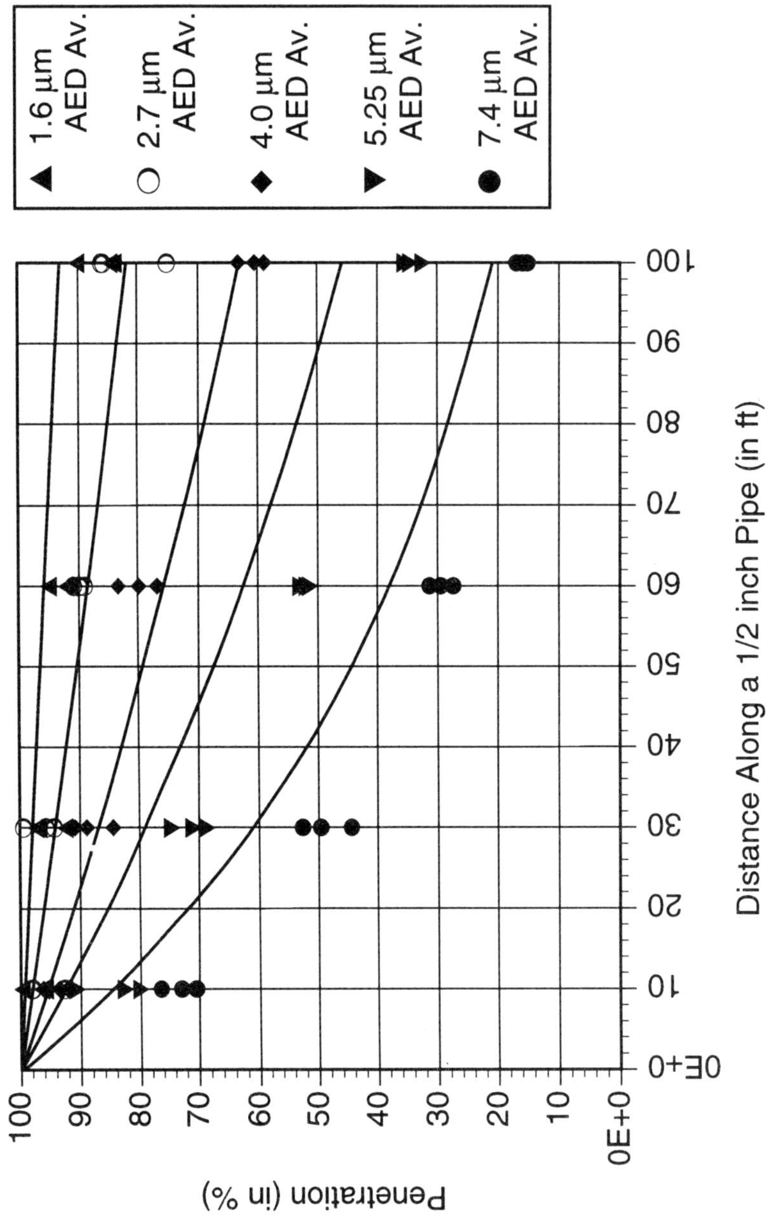

Figure 18-4. Penetration vs. distance with AED Av. as parameter. The symbols represent experimental data and the solid lines represent combined gravitational settling and turbulent diffusion model. (For Reynolds No. 3005.)

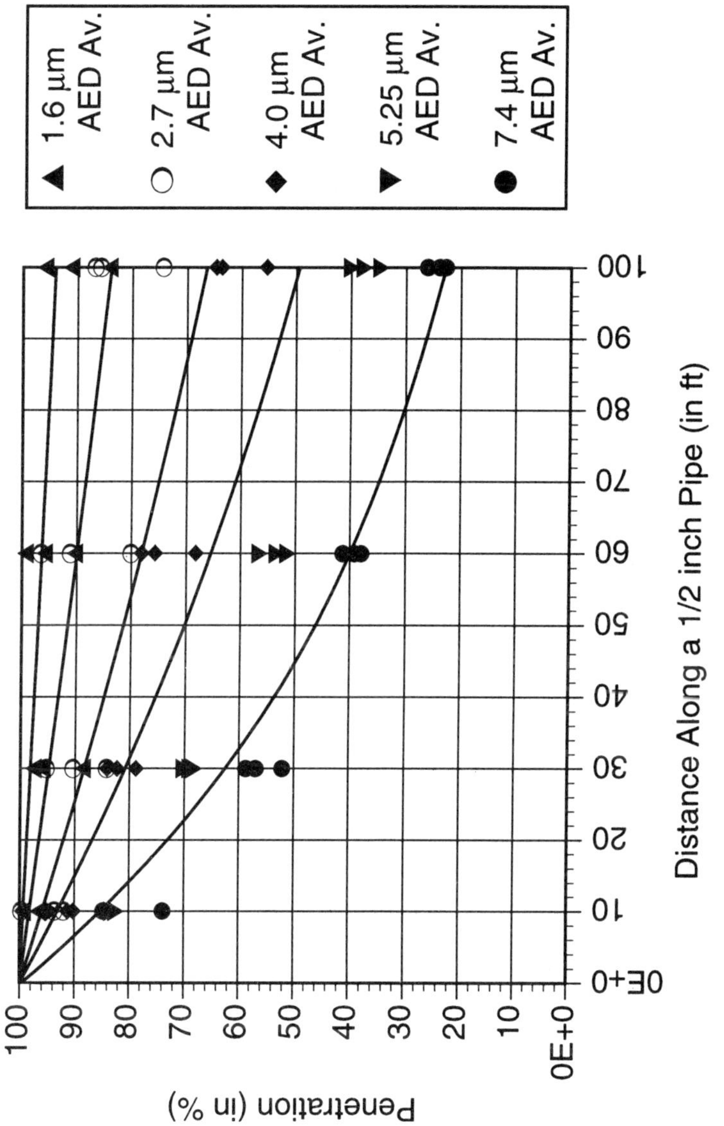

Figure 18-5. Penetration vs. distance with AED Av. as parameter. The symbols represent experimental data and the solid lines represent combined gravitational settling and turbulent diffusion model. (For Reynolds No. 3505.)

18-7 correspond to 18.3 m (60 ft) and 30.5 m (100 ft) distance along the supply pipe. For a particular pipe size if the flow rate is very low (low Reynolds number) the residence time of aerosol in the pipe is high and gravitational settling becomes the dominant mode for aerosol deposition in pipes. As the flow rate is increased, the residence time of the aerosol in the pipe is reduced because of higher flow velocity, i.e., the airborne particles are allowed less time to settle inside the pipe due to gravity. However, as the velocity of the flow is increased the turbulence also increases which results in higher turbulent deposition. For a particular diameter pipe there thus exists an optimum flow rate for maximum penetration, below which the penetration is low because of the dominant gravitational settling and above which the penetration decreases because of the dominant turbulent deposition. The evidence of decrease in penetration due to high turbulence and distance can be seen in Figures 18-6 and 18-7. Since the cascade impactor required a minimum 495.3 cm^3/s at STP flow (Reynolds number 2629), the test flow rates in the supply pipe could not be reduced below Reynolds number 2629. Experimental data points are thus not available below Reynolds number 2629 and the decrease in penetration at low Reynolds numbers did not show up clearly in Figures 18-6 and 18-7. However the computer program for the numerical model has been run for lower flow rates and the corresponding data for the penetration of oil mist in the supply pipe are plotted as a function of flow rate with impactor stage as the parameter in Figures 18-8 and 18-9. It has already

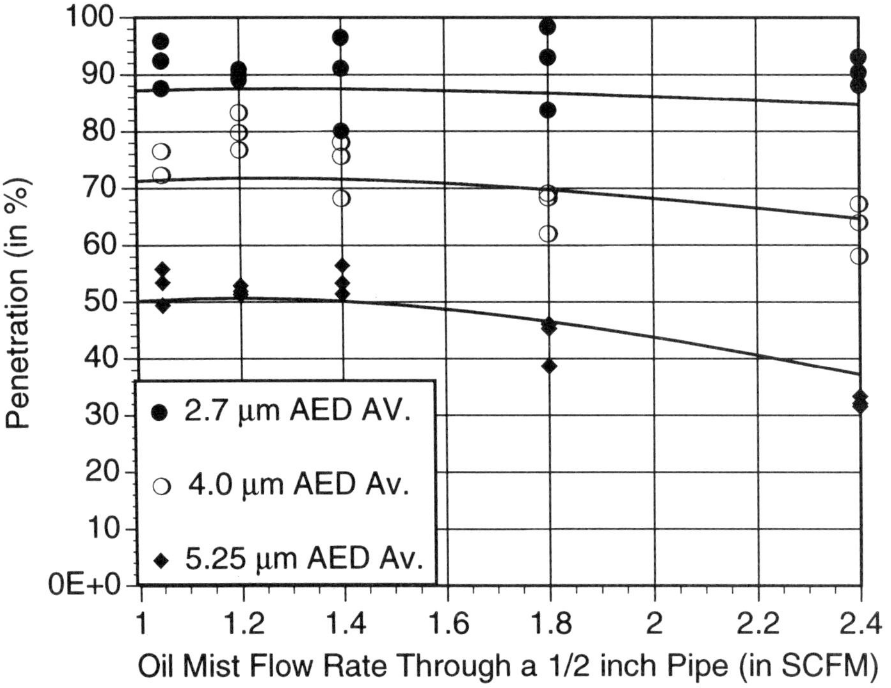

Figure 18-6. Variations of the penetration of oil mist with flow rate at 60 ft (18.3 m) distance along a 1/2-inch supply pipe. Here AED Av. is the average particle size for a size range shown in Figure 18-2. The solid lines represent combined gravitational settling and turbulent diffusion model and the symbols represent experimental data.

been shown that the numerical model for combined gravitational settling and turbulent diffusion performs well for low Reynolds numbers and these plots (Figures 18-8 and 18-9) for the model clearly show the reduction of penetration at lower Reynolds numbers due to excessive gravitational settling. It has been observed in Figures 18-3 through 18-9 that because of the decline in penetration, highly turbulent flows are not very useful in oil-mist lubrication system. It should be pointed out that the penetration analysis discussed here is generalized, i.e., independent of the type of oil used to produce oil mist.

It seems that the combined gravitational settling and turbulent diffusion model [36] for penetration of aerosol in pipes performed reasonably well in low Reynolds number ranges suitable for oil-mist flow through pipes. Using this model for low Reynolds number ranges it is possible to calculate a range of flow rate which is expected to provide the maximum penetration of the desirable larger particles. The larger particles are more desirable than the smaller particles because they do not contribute significantly to stray mist. Accordingly, the combined gravitational settling and turbulent diffusion model for aerosol flow through pipes may be used to determine the ideal flow rate of oil mist for a particular diameter pipe and for the overall design of the oil-mist supply line.

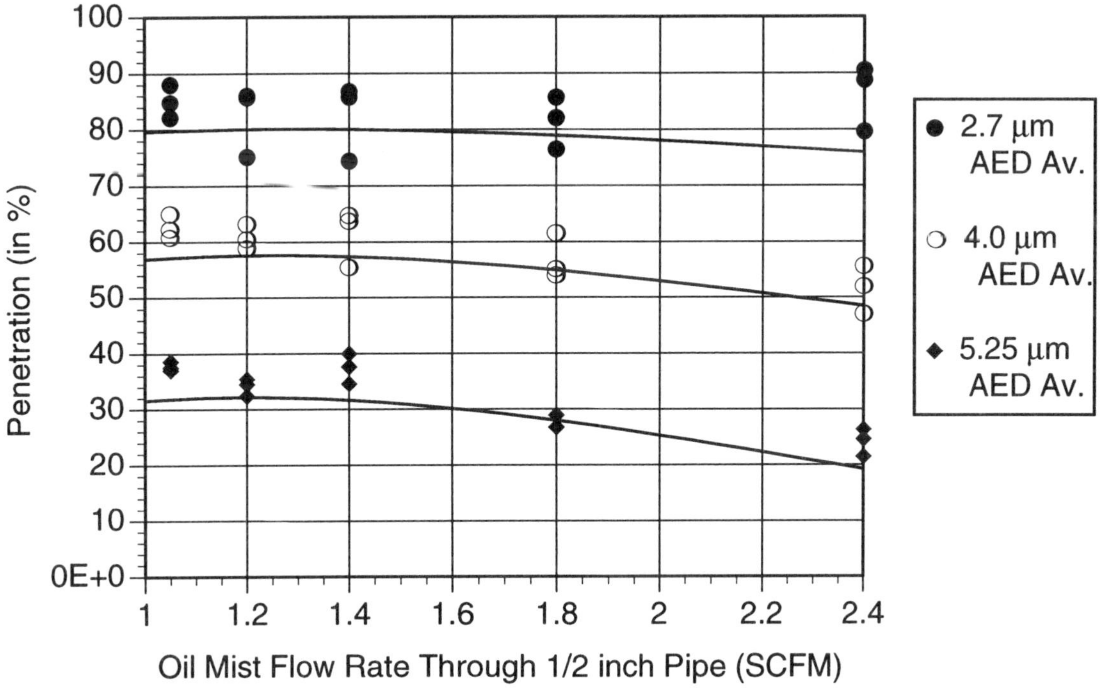

Figure 18-7. Variations of the penetration of oil mist with flow rate at 100 ft (30.5 m) distance along a 1/2-inch supply pipe. Here AED Av. is the average particle size for a size range shown in Figure 18-2. The solid lines represent combined gravitational settling and turbulent diffusion model and the symbols represent experimental data.

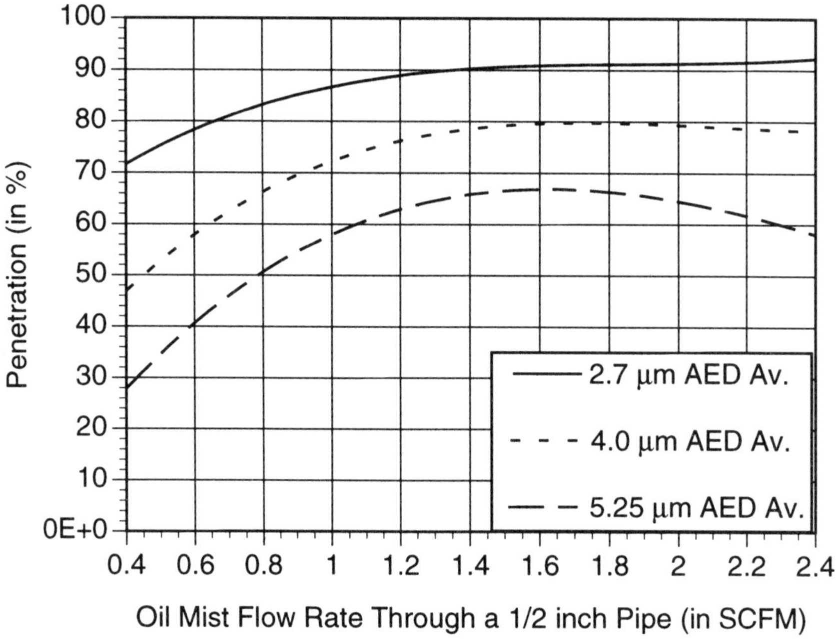

Figure 18-8. Variations of the penetration of oil mist with flow rate at 60 ft (18.3 m) distance along a 1/2-inch supply pipe, as predicted by the combined gravitational settling and turbulent diffusion model.

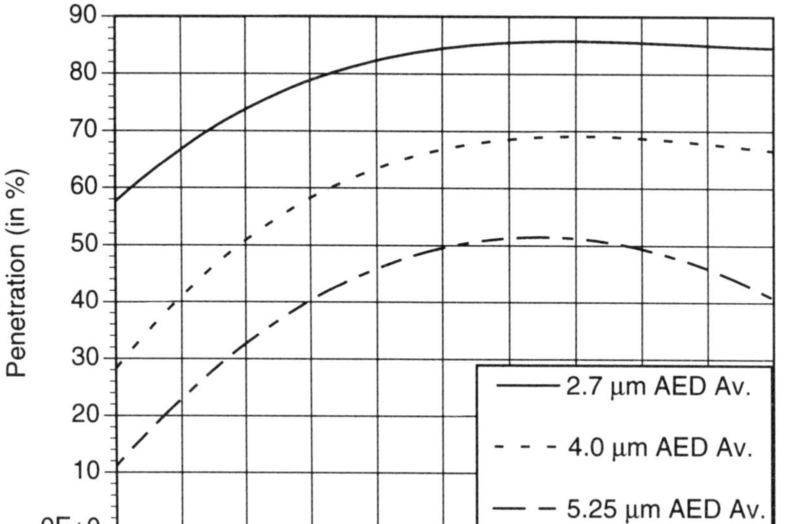

Figure 18-9. Variations of the penetration of oil mist with flow rate at 100 ft (30.5 m) distance along a 1/2-inch supply pipe, as predicted by the combined gravitational settling and turbulent diffusion model.

Chapter 19

Reclassification of Oil Mist

OPERATION OF RECLASSIFIERS

Before entering the point of lubrication oil mist flows through application fittings known as reclassifiers. The reclassifiers (Figure 19-1) meter flow of oil mist to the individual points of lubrication. Flow rates are controlled by the dimension of the restriction orifice at the end of the reclassifier and the supply pressure of oil mist. Some tribo-systems such as rolling element bearings can normally produce enough turbulence in oil mist as it flows through the bearing. This turbulence causes the oil droplets suspended in air to be deposited on the bearing surfaces to lubricate the bearing. Mist type reclassifiers are often used to lubricate bearings. They do not produce significant change in the droplet mass-size distribution. However, slow moving components such as chain and sprockets, gears, sliders etc. are unable to generate enough turbulence in and around themselves to cause the droplets in oil mist to be deposited on the operating surfaces. In these applications, reclassifiers which can induce turbulence in the flow such as the condensing or the spray type (Figure 19-1) reclassifiers are used.

The mechanisms of aerosol behavior which may be active in the operation of the reclassifiers are, (1) random turbulent coagulation, (2) kinematic coagulation at the exit and (3) turbulent deposition. The flow through the spray and the condensing type reclassifiers becomes very turbulent and the oil particles experience centrifugal force which tends to push them toward the reclassifier wall because of the swirling motion of the flow in the complex configuration of the swirl vanes. The turbulence in the flow produces eddies that can cause relative motion between particles. This relative motion can come from the velocity gradients or from the inertial projection of particles across eddies [35]. For monodispersed aerosol Fucs [43] gave the ratio of coagulation in turbulent flow to thermal coagulation as,

$$\frac{\mathbf{b}\,D_p^2}{64\,\pi\,D}\left[\frac{\rho_p\,\varepsilon}{\mu_g}\right]^{1/2}$$

Where, **b** is a constant of order 10, ε is the energy dissipation per unit mass of gas, D is the diffusion coefficient, μ_g is the viscosity of gas and ρ_g is the density of gas. The expression for coagulation due to random turbulence in polydispersed aerosol is much more complex and an explicit solution does not exist [38]. Kinematic coagulation may occur at the exit of the reclassifier. As the oil mist jet comes out of the reclassifier, the oil mist starts slowing down and the smaller particles slow down faster than the larger particle (due to inertia effects). This relative motion may cause the larger droplets to capture smaller droplets [35]. As oil mist flows through the narrow passages of the reclassifiers at high speed, the flow turbulence causes oil droplets to be deposited in the reclassifiers. The turbulent deposition mechanism may be the dominant mechanism in the operation of the spray and condensing type reclassifiers.

A model which combines all the mechanisms active in reclassifiers, which can account for the complex swirl vanes and the abrupt flow restrictions (the flow control orifice) and the polydispersed nature of oil mist, is not currently available.

Three commercially available reclassifiers have been tested [44]. These reclassifiers are known as the mist type reclassifier, the spray type reclassifier and the condensing type reclassifier. The cross-sectional views of these reclassifiers are shown in Figure 19-1. For slower moving equipment, reclassifiers which can induce turbulence in the flow, such as the condensing or the spray type reclassifiers, are often used. For rolling element bearings the mist type reclassifiers are normally used. In these tests the particle mass-size distributions of oil mist were measured before and after the test reclassifiers. For the mist reclassifier, the supply pressure of oil mist at the generator was maintained at 4.98 KPa (20 inch) water column pressure. For the condensing and spray type reclassifiers, the supply pressure was maintained at 8.71 kPa (35 inch) water column pressure. These supply pressures are normally used in the oil-mist lubrication system. A synthetic 100 (ISO VG 100) grade lubricating oil was used in the tests. The reclassifiers used in the experiments were rated for 84.9 cm^3/s (0.18 SCFM) flow rate and the reclassifiers had 1.2 mm (0.047 inch) diameter restriction orifices.

The fractions of the total mass of oil available in various droplet size ranges before and after the mist type reclassifier have been shown in Figure 19-2. The mass fraction is defined as,

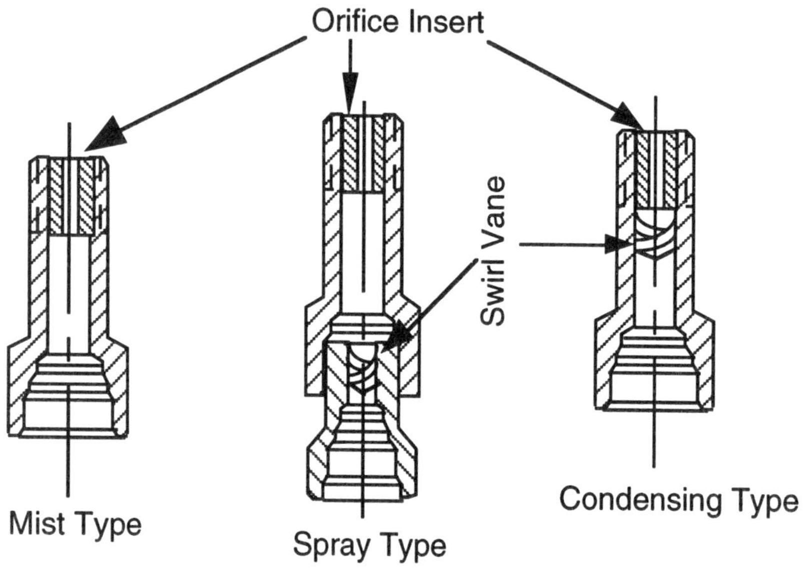

Figure 19-1. Cross-sectional views of reclassifiers.

$$\text{Mass Fraction} = \frac{\text{Mass of oil collected in an impactor stage}}{\text{Total mass of oil collected in the impactaor}} \times 100$$

The mass fractions shown in the plot represent the average of three measurements. There is negligible difference between the droplet mass-size distributions of oil mist before and after the mist type reclassifier. The maximum mass of oil is available in the 1.1 to 2.1 μm AED droplet size range. The mass fractions of oil mist in various size ranges have been shown in Figure 19-3 for spray and condensing type reclassifiers. For these reclassifiers there is a very sharp decline in the fraction of the total mass of oil available beyond the 1.1 to 2.1 μm AED size range after reclassification. However, for the condensing reclassifier the mass fraction of droplets in the size range 9.0 μm and above has increased. As oil mist flows through the spray or condensing type reclassifiers, it seems most of the large oil droplets are deposited in the reclassifiers. This deposited oil then drips out of the reclassifiers as large droplets due to gravity. The reclassifier exhaust jet often breaks these large droplets producing aerosol particles in the size range of 9.0 μm

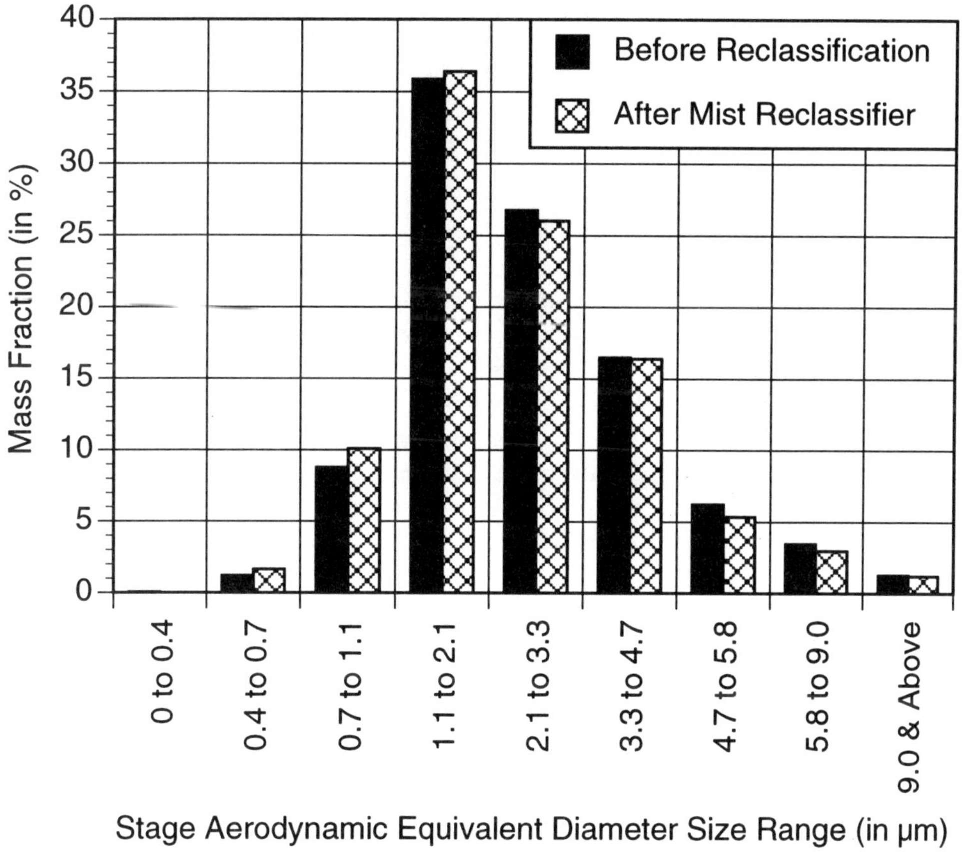

Figure 19-2. Mass-fractions of oil collected in various stages of the cascade impactor before and after the mist reclassifier. Here mass-fraction is the fraction of the total mass of oil in oil mist found in a size range.

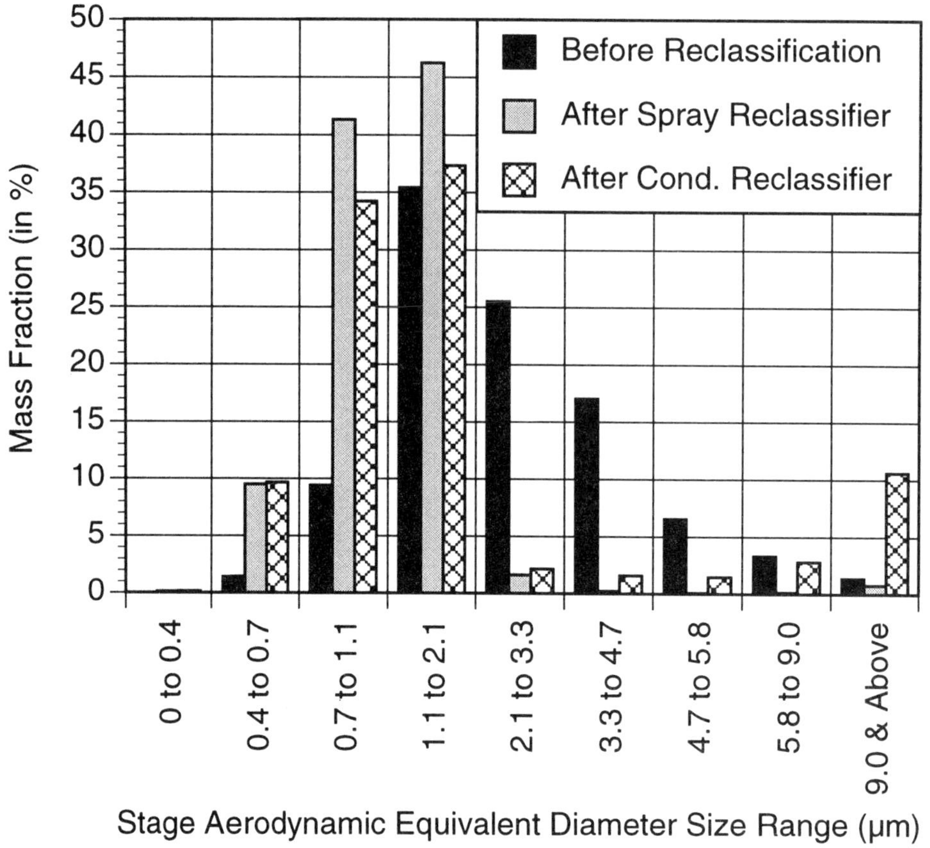

Figure 19-3. Mass-fractions collected in various stages of the cascade impactor before and after the condensing and spray type reclassifiers.

and above. Figure 19-4 shows the mass fractions of oil passing through the reclassifiers without deposition for various impactor stage size ranges. The stage size ranges are represented by the Effective Cutoff Diameter (ECD) for each stage which may be (approximately) considered as the average particle size for that stage. The mass fraction of oil passing through the reclassifier is defined as,

$$\text{Oil Passing through Reclassifier} = \frac{\text{Mass of oil at the reclassifier exit}}{\text{Mass of oil at the reclassifier inlet}} \times 100$$

The mist reclassifier seems to be capturing some large droplets (>1.5 μm AED) and reducing the mass of oil available in those size ranges slightly. However, the spray and condensing reclassifiers seem to be capturing almost all the oil droplets larger than 2.5 μm AED. The increase in the oil mass available in the largest size range is due to the breakup of the dripping oil from the reclassifiers by the exhaust jet. The oil deposited in the reclassifier as a percentage of the total mass of oil going into the reclassifier is shown in Figure 19-5. It seems that about 10% of oil in oil mist (by mass) is deposited in the mist

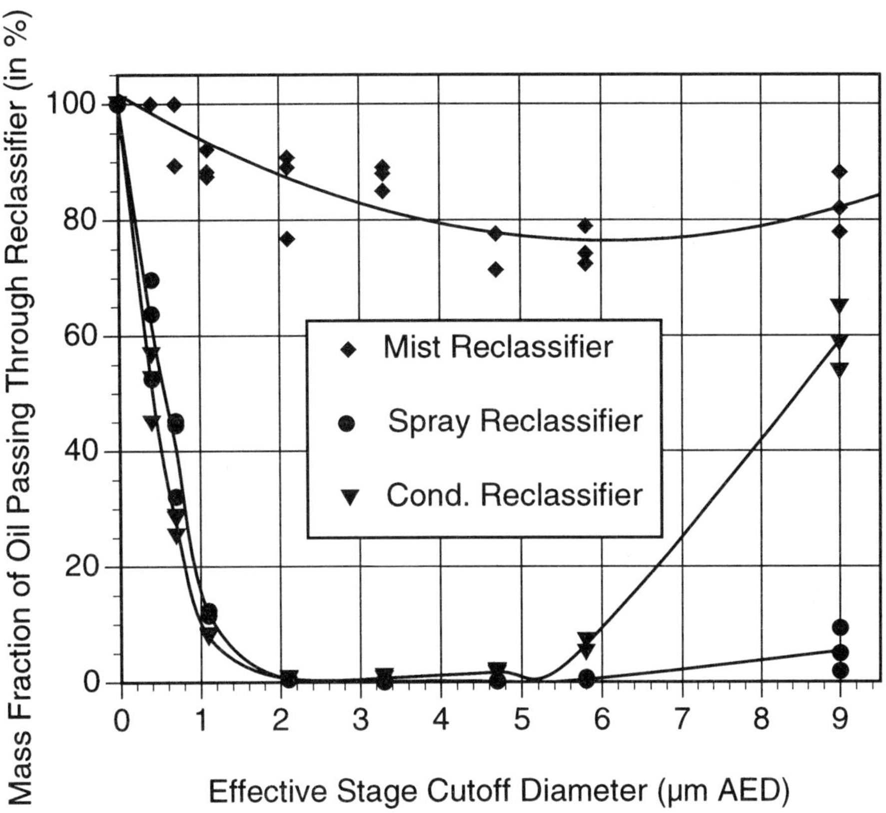

Figure 19-4. Mass-fractions of oil passing through the reclassifiers in various size ranges of the cascade impactor (see Figure 18-2). Here the particle size ranges have been expressed as effective stage cutoff diameters (see Figure 19-8) for the explanation of stage cutoff diameter).

reclassifier, 90% of the oil in oil mist is deposited in the spray reclassifier and about 92% of the oil in oil mist is deposited in the condensing reclassifier.

Experience shows that commercially available mist type reclassifiers cause negligible change in the mass-size distribution of oil mist. However, some larger beneficial oil particles are deposited in the mist reclassifier itself. The complex swirl vanes of the spray and condensing reclassifiers cause almost all the larger oil droplets (beyond about 2.0 μm AED) to be deposited in these reclassifiers. The deposited liquid oil drips out the reclassifiers due to gravity. The fine oil mist particles which pass through these reclassifiers mostly contribute to stray mist in an open loop oil-mist lubrication system.

Direction of Reclassified Oil-mist Jet

The oil mist comes out of the reclassifiers as aerosol jets. For the tests mentioned in the previous section, the oil mist coming out of the mist reclassifier had about 90 m/sec velocity. It has been shown that the width of a circular fluid jet is proportional to the axial

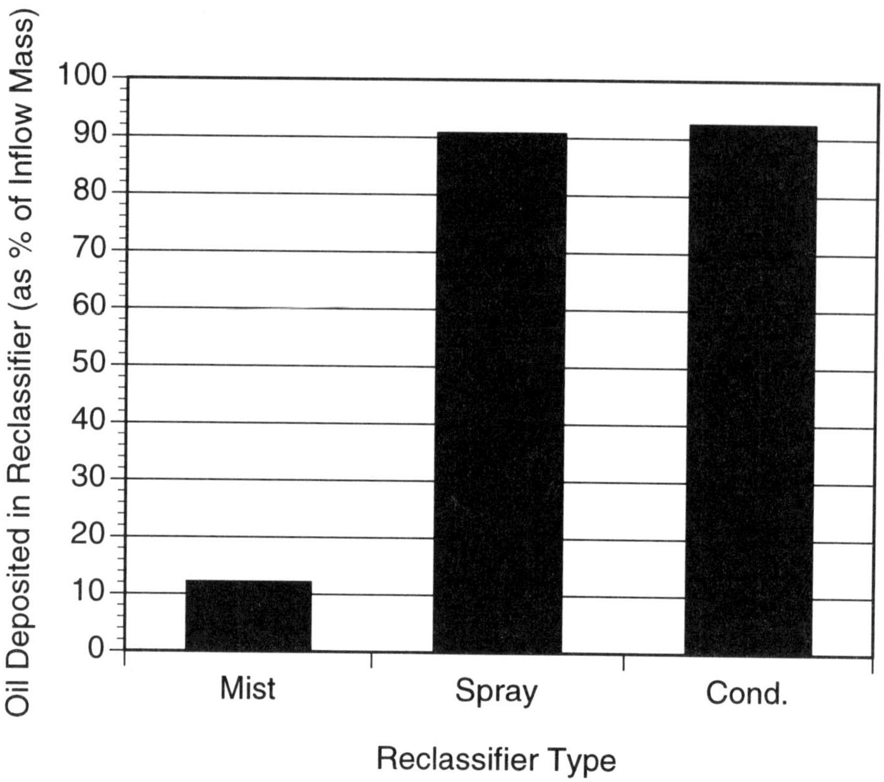

Figure 19-5. Oil deposited in a reclassifier as the percentage of total mass inflow of oil for three reclassifiers.

distance (x) from the origin of the jet and the center line velocity (which is the maximum velocity at any cross-section taken at 90 degrees to the center line or axis of the jet) is inversely proportional to the axial distance from the origin. A diagram of the pattern of streamlines in a circular turbulent free jet is shown in Figure 19-6. For a turbulent circular jet [45],

$$u = \frac{3K}{8\pi\varepsilon_o x}\left[\frac{1}{\left(1.0 + \frac{1}{4}\eta^2\right)^2}\right] \qquad (19\text{-}1)$$

$$u = \frac{1}{4}\left\{\frac{3}{\pi}\right\}^{1/2}\frac{K^{1/2}}{x}\left[\frac{\eta - \frac{1}{4}\eta^3}{\left(1.0 + \frac{1}{4}\eta^2\right)^2}\right] \qquad (19\text{-}2)$$

here, $$\eta = \frac{1}{4}\left\{\frac{3}{\pi}\right\}^{1/2}\frac{K^{1/2}}{\varepsilon_o}\frac{y}{x} \qquad (19\text{-}3)$$

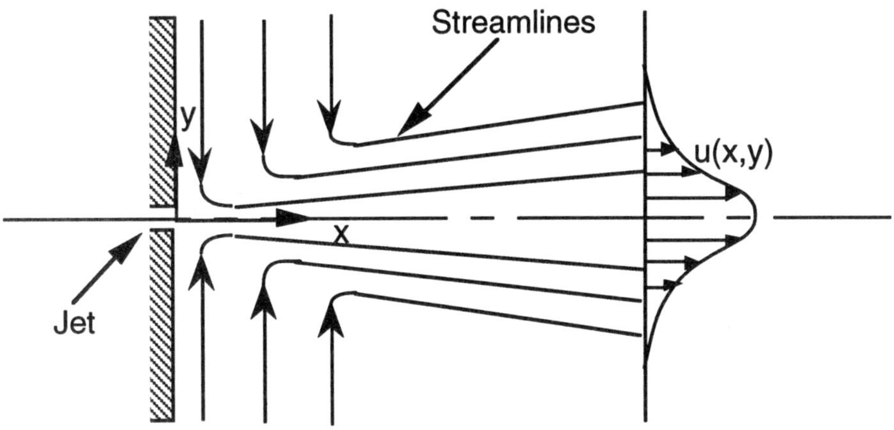

Figure 19-6. Patterns of streamlines for a circular jet.

$$\frac{\varepsilon_o}{K^{1/2}} = 0.0161 \tag{19-4}$$

$$K = 2\pi \int_0^{\alpha} u^2 y \, dy = \left(\frac{\pi D_0^2}{4} \right) U_o$$

here,

u = fluid velocity in x direction
v = fluid velocity in y direction
K = kinematic momentum of jet
ε_0 = virtual kinematic viscosity for turbulent flow
D_0 = diameter of orifice or nozzle = 0.047 inch = 0.1194 cm.
U_0 = exhaust velocity of orifice or nozzle = 90.7 m/s

It can be shown from the above equations that in order to reduce the speed of the oil-mist jet from 90.7 m/s to about 3 m/s, the jet needs to travel about 220 mm (or about 9 inch) distance. In most applications the oil-mist jet will impinge on the shaft or the casing (Figure 17-2) before traveling a distance of 220 mm (9 inches). The impingement of high speed aerosol on a surface causes particles (in particular, beneficial larger oil-mist particles) to be deposited on the impinged surface. This mechanism of aerosol particle deposition is known as the inertial impaction which is also the principle on which the Inertial Cascade Impactor works. With inertial impaction, aerosol is forced through a nozzle or orifice and the output jet is directed against a flat plate called the impaction or collection plate (as shown in Figure 19-7). The plate deflects the flow to form an abrupt 90 degree bend in the streamlines. Depending on jet velocity, the larger particles experience higher inertia forces compared to the drag force of the fluid which causes them to impact the flat plate as the streamlines bend. On the other hand, the smaller particles experience higher drag forces compared to the inertia force allowing them to stay in the streamline and flow with the fluid without impacting the flat plate [35,38]. The most important non-dimensional parameter for inertial impaction is the ratio of centrifugal force on the particle as it flows along a curved stream line to the drag force of the flowing fluid on the particle, which as explained before, is known as the Stokes number,

$$S_{tk} = \frac{C \rho_p D_p^2 U_o}{18 \mu_g R_c} \tag{19-5}$$

here, D_p = diameter of particle

C = Cunningham's correction factor which corrects for the slip of flowing fluid on particle surfaces when particle radius approaches mean free path of the carrier gas molecules [35].

$$= 1.0 + \frac{2\lambda}{D_p}\left[1.23 + 0.41 \exp\left(\frac{-0.55\, D_p}{\lambda}\right)\right]$$

λ = mean free path of gas molecule
ρ_p = density of particle
U_o = velocity of aerosol jet
μ_g = viscosity of gas
R_c = radius of curvature or radius of round jet

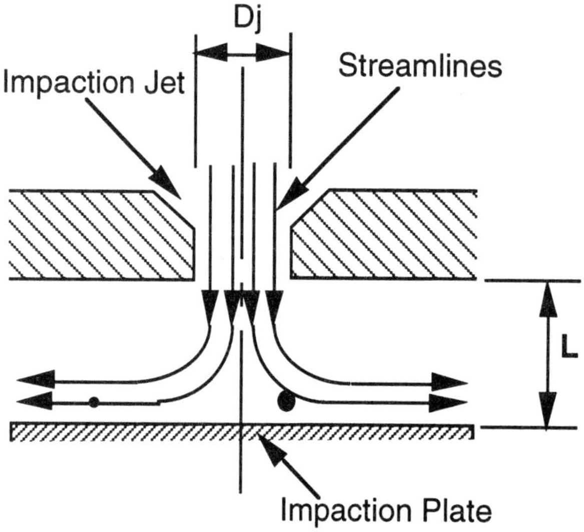

Figure 17-7. Diagram showing the principle of inertial impaction.

The particles above a certain aerodynamic size (ideal case) are collected on the impinged surface. The smaller particles remain air borne and flow out without deposition. This cutoff aerodynamic equivalent diameter (AED) depends on the jet velocity, the viscosity of the carrier gas and the diameter of the orifice or nozzle. The collection efficiency of the impaction is defined as,

$$\eta = \frac{\text{mass of aerosol collected}}{\text{mass of aerosol which enters the acceleration jet}}$$

A typical collection efficiency versus Stokes number curve is shown in Figure 19-8. The number which corresponds to 50% collection efficiency is known as the cut point Stokes number ($S_{tk0.5}$) for the stage and the particle diameter corresponding to that Stokes number is known as the cut point aerodynamic particle diameter (D_{cp}) for that stage [38],

$$D_{cp} = \left[\frac{9\,\mu_g\, D_j\, S_{tk0.5}}{C\,\rho_p\, U_o}\right]^{1/2} \tag{19-6}$$

In order to avoid deposition of larger beneficial oil droplets on non-load bearing surfaces, such as the casing or the shaft, whenever possible the reclassified oil-mist jet should be directed toward the load bearing rolling elements of a bearing.

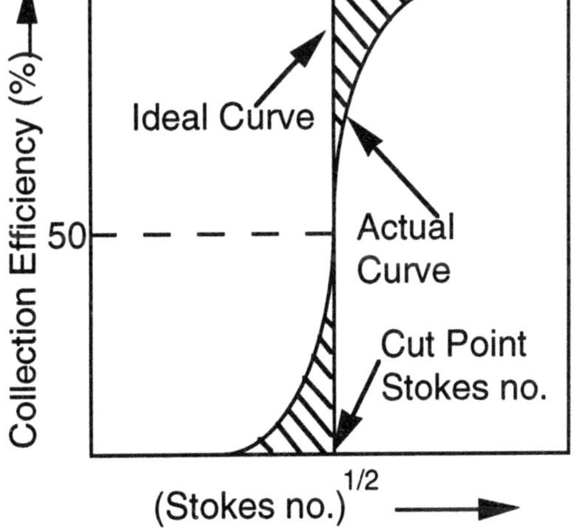

Figure 19-8. Collection efficiency vs. Stokes no. plot for an impactor stage.

Chapter 20

Deposition of Oil at the Point of Application and Stray Mist

DEPOSITION OF OIL AT THE POINT OF APPLICATION

After flowing through reclassifiers, oil mist enters the rolling element bearing to be lubricated. In general, mist type reclassifiers are used with rolling element bearings. While rotating, rolling element bearings can produce enough turbulence in the tortuous flow paths in and around the bearing to cause the droplets in oil mist to be deposited on the operating surfaces. Since in a typical rolling element bearing only the operating surfaces such as the inner race and the rolling elements rotate, oil mist is selectively deposited on the operating surfaces. On the other hand, the oil-sump lubrication splashes oil all over the casing and the bearing. This is the principal reason behind the ability of oil-mist lubrication to lubricate bearings with very little oil.

The deposition mechanisms of aerosol which may be active in the deposition of oil mist in rolling element bearings are impaction, interception, diffusion, turbulent coagulation and diffusion, and deposition due to the centrifugal force in the boundary layer of the spinning rolling elements. As oil-mist-flow stream lines bend around various obstructions such as the rolling elements, the bearing races and the rolling element retainers, some oil droplets are deposited on the obstruction surfaces due to inertia. This is known as the aerosol impaction deposition mechanism. As oil mist flows around an obstruction, some droplets may be deposited on the obstruction surface by coming in contact with the surface. This is known as the interception mechanism of aerosol deposition. Oil-mist particles may be diffused from the flow to the surfaces of bearings as oil mist flows over them. This is called the diffusion mechanism of aerosol deposition [35,38]. Also, as oil mist flows around the spinning rolling elements of the bearing, the oil-mist droplets in the boundary layer are thrown outward due to the centrifugal force caused by rotation of the rolling elements. These droplets are then deposited in the line of contact between the rolling elements, the inner race and the outer race, providing lubrication just where it is required. Models are available for the collection efficiency of aerosol on components with simple geometry such as spheres, cylinders etc. However work needs to be done to develop models for the deposition of oil mist on complex geometry components and flow fields around rotating bearings.

Tests have been conducted to investigate the deposition of oil droplets in oil mist in rolling element bearings [44]. In these experiments, the mass-size distributions were measured before oil mist entered the test bearing and after it flowed through the rotating bearing. Mist type reclassifiers rated for 85 cm^3/s (0.18 SCFM) were used in these tests.

The test bearing was a 60 mm bore diameter high precision angular contact ball bearing with 15 degree contact angle. The mist reclassifier was connected radially to the load side of the bearing at the oil mist inlet as shown in earlier Figure 17-2. The exhaust oil mist from the bearing went through a radial vent hole on the support side of the bearing. The vent hole was placed on the diametrically opposite side of the inlet hole. The test bearing speeds were 200, 1200, 2200 and 3200 rpm which corresponded to 0.81, 4.87, 8.93 and 12.98 m/s linear speed at the pitch diameter of the test bearing.

In Figure 20-1, the overall oil collection efficiency of the bearing has been plotted as a function of the linear speed of the bearing at pitch diameter. The overall collection efficiency is defined as,

$$\text{Overall Collection Efficiency} = \frac{\text{Total mass of oil collected in the bearing}}{\text{Total mass of oil entering the casing}}$$

The solid line represents a second degree polynomial fit of the experimental data (R-square >0.95). The bearing speed seems to have a marginal influence on the overall oil collection efficiency. At very low speeds, the collection efficiency is still significant. A sixteen-fold increase in bearing speed caused the overall collection efficiency to increase from about 40% to about 60%. Figure 20-2 shows the variation of collection efficiency of

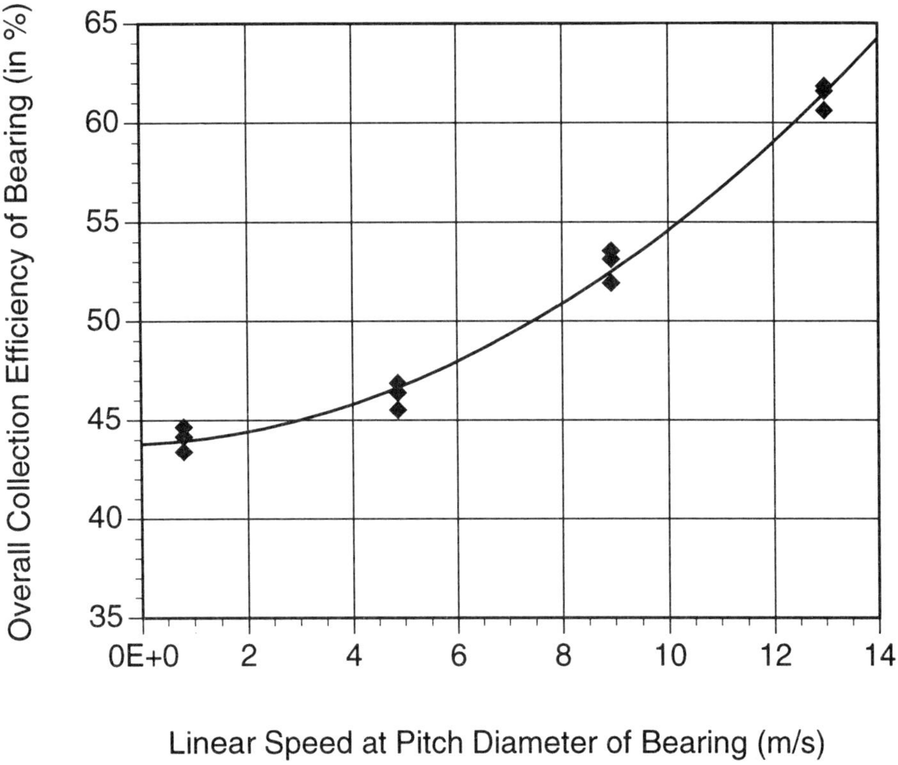

Figure 20-1. Variation of the overall oil collection efficiency of the test bearing with variation of bearing speed. Here oil collection efficiency is the percentage of oil entering the bearing housing being deposited in the bearing.

oil droplets in four different size ranges with variations of the bearing speed. Here, the collection efficiency is defined as,

$$\text{Collection Efficiency} = \frac{\text{Mass of oil collected in the bearing in a size range}}{\text{Mass of oil entering the casing in the same size range}}$$

The experimental data has been fitted to second degree polynomials which are shown as solid lines in the plot (R-square > 0.90). As speed increases, the collection efficiency of the large droplets should increase faster than the smaller droplets, since larger droplets are more prone to turbulent deposition caused by the increase in speed. However, the rate of increase of the collection efficiency with bearing speed is higher for smaller droplets than that of larger droplets in Figure 20-2. This is due to random turbulent coagulation among the smaller droplets which produces large droplets that show up in the exhaust. This may cause the calculated collection efficiency for the larger droplets to be lower than their actual value. The variations of the collection efficiency of the bearing are shown as functions of droplet size ranges for two bearing speeds in Figure 20-3. The droplet size ranges are represented as the effective cutoff diameters for the various stages of the impactor. The solid lines in the plot represent the best fit (spline fit) of the experimental data. Most droplets with less than 1.0 μm AED size seem to be going out of

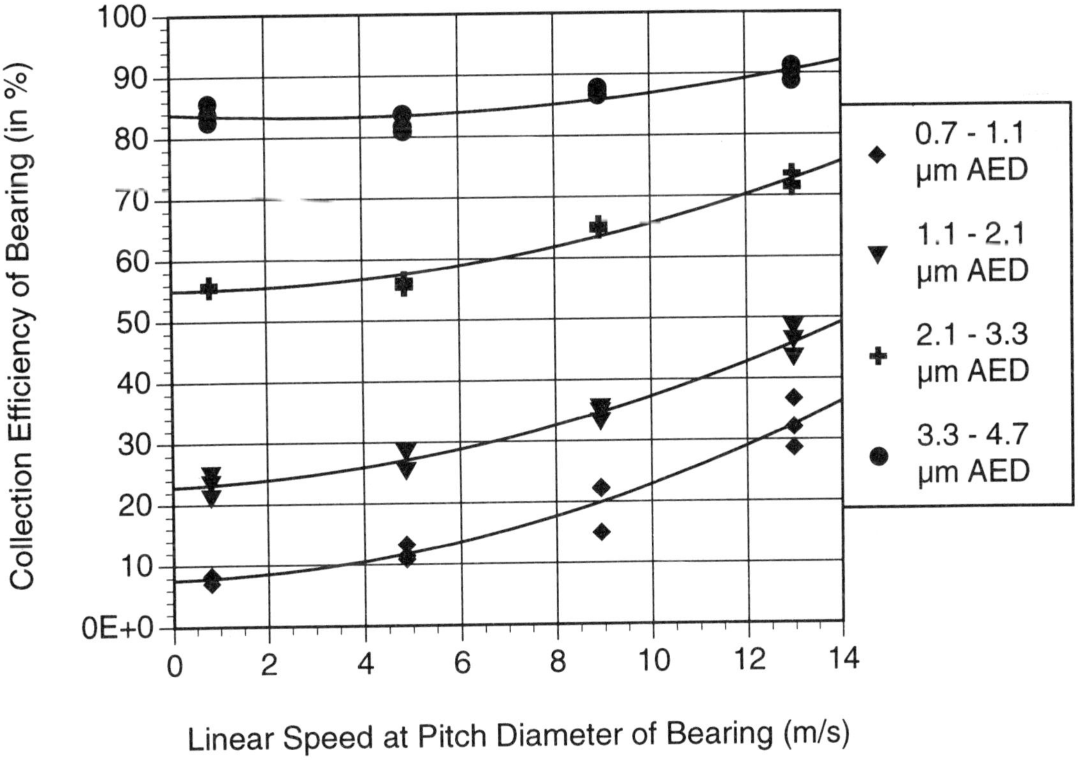

Figure 20-2. Variations of the oil collection efficiency of the test bearing speed for four oil droplet size ranges.

the bearing without deposition. Between 1.5 μm AED and 3.5 μm AED the rate of increase in the collection efficiency is very sharp. By increasing the droplet size from 1.5 μm AED to 2.5 μm AED, the collection efficiency can be almost doubled. If the droplet mass-size distribution of oil mist is changed at the generator (by shifting the peak of the mass-size distribution to a higher size range), the penetration of oil mist in the supply line will decrease slightly [44]. However, it should be pointed out that the decrease in penetration will be compensated by the increase in the collection efficiency.

The overall amount of oil reaching the bearing will be less, but since the collection efficiency is higher, most of the oil will be deposited. This should result in the same or better lubrication and a lower amount of oil in the exhaust. Figure 20-3 may be used as a guideline in designing reclassifiers to produce droplets most suitable for deposition in bearings. The mass-size distributions of oil mist before and after the test bearing are shown in Figure 20-4. This plot shows the change in the mass fraction as oil mist flows through the test bearing. The number size distributions of oil mist before reclassification, after reclassification and after the test bearing (operating at 3200 rpm) are shown in Figure 20-5. The number fraction is defined as,

$$\text{Number Fraction} = \frac{\text{Number of particles collected in a stage}}{\text{Total number of particles collected in the impactor}}$$

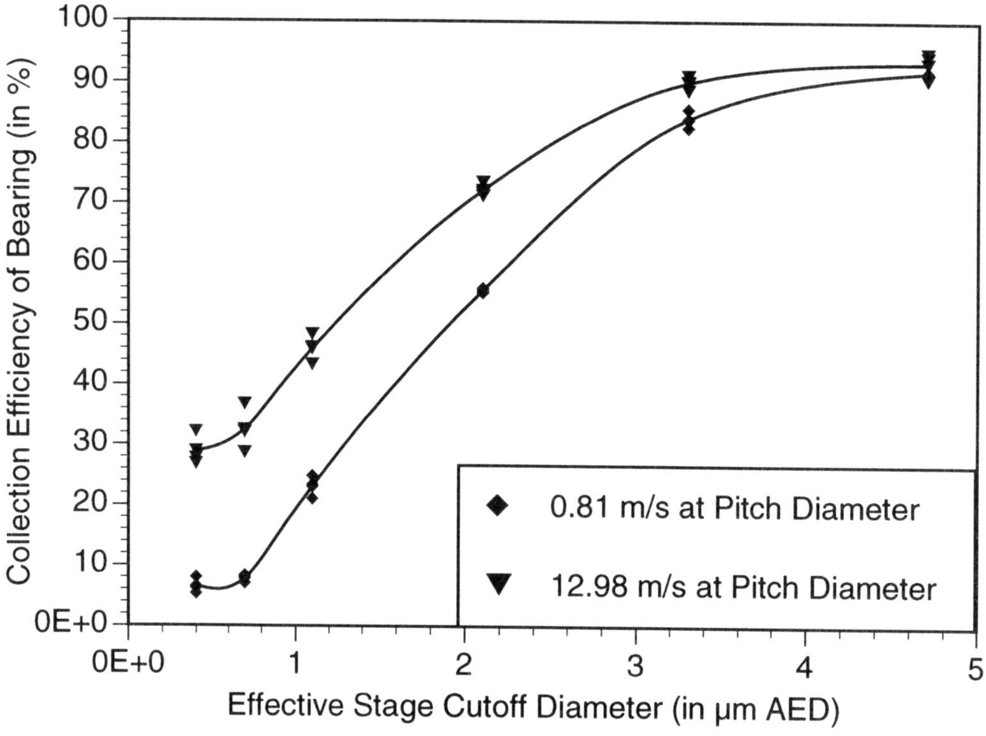

Figure 20-3. Variation of the oil collection efficiency of the test bearing with variation of oil droplet size for two bearing speeds. Here oil collection efficiency is the percentage of oil entering the bearing housing being deposited in bearing.

As shown in Figure 20-4, only about 2% of the oil mist by mass has been found in the 0.4 to 0.7 μm AED size range after the bearing. However, by number, this oil mist represents about 30% of the total number of droplets in stray mist as shown in Figure 20-4.

It seems that about half of the oil in oil mist (by mass) may leave a bearing without deposition. About 70% to 90% of the mass of oil available below 1.0 μm AED leaves the bearing without deposition. However, the actual mass of this oil leaving the bearing is very small because the oil-mist lubrication system consumes very little oil. It should again be mentioned that all applications of the oil-mist lubrication system have easily met the United States OSHA requirement of maximum allowable 5 mg of oil per cubic meter of air. It has been observed that the operating speed of bearings has marginal influence on the deposition of oil droplets in bearings. Oil droplets in the size range of 2.0 to 4.0 μm AED were found to be ideal for an oil-mist lubrication system. These particles provide high oil collection efficiency in the bearing and reasonable penetration in the supply line. Oil droplets smaller than about 1.5 μm AED do not lubricate bearings. Only a small percentage of oil by mass has been found to be present in the sub-micrometer size ranges. However, by number this oil represents a very large percentage of the oil droplets. These particles cause the oil mist to appear dense. Moreover, these particles primarily contribute to stray mist in an open loop oil-mist lubrication system and should be avoided.

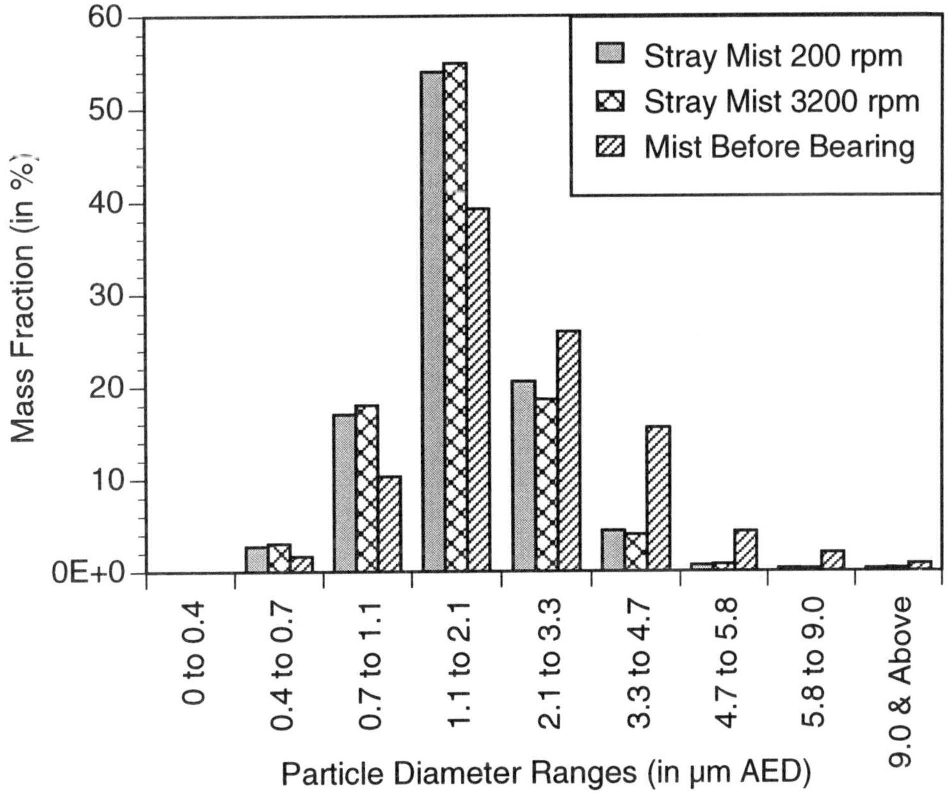

Figure 20-4. Mass-size distribution of oil mist before and after the test bearing.

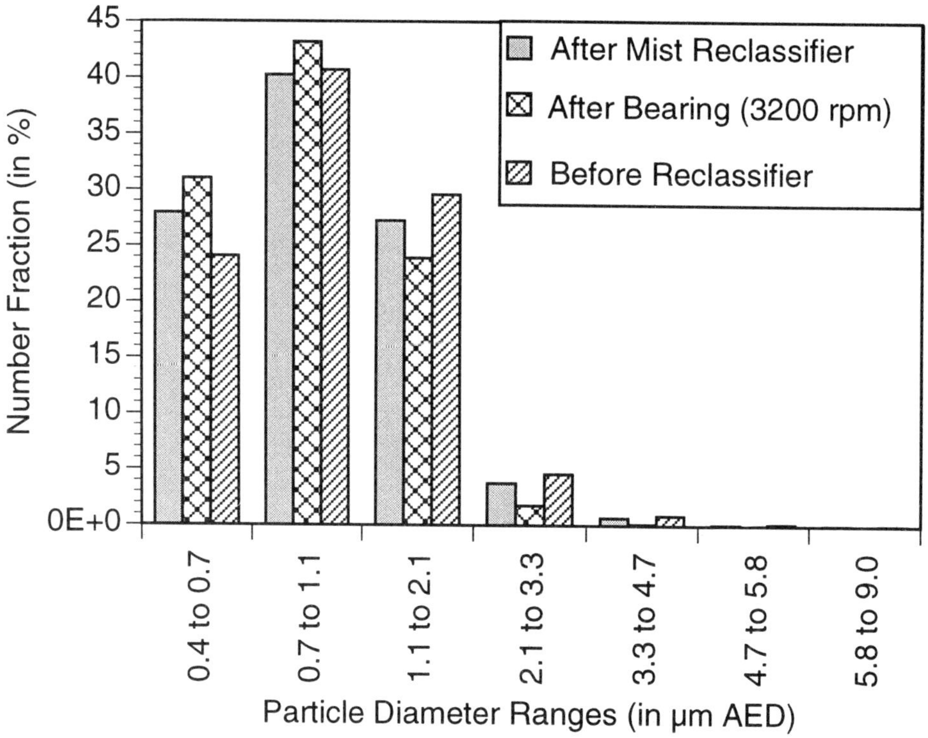

Figure 20-5. Number-size distributions of oil mist before reclassification, after reclassification and after test bearing. Here number-fraction is the fraction of the total number of particles found in a size range.

Stray Mist

Open loop oil-mist lubrication, which allows the used oil mist containing small amounts of lubricating oil to escape into the atmosphere, is primarily used in outdoor applications. The base lubricating oils typically present few hazards to humans from acute exposure. The additives and the contaminants in used lubricating oils however, may be toxic [47]. Since the stray mist contains a small mass of oil, applications can easily meet the OSHA requirement of maximum allowable 5 mg of oil per m^3 of air. This requirement was still on the books when this text was prepared in 1997. However, it has been found that the stray mist often contains large numbers of sub-micrometer oil particles [46]. Relatively large aerosol particles (about 2.5 μm Aerodynamic Equivalent Diameter or larger) can be filtered out by the protective filtering mechanisms of the human respiratory system. The Aerodynamic Equivalent Diameter (AED) is defined as the diameter of the unit density (1 g/cm^3) sphere that has the same terminal settling velocity due to gravity as the particle in quiescent air. Aerosol particles having approximately 1μm AED or less can penetrate sensitive regions of the lung [35] as shown in Figure 20-6

Several oils containing special additives have been studied [46]. The study was focused on reducing sub-micrometer particles in the stray mist without adversely affecting the performance of the oil-mist lubrication system. In the initial phase of this study several commercially available oils were tested for their stray mist characteristics. These oils are code named as Lube C, Lube J, Mist Lube 100, and Mineral Oil 46, and two variations of Lube C, identified as Lube A and Lube B. Later, five more variations of Lube C were studied, considering the performance of Lube C as the "bench mark." These oils are identified as Lube D, Lube E, Lube F, Lube G and Lube H. Some general information on these oils is given in Chapter 17 of this text. The purpose of this discussion is to demonstrate the effectiveness of additives in the reduction of fine particles in the stray mist.

In these tests for each oil the particle mass-size distributions of oil mist were measured before entering the test bearing and after the test bearing (stray mist). A diagram of the test bearing setup has been shown in Figure 17-2. For these tests the 77.5 mm pitch diameter angular contact ball bearing was operated with about 125 kg load at 2400 rpm. A mist type reclassifier (0.09 standard cubic ft. per minute or 43 cm^3/s) was used to lubricate the bearing.

Figure 20-7 shows the results from the first phase of this study, where four different oils along with two variations of Lube C, identified as Lube A and Lube B, were tested. Lube J, Mist Lube 100, Lube A and Lube B oils had less sub-micrometer particles in the stray mist than Lube C and Mineral Oil 46. However, the oils with less fine particles in the stray mist also deposited less oil in the bearing as compared to the oil deposited in the

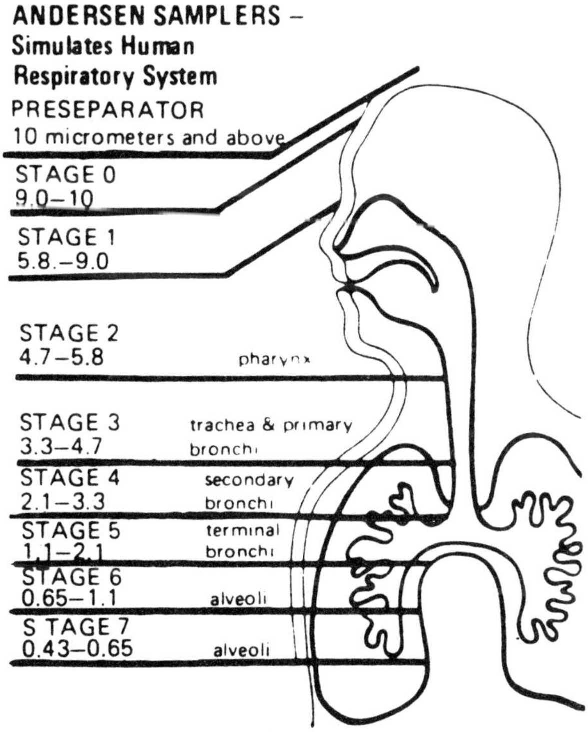

Figure 20-6. Human respiratory system as simulated by the inertial cascade impactor used in tests to measure particle mass-size distribution of oil mist (Andersen Instruments, Inc., Atlanta, GA 30336).

bearing by Lube C. Figures 20-8 through 20-11 show the results from the second phase of this study in which the stray mist characteristics and the performance of several variations of Lube C have been compared. The plots show all seven variations of Lube C along with the data for Lube C. The mass-size distributions of stray mist for the eight test oils are shown in Figure 20-8. The total mass of oil collected in the test bearing per one cubic foot of oil-mist flow through the bearing is shown in Figure 20-9 for the eight lubricating oils. The Lube A and Lube G oils deposited about 4% more oil in the bearing than Lube C. All other oils deposited less oil in the bearing than Lube C. The collection efficiency i.e., the fraction of the total mass of oil deposited in the bearing, is shown in Figure 20-10 for the eight lubricating oils. This plot shows that in terms of the collection efficiency all except Lube F performed better than Lube C. Lube D, Lube E and Lube G provided about 60% collection efficiency. Lube A gave about 5% higher efficiency than Lube C, which yielded a collection efficiency of about 54%. Lube F gave the least satisfactory performance in terms of collection efficiency. In Figure 20-11 the number of particles in the stray mist for any of the eight oils in any size range is shown relative to every 100 particles in the 0.4 to 0.7 μm AED size range stray mist for Lube C. Figure 20-11 shows the comparison of the appearance of stray mist for the eight lubricating oils. Among the eight test oils, Lube C had the maximum number of sub-micrometer particles in the stray mist. Lube A and Lube G both had significantly less visible stray mist than Lube C. In the stray mist, the Lube A and Lube G oils had about 75% less sub-micrometer particles than Lube C.

It is clear from the above discussion that the amount of useless sub-micrometer particles in the stray mist can be reduced by the use of appropriate additives in the lubricating oil without sacrificing the performance of the oil-mist lubrication system.

In recent years considerable advances have been made in designing devices to remove oil particles from the stray mist. These devices typically use the power of an electric motor to force oil mist through filters which remove oil particles from the spent oil mist (Figure 20-11a). This filter has to be replaced periodically for peak performance and less wastage of power. These systems are used with closed loop oil-mist lubrication system, in which spent oil mist is collected and passed through these filtering devices [48,49]. It has been shown that these systems can significantly reduce the amount of oil and visible oil particles in stray mist and help meet and exceed the OSHA requirement of maximum 5 mg/m^3 of oil in air [50]. However, as discussed earlier, from the human health point of view removing sub-micrometer oil particles from the spent oil mist is more important than removing larger and more visible particles. Mist collecting devices should be studied to check for the presence of sub-micrometer particles in the filtered air, since sub-micrometer particles are typically difficult to filter out using conventional filtering systems. Fiber bed filters have also been successfully used to remove oil particles from the stray mist [51].

A filtering system using the cyclone principle of particle filtration may be more suitable in filtering oil particles from the spent oil mist [35]. The cyclone system is more suitable for liquid particle filtration because unlike solid particles, oil particles cannot bounce back into the air flow from the walls of the cyclone filtration system. These systems have been developed under the auspices of the Environmental Protection Agency to collect inhalable airborne particles. Cyclone systems require a pressure drop to operate; however, they do not generally require a filter.

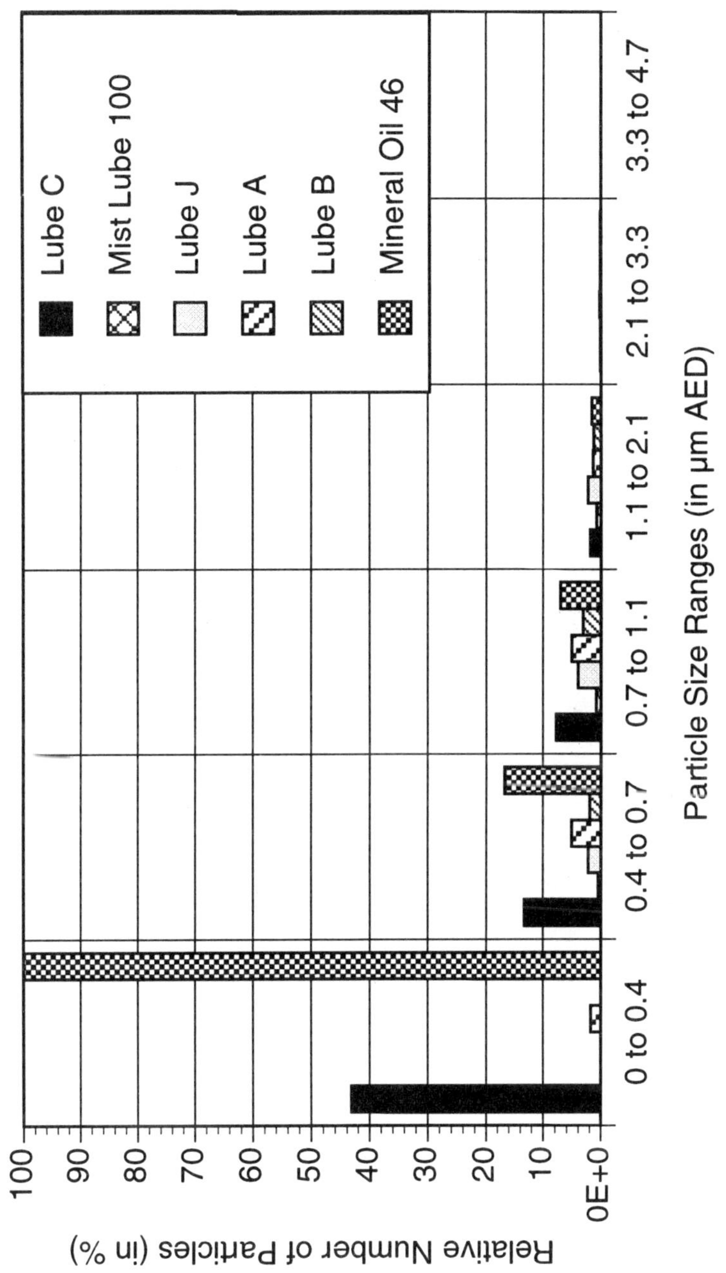

Figure 20-7. Relative number-size distribution of stray mist for several lubricating oils. The numbers of particles shown in various size ranges are for every 100 particles in the 0 to 0.4 µm AED size range for the Mineral Oil 46.

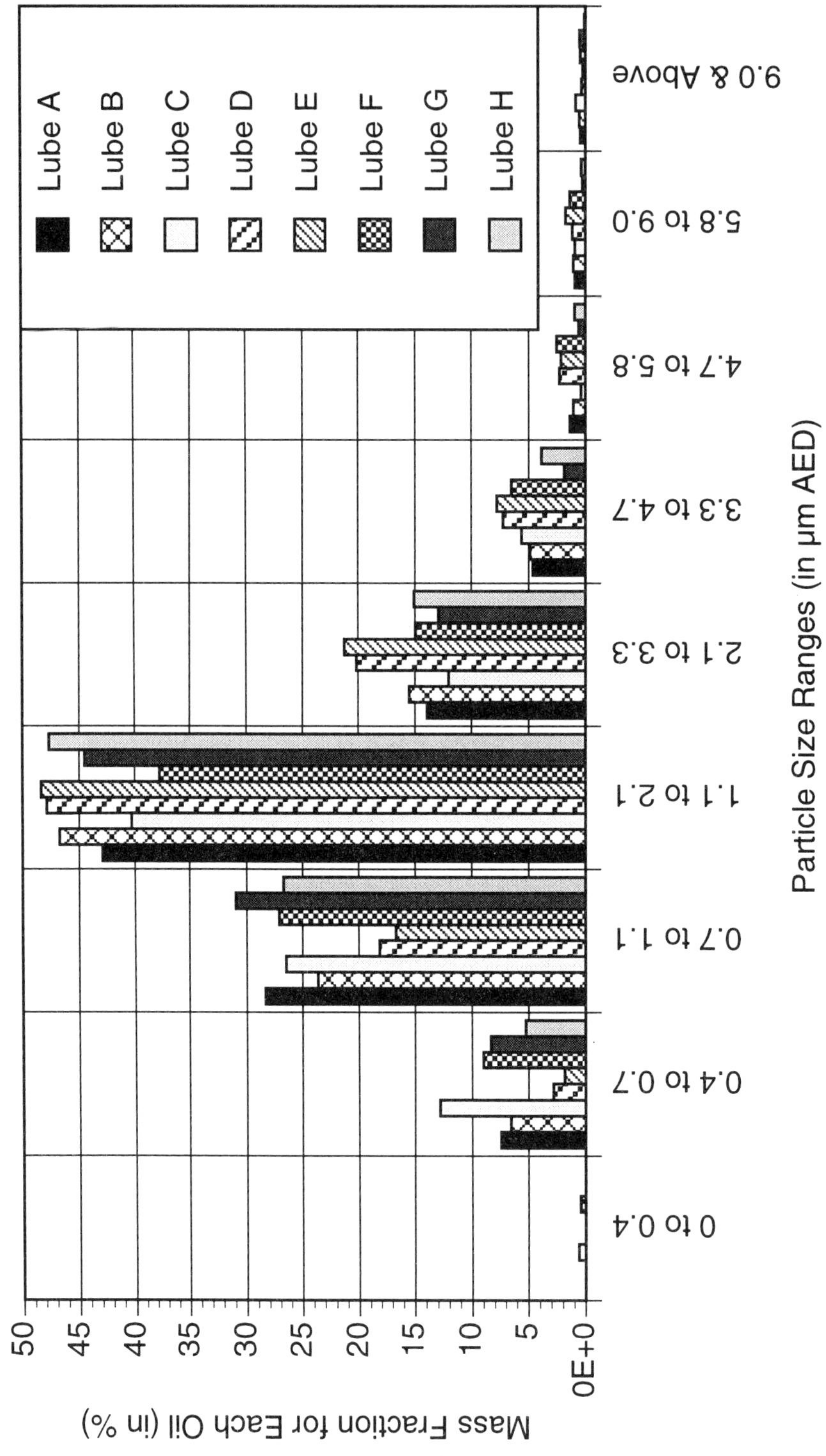

Figure 20-8. Mass-size distribution of stray oil mist for several lubricating oils. Here mass fraction is the fraction of the total mass of oil found in a size range for a particular oil.

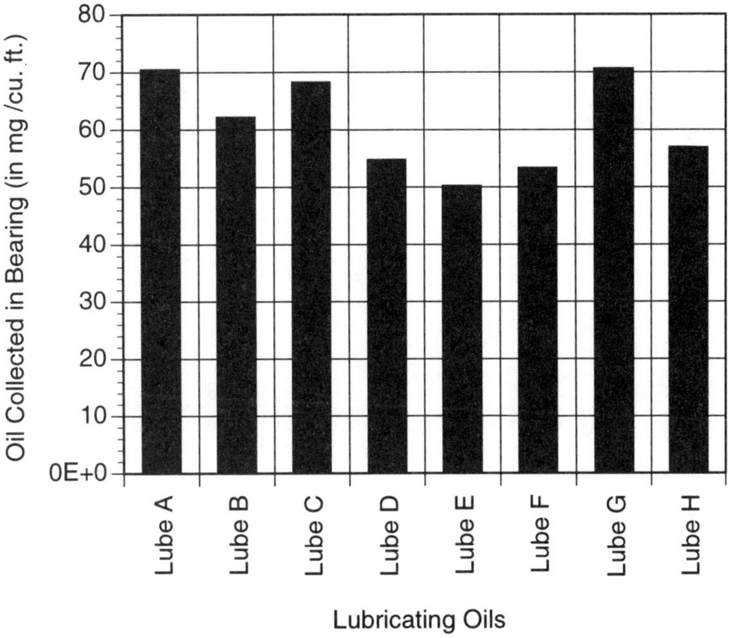

Figure 20-9. Total mass of oil collected in the test bearing for various lubricating oils per 1 cu. ft. of oil mist flow.

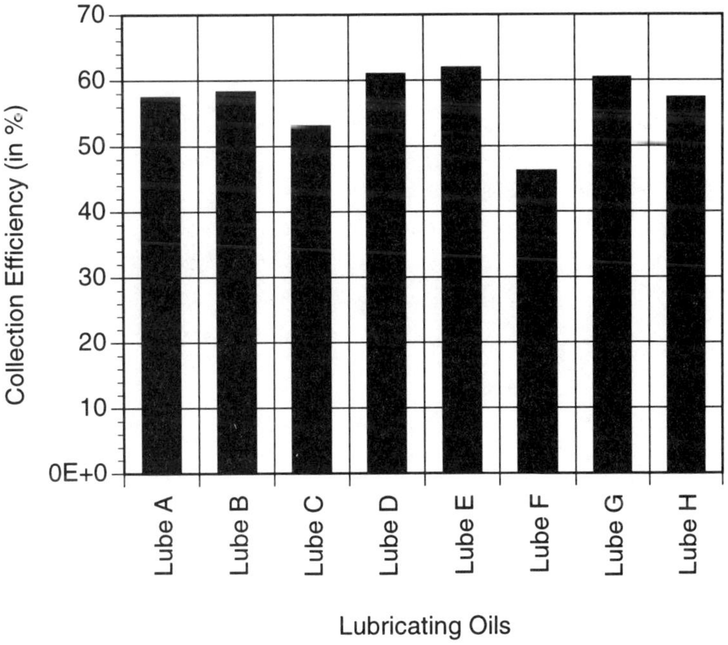

Figure 20-10. Overall oil collection efficiency for various lubricating oils. Here oil collection efficiency is the fraction of the total mass of oil (available in oil mist) collected in the test bearing.

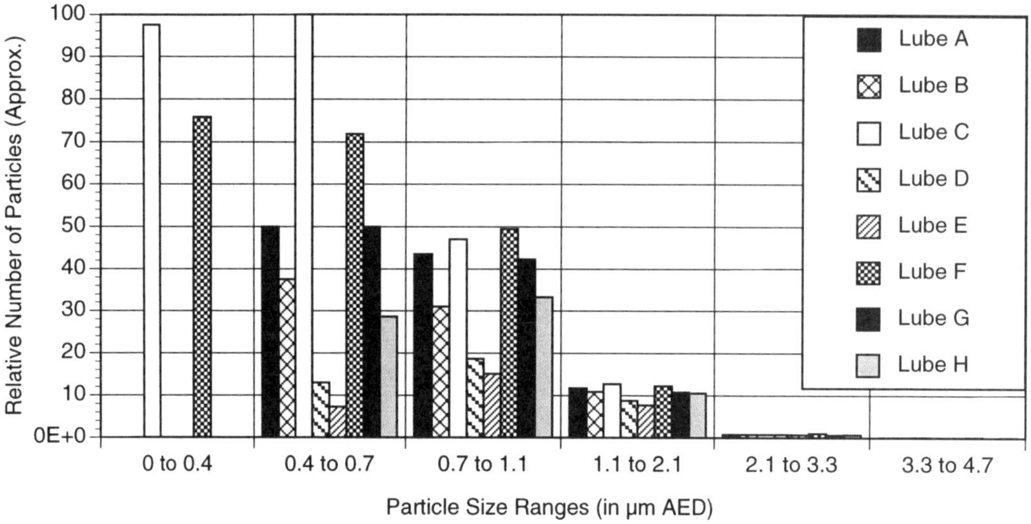

Figure 20-11. Relative number-size distribution of stray mist for several lubricating oils. The numbers of particles shown in various size ranges are for every 100 particles in the 0.4 to 0.7 µm AED size range for the Lube C.

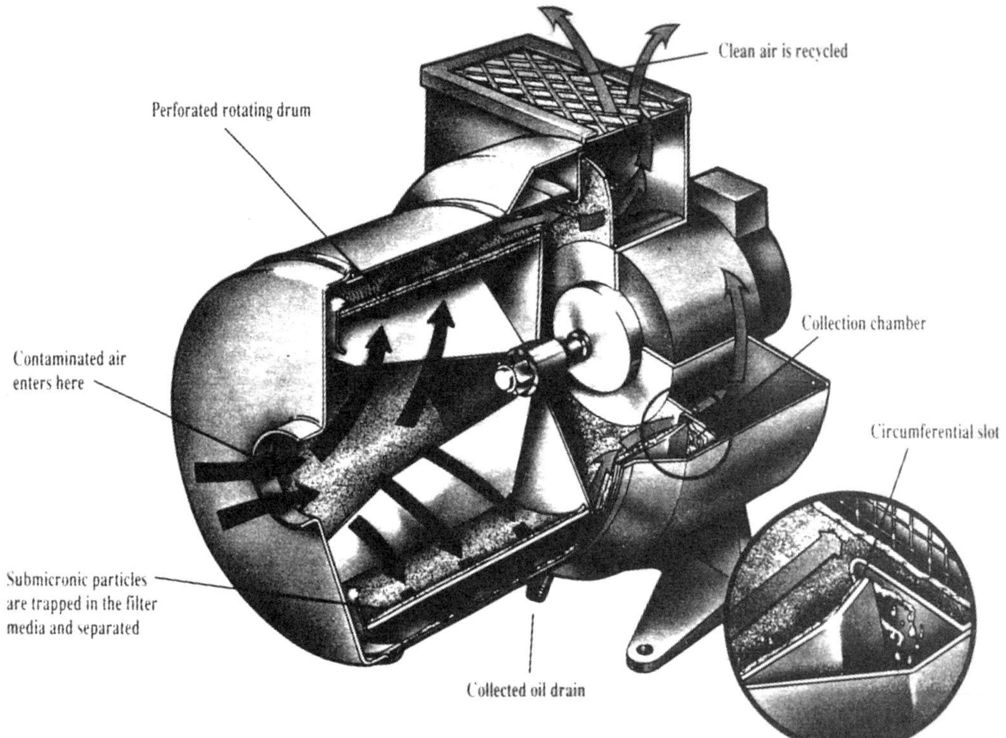

Figure 20-11a. Stray mist filtering system commonly used with closed loop oil mist lubrication system (Aercology, Inc.)

Chapter 21

Tribological Performance of Oil-mist Lubrication

Although oil-mist lubrication has been available for a long time, it is only now gaining more widespread acceptance in many industries. There used to be a very sharp difference in opinion about the performance of oil-mist lubrication among practicing engineers. The manufacturers and users of oil-mist lubrication have always been highly impressed by its performance [30,14&29]. In one application [29], it has been reported that the bearing failure rate went down by as much as 90% when conventional sump lubrication was replaced by oil-mist lubrication. However, nonusers and many other practicing engineers were often skeptical about oil-mist lubrication. A scarcity of published scientific data and little, if any independent research findings existed on how and why oil-mist lubrication systems work. This may be the principal reason for the earlier skepticism about oil-mist lubrication. Most of the knowledge about the performance of oil-mist lubrication systems has been obtained from practical industrial applications, where often little control could be exercised over the elements which influence the performance of tribo-systems.

Performance Tests

In 1992-93, a study was conducted to evaluate the tribological performance of the oil-mist lubrication system [32, 53]. In the first phase of this study the performance of oil-mist lubrication was compared to the performance of oil-sump lubrication in terms of operating temperature and friction. The most important characteristic of oil-mist lubrication which influences the operating temperature and friction of rolling element bearings is that oil-mist lubrication does not require an oil sump in the bearing housing. With oil-mist lubrication the rotating bearing does not churn oil; this results in lower friction (hence lower energy loss) and lower operating temperature. Also, as the oil mist at room temperature flows through the narrow passages and over the hot operating surfaces of the bearings, some heat is removed by forced convection from the operating surfaces of bearing which results in lower operating temperatures. In this part of the study the influence of the two lubrication systems on the performance of the test bearings was investigated with variations of load and speed. Statistical techniques were used to design an efficient experimental plan. The experiment was statistically balanced, randomized and blocked. For each experimental run, after applying load and speed, the test bearings were allowed to achieve steady state temperature. At steady state condition the temperature and frictional loads were measured. In this part of the investigation four independent parameters, lubrication system, bearings, load and speed were chosen to be varied. Oil-mist and oil-sump lubrication systems were used in this experiment. Synthetic diester base lubricant no. 32 (ISO VG 32) was used in both lubrication systems. The lubricating oil level in the oil sump was maintained at one half the height of the lowest ball in the test bearing at stand still condition as recommended by the bearing manufacturer [54]. Two randomly chosen new 60 mm bore diameter high precision angular contact ball bearings (B7012C.TPA.P4.K5.UL, FAG) were used in this experiment. The thrust load on the test bearings was varied from 0.56 KN (125 lbs.) to 5.0 KN (1125 lbs.). The load range was chosen to approximate industrial application conditions. The maximum and minimum test speeds at the pitch diameter were chosen as 8.12 m/s (2000 rpm) and 4.06 m/s (1000 rpm) respectively. The maximum speed and load combination was limited by the requirement that a minimum oil viscosity of 13 mm^2/s (approximately 70 SSU) has to be maintained at the operating temperature as recommended by the bearing manufacturer.

The dependent variables chosen to be monitored in this experiment were,

a) operating temperature and
b) frictional load.

Some assumptions were made about the variables in the design of the experiment, a) in the course of the experiment the surface characteristics of the test bearings did not change significantly to influence the operating temperature and friction, b) in the course of the experiment the oil seal characteristics of the test bearing casing did not change significantly to influence the operating temperature and friction, c) the auxiliary bearings (Figure 17-2) which keep the test bearing casing floating and isolated from the rigid thrust load applicators, the sensor wires and oil-mist supply tubes caused negligible or approximately constant effect on the frictional load measurement. Analysis of Variance (ANOVA) tests were run on the data using Statistical Analysis Software (SAS). It was found that in these tests the bearings and the test shafts had negligible influence on both operating temperature and friction. Also as expected, the type of lubrication system, load, and speed had strong influences on the operating temperature and friction.

Next, General Linear Model (GLM) procedure, which estimates and tests hypotheses about linear models, was used to model operating temperature and frictional torque as functions of the independent variables which have strong influence on the operating temperature as identified in the ANOVA test. For the operating temperature, the log-log model gave the highest regression coefficient (0.98). This model is plotted in Figure 21-1a and 21-1b to show the variation of the operating temperature with variation of load and speed. For frictional torque the linear model gave the best R-square value (0.83). This model is plotted in Figure 21-2a and 21-2b to show the variation of friction with variation of load and speed. In both cases the independent variables had good signal to noise ratio.

It can be seen from the plots for the models of temperature and friction that in general the oil-mist lubricated bearings operate at lower temperatures (8 to 10°C) and friction (20 to 25%). It should be pointed out that the contrast between oil-mist and oil-sump lubrication is expected to be more dramatic when regular bearings are used instead of high precision bearings. Regular bearings, in general, have larger balls than high precision bearings, so they require a higher level of oil in the casing for oil-sump lubrication. Since a regular bearing with the same inner diameter has to churn more oil, it runs hotter than high precision bearings with oil-sump lubrication. Morrison et al. [13] have observed in their experi-

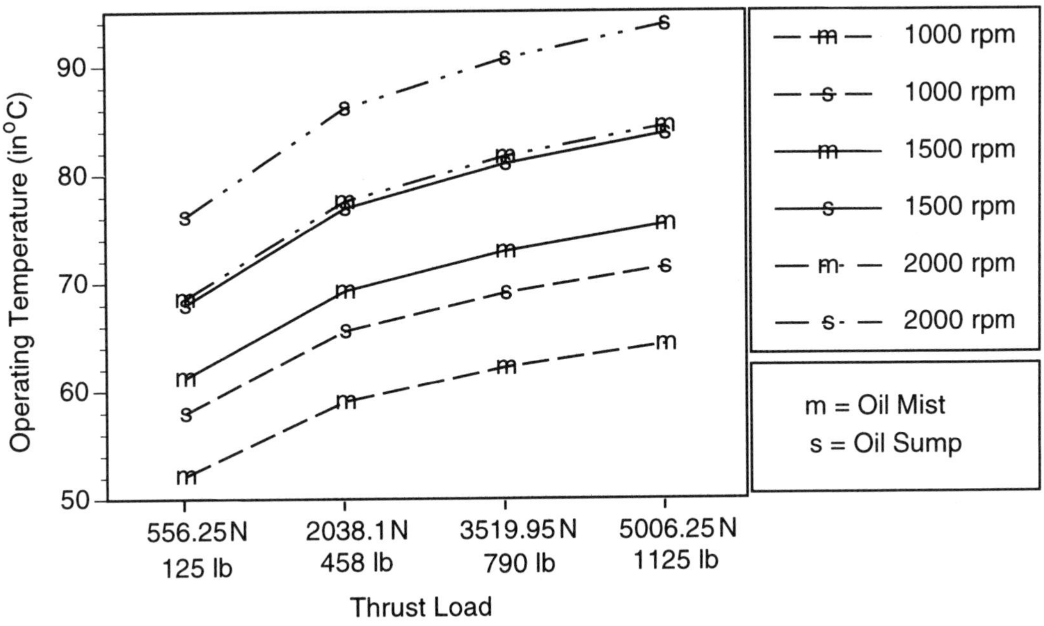

Figure 21-1a. Graphical representation of the statistical model describing the variation of the test bearing operating temperature with load.

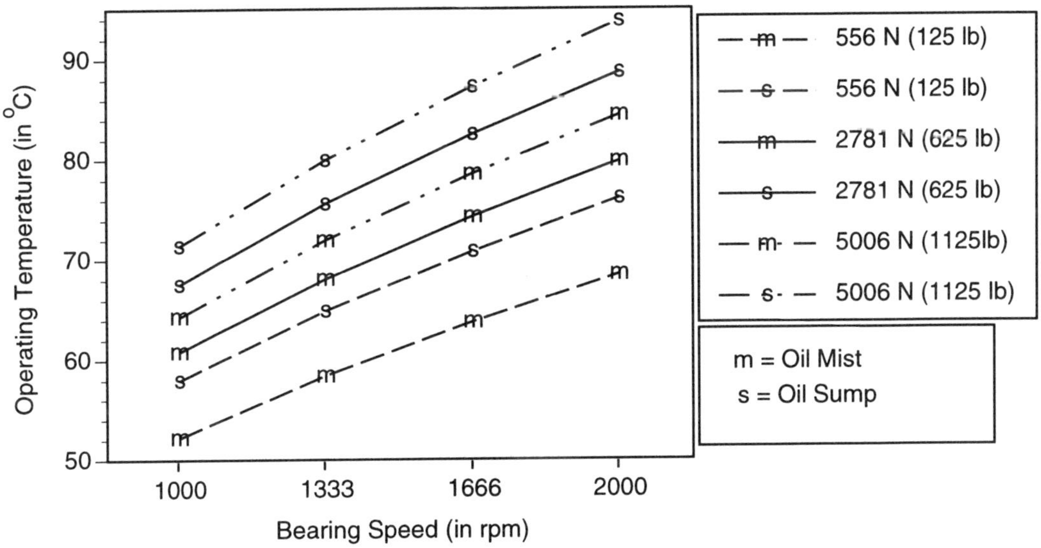

Figure 21-1b. Graphical representation of the statistical model describing the variation of the test bearing operating temperature with speed for several load conditions.

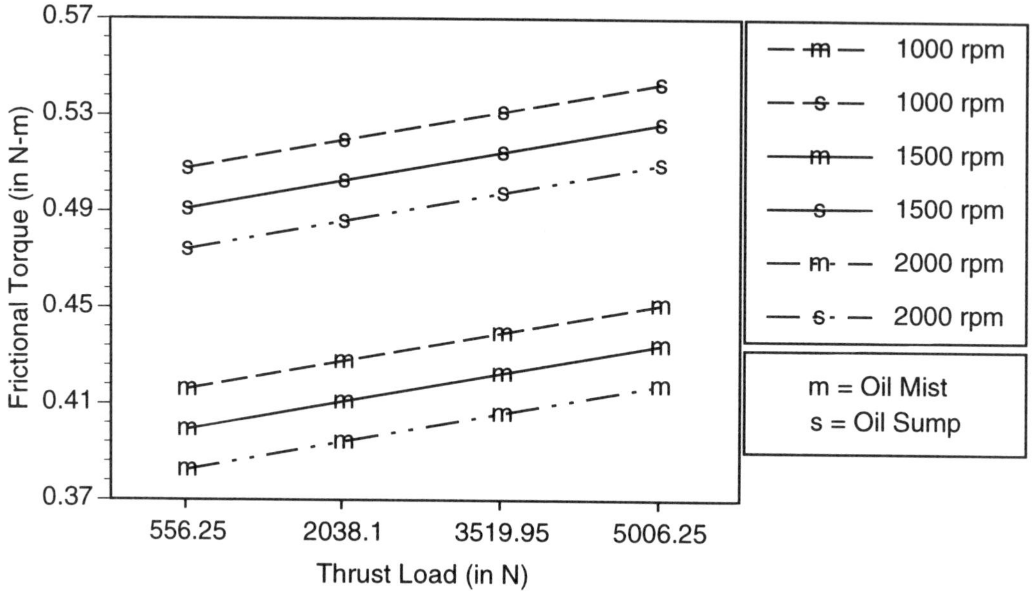

Figure 21-2a. Graphical representation of the statistical model describing the variation of the test bearing frictional torque with load.

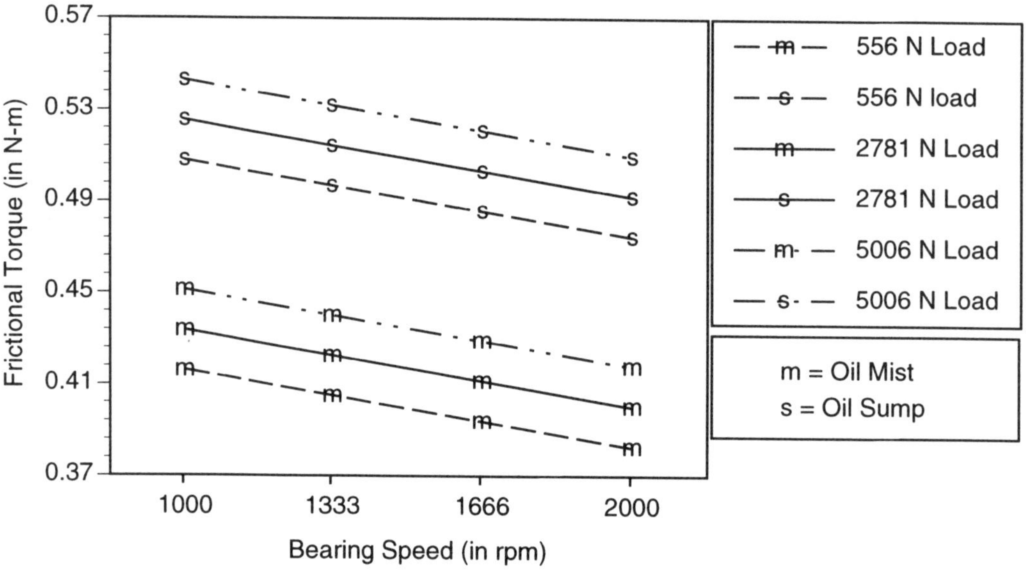

Figure 21-2b. Graphical representation of the statistical model describing the variation of the test bearing frictional torque with speed.

ments that 45 mm diameter regular angular contact ball bearings with 8.9 KN (2000 lbs.) load and 3000 rpm, ran 12°C to 15°C cooler with oil-mist lubrication than with oil-sump lubrication (for synthetic 32 oil).

Extended Performance Tests

In the second part of this performance study [52], two extended comparative performance tests were conducted under endurance test conditions. In each of these tests one test bearing was operated with oil mist and another test bearing was run with oil-sump lubrication, under identical load, speed, lubricant and exterior condition, until one of the test bearings failed. The operating temperature, frictional load and vibration of the test bearings were monitored throughout the duration of the experiments. The objective was to provide a direct comparison between the performances of oil-mist and oil-sump lubrication systems and to observe if their performances change over time. The oil-mist and oil-sump lubrication systems were alternated between two identical test shafts of the test rig in the two tests. Four new test bearings were used in these tests.

For the extended performance tests a synthetic diester base lubricant (ISO VG 68) was used in both lubrication systems. In these tests 60 mm bore diameter high precision angular contact bearings (B7012C.TPA.P4.K5.UL,FAG) were operated with 18.9 KN (4250 lbs.) thrust load at 2400 rpm (endurance test conditions). The variables continuously monitored in this experiment were, a) operating temperature, b) frictional load and c) vibration. Based upon previous experience, an increase of 5°C above the normal operating temperature was used as the failure criterion for the test bearings. This failure criterion was used as a trigger to activate the shut down procedure for the test rig, the data acquisition system and the oil-mist supply system. An increase of 50% in the frictional load over the normal value was also taken as a failure criterion for the test bearing. Accelerometers mounted on the bearing housings measured the vibration acceleration. The 0 to 10 Khz vibration signal was divided into three band widths (0 to 3 Khz, 3 to 6 Khz and 6 to 10 Khz) using the band pass, low pass and high pass filters of the LabView data acquisition software. It has been shown that early damage in rolling element bearings shows up as an amplitude excursion in the lower band widths. As the damage progresses it shows up in progressively higher band widths [55]. Next the root mean square values of all three filtered acceleration signals were calculated and stored. The RMS value can indicate bearing damage when it is compared to the RMS value of the same bearing operating at identical conditions when the bearing is undamaged [56].

The characteristics of the two test shafts (identical in design) of the experimental setup were assumed approximately identical or having small differences which have negligible effects on the lives of the test bearings. Before starting the tests the performance parameters such as the operating temperature, friction and vibration of the test bearings in the two test shafts were measured and compared under identical load, speed, lubrication and external conditions. The characteristics of the two test shafts were found to be virtually identical. Also, it was assumed that the operating characteristics of new bearings are approximately identical.

The first extended performance test was terminated after 212 hours (or 30.528×10^6 revolutions) and the second extended performance test was terminated after 62.5 hours (or 9.0×10^6 revolutions). The tests were terminated because of the failure of test bearings lubricated by oil-sump lubrication. The variations of operating temperatures for oil-mist and oil-sump lubricated test bearings are shown as a function of time in Figure 21-3a and 21-3b. The variations of the frictional torques with time are shown in Figure 21-4a and 21-4b, for the two extended performance tests. The RMS values of the vibration acceleration of the test bearings for the 0 to 3 Khz band width are shown in Figure 21-5a and 21-5b as functions of time.

Extended performance tests provided a direct comparison between the performance of oil-mist and oil-sump lubrication. In both of these extended performance tests the oil-mist lubricated bearing ran about 10°C cooler than the oil-sump lubricated bearing. Also, the oil-mist lubricated bearing ran with about 25% less friction than the oil-sump lubricated bearing. In both tests the oil-sump lubricated bearings failed earlier. The differences in temperature and friction between the two lubrication systems remained approximately constant throughout the tests until the test bearings lubricated by the oil sump failed.

Endurance Tests

In this experimental study the L_{10} life of bearings lubricated by oil-mist lubrication was determined and compared to the L_{10} life for oil-sump lubrication calculated for the

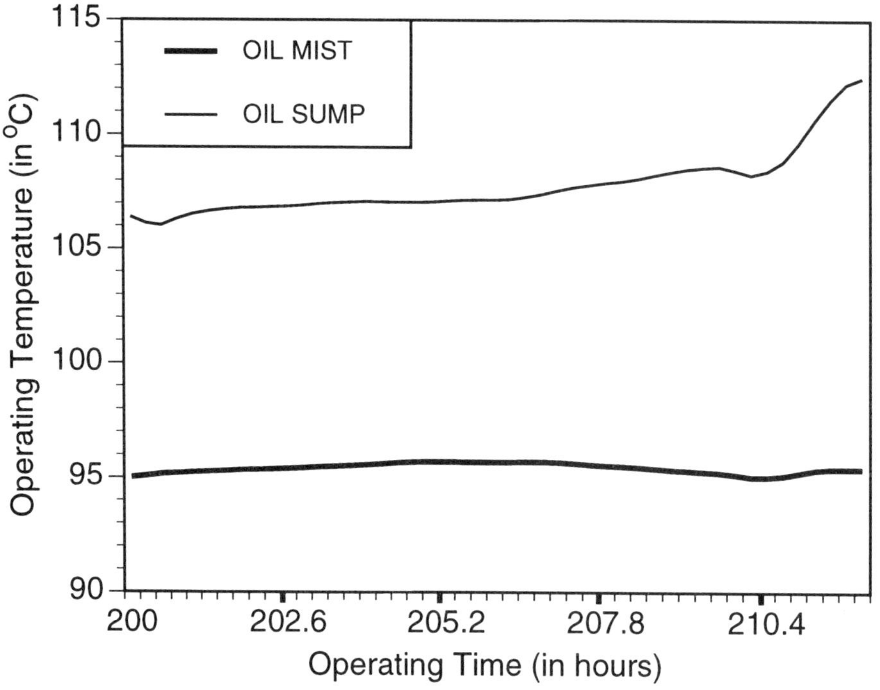

Figure. 21-3a. Variation of the test bearing operating temperature with time for the expended performance test number 1.

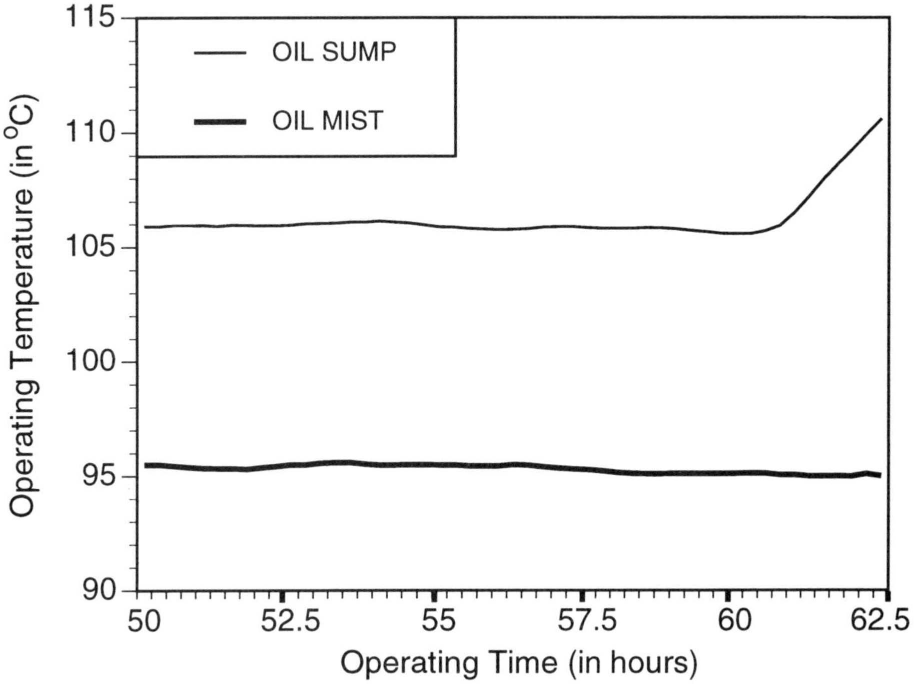

Figure 21-3b. Variation of the test bearing operating temperature with time for the expended performance test number 2.

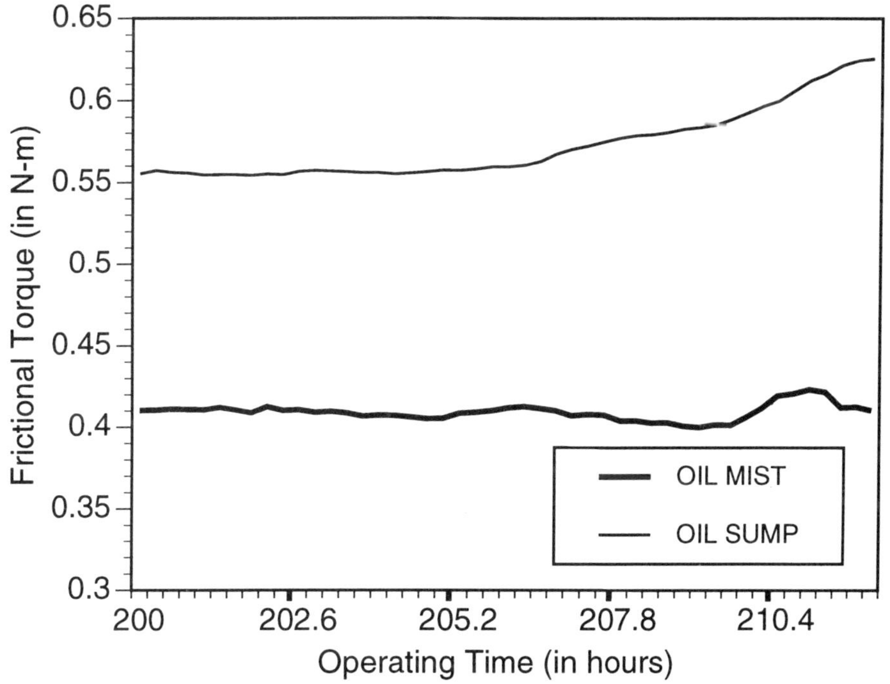

Figure 21-4a. Variation of the test bearing frictional torque with time for the expended performance test number 1.

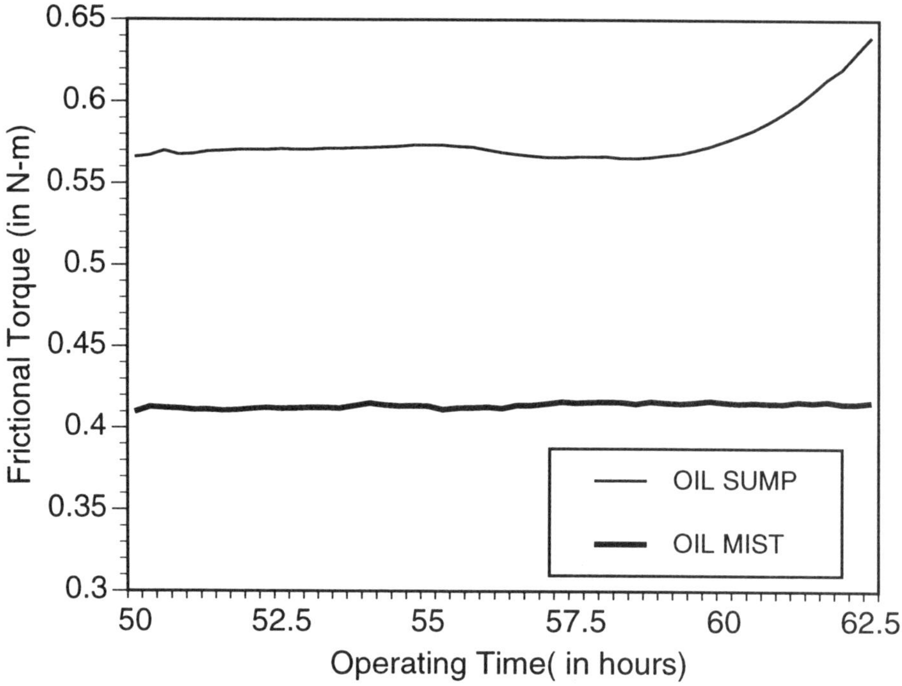

Figure 21-4b. Variation of the test bearing frictional torque with time for the expended performance test number 2.

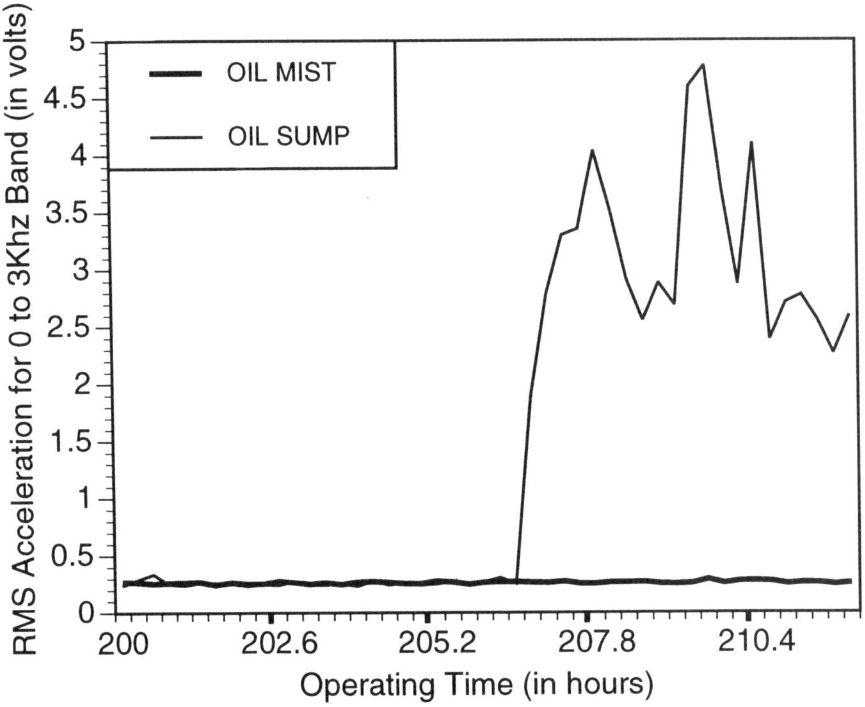

Figure 21-5a. Variation of the RMS acceleration (for 0 to 3 Khz band) with time for the expended performance test number 1. (Pizoelectric accelerometer sensitivity 10 mv/g)

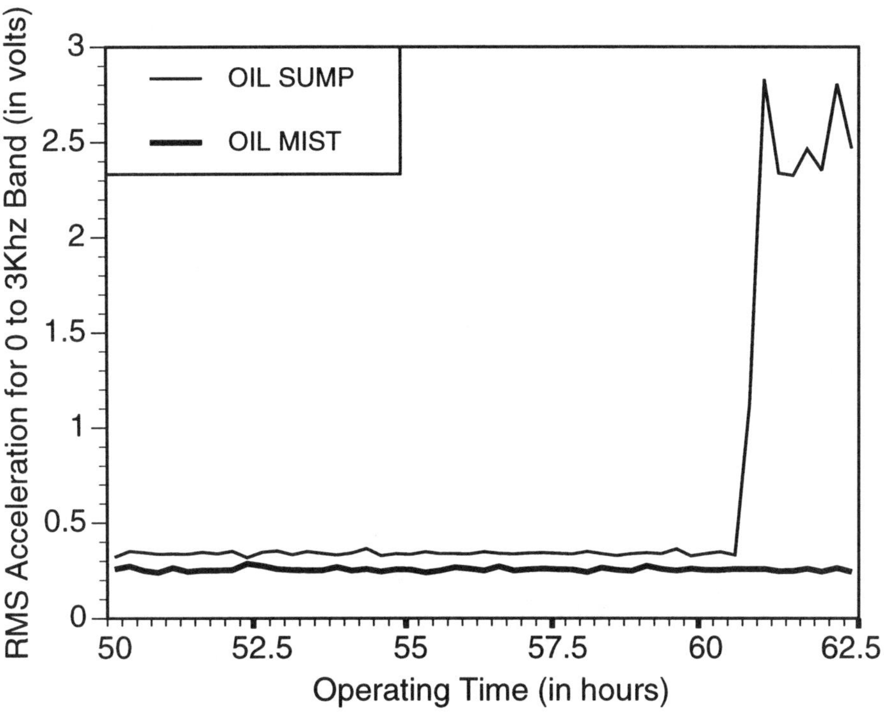

Figure 21-5b. Variation of the RMS acceleration (for 0 to 3 Khz band) with time for the expended performance test number 2. (Pizoelectric accelerometer sensitivity 10 mv/g)

Table 21-1. Results of the bearing endurance test (with oil mist lubrication).

Test bearing number	Test shaft number	Duration of test: Hours	Duration of test: $\times 10^6$ rev.	Failure status
1	1	315	45.36	Failed
2	2	551	79.34	Failed
3	1	768	110.59	No failure
4	2	768	110.59	No failure
5	1	657	94.61	Failed
6	2	768	110.59	No Failure
7	1	768	110.59	No failure
8	2	209	30.10	Failed
9	2	768	110.59	No failure
10	1	271	30.02	Failed

same bearings operating under identical load, speed and external condition [52]. The lubricating oil used in the endurance tests with oil-mist lubrication was synthetic ISO Grade 68 oil. However, the L_{10} life for oil-sump lubrication was calculated for mineral ISO Grade oil. This was done because often in industry, conventional sump lubrication with mineral oil is replaced by oil-mist lubrication with synthetic oil; also the data required to calculate the correction factor for the effect of lubricant on the L_{10} life are not readily available for synthetic oils. For the endurance tests the test bearings were subjected to 18.9 KN (4250 lbs.) of thrust load and they were operated at 2400 rpm. The load and speed for this experiment were chosen considering, (a) approximate simulation of industrial application, (b) acceptable operating temperature and (c) reasonably short duration of the experiment. An experiment adequately reflects the operating condition of a bearing if the linear speeds at the pitch diameters are the same [13]. The endurance tests approximately simulated 40 mm bore diameter bearings operating at 3600 rpm. In order to increase the rate of heat removal from the test bearings, blowers had to be used to blow air over the test bearing housings. The room temperature, as well as the position and speed of the blowers were kept identical throughout the experiment to ensure identical external conditions for all tests. A group of ten randomly selected 60 mm bore diameter, high precision, angular contact (15° contact angle) ball bearings (B7012C.TPA.P4.K5.UL, FAG) were tested for endurance. According to the type II censored endurance test procedure [57] endurance tests were continued until five fatigue failures were obtained. The results of the bearing endurance test with oil-mist lubrication are given in Table 21-1.

The fatigue life of rolling element bearings is a random variable widely accepted to follow a Weibull probability distribution.

$$
\begin{aligned}
F(L) &= 1.0 - \exp\left[-\left\{\frac{L - L_o}{a}\right\}^b\right] ; L > L_o \\
&= 0 \qquad ; L > L_o
\end{aligned}
\qquad (21\text{-}1)
$$

Where, L = bearing life
L_o = location parameter
a = scale parameter or characteristic life and
b = shape parameter or Weibull slope
$F(L)$ = cumulative distribution function or hazard rank

In the analysis of the endurance test data, a widely accepted and well established procedure for hazard rank calculation was used. This procedure has been described by Johnson [58,59]. The hazard rank and bearing failure data are plotted in Figure 21-6. The straight line in Figure 21-6 represents the least square curve fit of the endurance test data in the Weibull probability distribution. From the least square curve fit of the endurance test data,

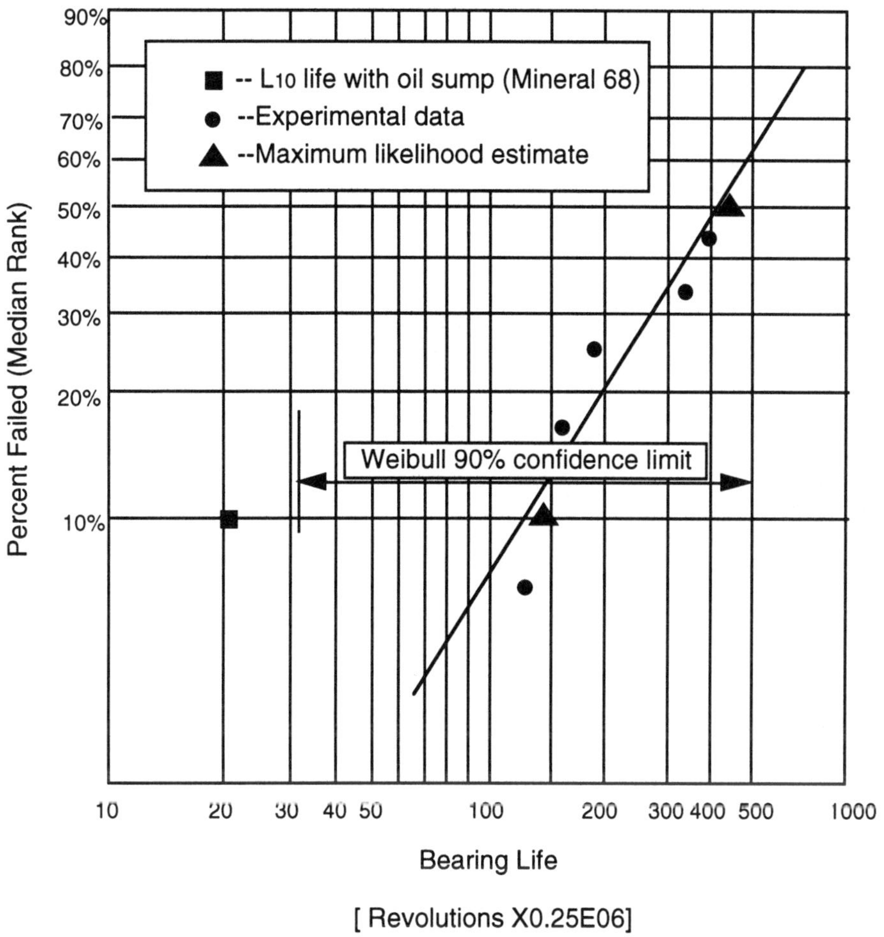

Figure 21-6. Weibull plot of the bearing endurance test data.

Weibull slope (b_{wb}) = 1.6509
Characteristic life (a_{wb}) = 121.818 million revolutions
L_{10} life (L_{10wb}) = 31.168 million revolutions
Regression coefficient for the least square curve fit = 0.93

Another well accepted technique for the L_{10} life calculation is the maximum likelihood technique. It has been shown that if in an endurance test of n bearings which leads to the failure of r of the bearings at lives L_1, L_2, L_3.........L_r then the maximum likelihood estimate of the Weibull slope is the positive solution b_{ml} of the following transcendental equation [57],

$$\frac{1}{b_{ml}} + \frac{1}{r}\left[\sum_{i=1}^{r} \mathrm{Log}_e(L_i)\right] - \frac{\sum_{i=1}^{n} (L_i)^{b_{ml}} \mathrm{Log}_e(L_i)}{\sum_{i=1}^{n} (L_i)^{b_{ml}}} = 0 \qquad (21\text{-}2)$$

Using the endurance test data from Table 21-1, equation (21-2) has been solved using the Newton-Raphson technique for the value of the maximum likelihood estimate of the Weibull slope. According to the calculation procedure described by McCool [57],

Maximum likelihood estimate of Weibull slope (b_{ml}) =1.62
Maximum likelihood estimate of characteristic life (a_{ml}) = 134.951 million rev.
Maximum likelihood estimate of L_{10} life (L_{10ml}) = 33.643 million rev.

It has been found that with oil-sump lubrication (for mineral 68 oil) when the test bearings are subjected to the same load, speed and external condition as the endurance test with oil-mist lubrication, they operate at an average temperature of 107°C. This temperature was used to calculate the correction factor for the oil viscosity. The correction factors for cleanliness and bearing material were calculated, assuming standard bearing material and moderate to high cleanliness in the lubricating gap. Using the FAG Rolling Bearings Catalog No. WL41510/2ED (1989), L_{10sump} was found to be $5.24(10^6)$ revolutions.

According to the maximum likelihood technique the 90% confidence interval may be calculated as,

$$\left[\exp\left\{\frac{-u_{0.95}}{b_{ml}}\right\}\right] \times L_{10ml} \le L_{10} \le \left[\exp\left\{\frac{-u_{0.05}}{b_{ml}}\right\}\right] \times L_{10ml} \qquad (21\text{-}3)$$

For n = 10 and r = 5, $u_{0.95}$ = 3.78 and $u_{0.05}$ = -0.918 [57]. The 90% confidence interval on L_{10} life by the maximum likelihood technique is, thus:

$3.26 \le L_{10} \le 59.28$ $[10^6$ revolutions]

However, a conservative estimate of 90% confidence interval on the L_{10} life determined from the Weibull plot is given as [59],

$0.25L_{10wb} \le L_{10} < 4L_{10wb}$
i.e., $7.792 \le L_{10} \le 124.67$ $[10^6$ revolutions]

In the statistical hypothesis testing, the null hypothesis is that there is no significant difference between the calculated L_{10} life with oil-mist lubrication and the L_{10} life with oil-sump lubrication for the test bearings.

i.e., H_0: $L_{10mist} = L_{10sump} = 5.24(10^6)$ revolutions

The alternative hypothesis is that the L_{10} life with oil-mist lubrication is higher than the L_{10} life with oil-sump lubrication,

i.e., H_1:$L_{10mist} > L_{10sump}$

The significance level for this test was taken as $\alpha = 0.10$. Using the maximum likelihood technique, the null hypothesis is accepted if,

$$b_{ml} \, \text{Log}_e \left[\frac{L_{10ml\,(mist)}}{L_{10\,sump}} \right] < u_{0.90} \tag{21-4}$$

The left hand side of equation (21-4) is 3.012 and $u_{0.90}$ is 2.64 [57]. As 3.012 > 2.64 we reject the null hypothesis i.e., we accept that L_{10} life for oil-mist lubrication is higher than the L_{10} life for oil-sump lubrication at the 10% significance level.

As was brought out earlier, under identical load, speed, lubricant and external conditions oil-mist lubricated bearings were found to run cooler by about 10°C compared to oil-sump lubricated bearings. The oil-mist lubricated bearings also ran with about 25% less friction than oil-sump lubricated bearings. It has also been found that the difference in operating temperature and friction between oil-mist and oil-sump lubricated bearings remain approximately constant with variation of load and speed. Moreover, they do not change appreciably with time. In the ten bearing endurance test it has been found that under identical load, speed and external condition the L_{10} life of a 60 mm bore diameter high precision angular contact ball bearing with oil-mist lubrication using synthetic 68 oil is about 6 times the L_{10} life calculated for oil-sump lubrication using mineral 68 oil. The L_{10} life for sump lubrication fell below the lower limit of the 90% confidence interval on the L_{10} life with oil-mist lubrication using the Weibull technique, indicating better wear and fatigue protection capability of oil-mist lubrication. For the maximum likelihood technique the L_{10} life for sump lubrication was found to be very close to the lower 90% confidence limit but inside the interval for the L_{10} life with oil-mist lubrication. However, in the hypothesis test for the 10% significance level the null hypothesis $L_{10mist} = L_{10sump}$ was rejected in favor of $L_{10mist} > L_{10sump}$. It should be pointed out that the corrosive wear protection capability of oil-mist lubrication is not reflected in the results since it could not be simulated in the accelerated endurance tests.

As shown by Morrison et al. [13], a 25% reduction in friction per bearing can lead to significant savings in energy when elements of a total petrochemical complex are considered. Also, through better wear and fatigue protection, oil-mist lubrication can reduce maintenance requirements and downtime of expensive equipment, leading to savings in resources.

Combined Solid and Liquid Lubrication with Oil Mist

During the research effort involving the endurance testing of ball bearings with oil-mist lubrication, it has been discovered that with certain lubricating oils it is possible to form and sustain a significant continuous solid carbonaceous layer along with a liquid lubricating layer on the surfaces of bearings operating at or beyond 80°C temperature, without any deterioration of the lubricating oil. However, with the conventional sump lubrication the same lubricant, under the same load and speed condition could not produce any carbonaceous layers on the bearing surfaces (Figure 21-7).

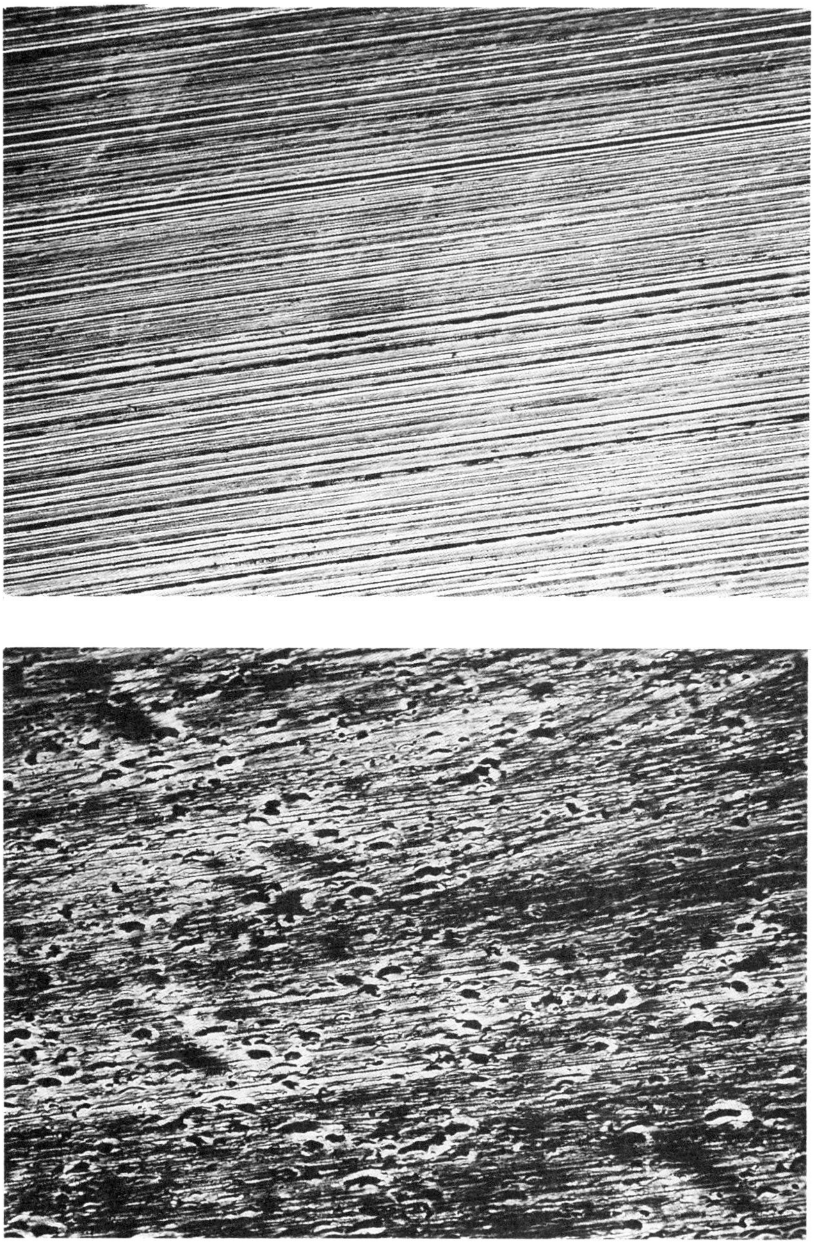

Figure 21-7. (a) Scanning Electron Photomicrograph of a bearing surface lubricated by the oil mist lubrication and (b) a bearing surface lubricated by the conventional sump lubrication. The same oil and the same operating conditions were used for both the bearings. Only the surface shown in the top photomicrograph contains a carbonaceous layer. The magnification level is 500.

In conventional liquid lubrication excessive loading, cyclic loading or high temperature operation often lead to breakdown in the liquid lubricant layer allowing direct metal to metal contact. This causes excessive wear and reduction in the operating life of tribo-mechanical systems (bearings, automotive engines etc.). As shown in Figure 21-8, if a continuous solid lubricating layer is maintained under the liquid oil film, the metal to metal contact can be almost completely eliminated resulting in a dramatic decline in friction and wear [60]. The additional protection provided by the solid lubricating layer is particularly significant for today's high speed, heavily loaded and high temperature rotating equipment. In conventional lubrication the brittle solid lubricant layer can not be sustained and in the course of time this layer is removed by the loaded and operating system.

A comprehensive literature survey indicated that, to date research has been conducted on two techniques for lubricating tribo-mechanical systems with carbonaceous layers:

(a) Solid lubrication by oil in vapor phase: Lubricating oil is supplied to the hot operating surfaces in the form of vapor carried by hot inert gases. The oil vapor dissociates on the operating surfaces (influenced by the catalytic action of iron) and leaves a lubricating carbonaceous layer [61].

(b) Solid lubrication by carbonaceous gases: Diluted high temperature hydrocarbon gases are supplied to the hot operating surfaces. By hydrocarbon acid-catalytic cracking or by pyrolysis these gases form lubricating carbonaceous layers on hot bearing surfaces [62,63].

The important aspects of vapor, gas and oil-mist lubrication may be summarized as follows:

(i) All of these methods, including oil-mist lubrication, are continuous flow systems which continuously replenish the brittle solid lubricating layer.

(ii) Vapor phase and hydrocarbon gas lubrication are for specialized equipment such as adiabatic or ceramic engines. They require between 300 to 500°C surface temperature. Oil-mist lubrication currently requires only about 80°C to form a carbonaceous layer. Hence, this can be applied to conventional tribo-mechanical systems.

(iii) The vapor phase and gas lubrication techniques can produce only solid carbonaceous

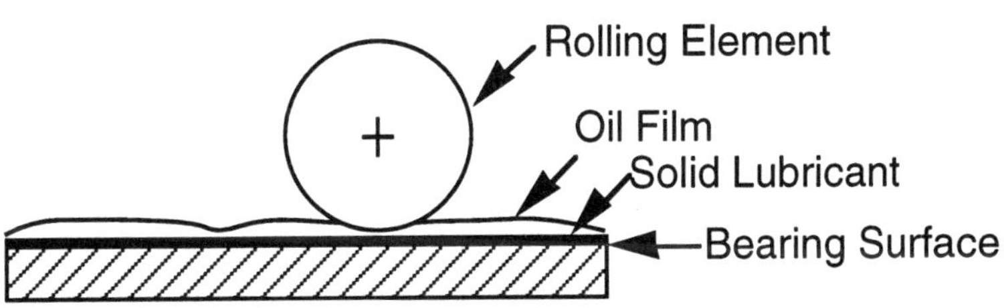

Figure 21-8. Schematic diagram of combined solid and liquid lubrication.

lubricating layers. Oil-mist lubrication can provide combined solid and liquid lubrication.

(iv) Vapor and gas lubrication require the high temperature gases to be carried to the point of lubrication. However, in oil-mist lubrication the oil mist is supplied to the points of lubrication at room temperature.

Need for Research

Research work is needed to investigate the phenomenon described above.

(i) It solves a Significant Lubrication Problem: Technology moves toward high speed, high efficiency and heavily loaded rotating equipment. The tribo-mechanical systems associated with this rotating equipment invariably operate at high temperatures. However, at temperatures beyond 80 or 100°C, due to low viscosity, most lubricating oils provide poor elastohydrodynamic films [64]. In these cases combined elastohydrodynamic film and a continuously replenished solid lubricant layer can provide the ideal solution.

(ii) Savings in Energy and Resources Will Accrue: The savings provided by lower maintenance and higher reliability of rotating equipment can be very significant. The combined solid and liquid lubrication will protect high temperature surfaces primarily by forming solid carbonaceous layers. The task of forming a carbonaceous layer can be accomplished by relatively low viscosity lubricating oils. In this method low operating viscosity oils can thus be used to lubricate relatively high temperature tribo-mechanical systems. It has been shown that the power loss in rolling element bearings can be reduced by 28% if ISO VG 68 oil is replaced with ISO VG 32 oil [13]. Considering the large number of equipment where the combined solid and liquid lubrication with oil mist can be used, the overall savings in energy can be very significant [65].

A number of bearing tests have been run which duplicated and verified the phenomenon of carbonaceous solid layer formation with oil mist using the bearing endurance tester discussed earlier [40]. The solid layer has been found to be brownish-semi-transparent in appearance and strongly bonded to the bearing surfaces. After a large number of failed attempts, a special experimental setup, as shown in Figure 21-9 has been built which simulates this phenomenon on heated flat steel plates. It appears that after the oil mist enters a hot bearing, part of the lubricating oil evaporates. This evaporated oil is then forced by the turbulence caused by the rotating bearing to be deposited on the hot bearing surfaces and this forms the solid carbonaceous layer on the surfaces.

More specifically, we believe that research should be performed:

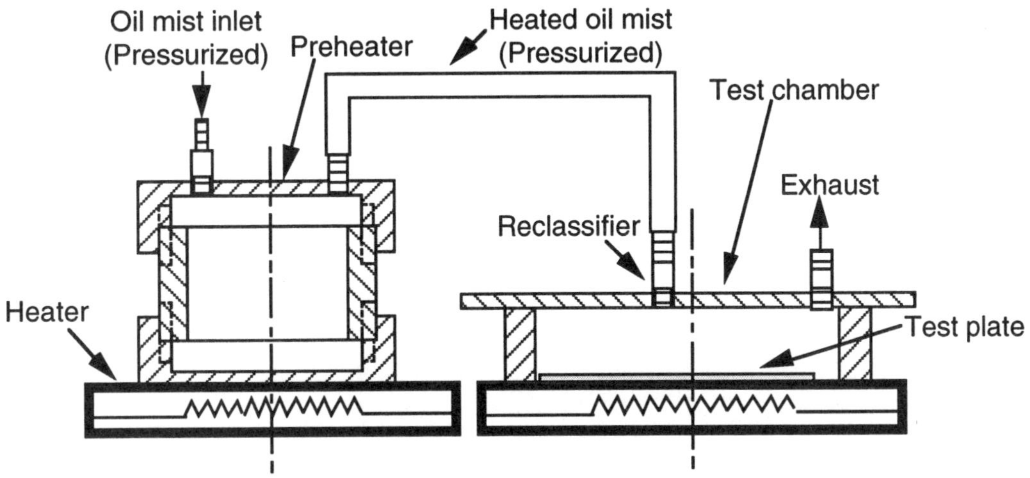

Figure 21-9. Cross-sectional sketch of the experimental setup which simulates carbonaceous layer formation on flat steel plates with oil mist.

(i) to identify the chemical composition of the solid carbonaceous layer and the component of the lubricating oil causing this phenomenon,

(ii) to investigate the control and optimization of this phenomenon by controlling the type and quantity of the component of the lubricating oil responsible for this phenomenon, and

(iii) to investigate the capability of this solid layer to protect surfaces against wear.

Closure

The main impediment in the way of widespread acceptance of oil-mist lubrication has been convincing engineers that liquid oil splashing all over a high speed and heavily loaded rolling element bearing can be replaced by oil mist which looks like "puffs of dry smoke." A safety-conscious engineer will, at a first glance, raise questions about oil-mist lubrication. The objective of Part 2 of this text has been to explain as to why and how an oil-mist lubrication system works. The experimental and analytical work discussed in Part 2 of this text was performed at Texas A&M University, College Station under the guidance of faculty members of the Mechanical Engineering Department. Details of the experiments and analysis have been shared to allow the reader to judge the scope of these endeavors. It is evident from the work presented here that more investigations need to be conducted on several aspects of oil-mist lubrication. However, the authors hope that the information presented will give practicing engineers and reliability professionals previously unavailable insight into the operating principles of the oil-mist lubrication system.

Appendices

Appendix A

Why and How to Develop a Specification for an Oil-mist Lubrication System

Before a specification is being prepared, the diligent reliability professional will have assisted plant designers in scoping studies aimed at defining the economics of oil mist lubrication. Our text has highlighted the savings documented at plants located in the US gulf coast region. These cost calculations indicate that regardless of systems size, implementation decisions are quite obviously influenced by a number of tangible factors.

But, there are also some more elusive considerations. First and foremost among these is the one we might collectively call management philosophy. If it is management's opinion that operators should not perform lubrication-related jobs, the entire responsibility for implementing the various and sundry associated tasks will rest on the shoulders of either a resident, or a contract maintenance workforce. On the other hand, in a Total Productive Maintenance (TPM) environment, many of the routine tasks associated with conventional lubrication are being handled by a competent, cross-trained group of technician-operators.

Total Productive Maintenance (TPM) is a system based on maintenance activities carried out by all employees in an organization. At its core, TPM encourages *knowledge-based* maintenance. To that end, TPM has a much broader scope and includes maintenance prevention/deferral, preventive maintenance, and reliability and maintainability improvement. The broad objectives of TPM are:

1. Maximizing maintenance effectiveness (overall efficiency)
2. Developing a system of maintenance for the life of the equipment
3. Involving all departments that plan, specify, design, use or maintain equipment in implementing TPM. Thus, it includes the technical, engineering, design, production & maintenance work force members
4. An involvement of all personnel from top management to the shop and production floor worker

5. The promotion of TPM through small group activities

Once in action, TPM focuses on the elimination of six big losses which are endemic to all organizations:

1. Breakdown-related losses
2. Setup and adjustment losses
3. Speed losses
4. Idling and minor stoppage losses
5. Quality defects and rework losses

In working towards reducing these losses, TPM encourages all involved to gain "*Total Knowledge*" of production machinery to effectively use and maintain them. This entails understanding how the various machines respond to different production cycles, their stable operating ranges, capabilities and failure modes. Once this knowledge permeates the work force, system deficiencies become clearer. Skill-based training programs can now be devised to enable all employees to participate in TPM. Autonomous maintenance, the real core of TPM, depends on the development of skills, standards, and judgment of all employees.

Fundamentals of *autonomous maintenance* are:

1. Plant and equipment cleaning
2. Reacting to problems at their source and at the point of control
3. Plant lubrication and development of lubrication standards
4. Value-added and action-oriented inspection (standards!)
5. Continuous improvement toward zero defects and no breakdowns

As can be seen, TPM implementation would require a major departure from the existing, entrenched work practices, attitudes and prevailing cultures. It would require commitment of personnel and monetary resources from an organization and will typically take 3-6 years to bear solid, tangible results.

Without even attempting to sort out which approach is the preferred one, we are simply asking such questions as who will be in charge and for how long? What will be the cost of conventional approaches to the assignment of your own, or the contractor's maintenance workforce from the time of plant commissioning until TPM is firmly established? What will be the administrative and labor relations logistics associated with the various options that present themselves? What would be the real benefit of an oil-mist installation?

Once these issues have been considered, and the decision made to implement an oil mist lubrication system, we firmly believe that a good specification needs to be developed.

To the superficial observer, the job of specifying an oil mist system would seem rather routine. Some may leave the task to the design contractor, installation contractor, or systems vendor. However, we recommend that the equipment owner/operator to take a lead role in specification development and also in verifying that the actual field installation is in full compliance with the specification.

We are now ready to consider a typical sample specification.

Sample Specification for Plant-wide Oil-mist Lubrication Systems

1.0 Scope

1.1 The vendor shall design and furnish a complete oil mist lubrication system as required by the following specification. The vendor must also be able to provide, under a separate agreement and at a later date, a service representative for technical assistance if, and when, required by the purchaser. This separate agreement shall be contractually binding.

*1.2 All centrally located on-site centrifugal pumps, steam turbine drivers, electric motor drivers, gear boxes, and cooling tower gear speed reducer with antifriction or sleeve bearings shall be lubricated by means of pure mist or purge mist as will be specified by the user in a separate listing. This listing shall identify these machinery items and shall state the required lubrication method, using sketches as needed.

1.3 Unless otherwise specified, all tasks described in this specification are to be executed by the oil-mist system vendor.

2.0 Definitions

2.1 An *oil-mist lubrication system* includes the mist generator console, distribution piping, distribution manifold, application fittings, and lubricant supply tank. See Figures A-1 and A-2 for system schematics. (Reader please note: Be sure to provide figures/sketches that accurately represent your requirements.)

2.2 The *mist generator console* is to include the oil-mist generator, generator oil-supply system, air-supply moisture separator/filtering system, and attendant controls and instrumentation.

*2.3 The *distribution piping* includes console oil-supply piping, distribution header to distribution manifolds, application fittings, and stainless steel tubing from application fittings to equipment lubrication points. Alternatively, the purchaser may specify application fittings to be located directly at the lubrication points.

2.4 *Distribution manifolds* are multiported terminal blocks at the end of pipe drops. They divert oil mist to individual equipment lubrication points.

*2.5 *Application fittings (reclassifiers)* meter oil mist to individual equipment lubrication points. The purchaser shall specify the location of these fittings (see Paragraph 2.3).

*Purchaser to separately state specific requirements on items identified with asterisk.

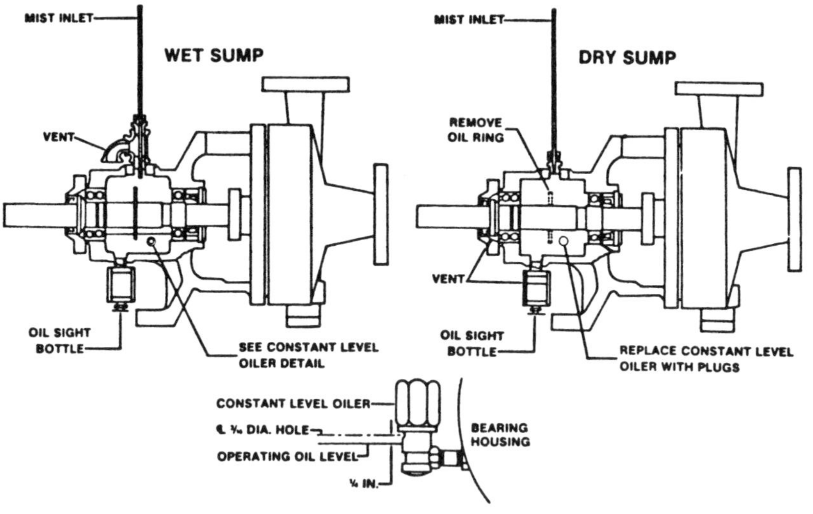

CONSTANT LEVEL OILER DETAIL

DRY SUMP APPLICATIONS

This is sometimes referred to as pure mist. No oil level is maintained, oil rings and constant level oilers are removed. Venting through each bearing is essential. Lip type seals will prevent proper venting, and must be modified. Proper venting can be accomplished through labyrinth type seals. Housing drains must be provided. One method of accomplishing this is by the use of a vented oil sight bottle as shown in the above drawing.

WET SUMP APPLICATIONS

This is sometimes referred to as purge mist. The oil level is maintained in the reservoir. The oil mist is induced into the bearing housing to exclude contaminants and to insure proper lubrication. The housing must be vented from the top of the bearing housing.

The constant level oiler must be modified to provide an oil overflow. This is done by drilling a 3/16" hole in the vertical internal tube 1/4" above the oil level. The oil will have a tendency to rise due to the turbulence in the housing. Raising the over flow point will insure that the proper level is maintained.

Figure A-1. Typical pump hookup.

2.6 Oil-mist lubrication shall be applied in one of the following two ways (purchaser to specify per Paragraph 1.2):

A. *"Pure mist"* or dry-sump lubrication—an arrangement in which no oil level is maintained in the bearing housing and lubrication is accomplished by condensation of the oil mist on the rolling elements within the bearing housing. Adequate venting shall be provided by equipment supplier or user. (Note to reader: It is recommended that appropriate sketches be made part of this specification.)

B. *"Purge mist"* or wet-sump lubrication—an arrangement in which normal bearing housing oil levels are maintained by introducing oil mist. An adequate vent orifice and a method of maintaining a constant oil sump level shall be provided by oil-mist system vendor. (Note to reader: Please provide appropriate sketch.)

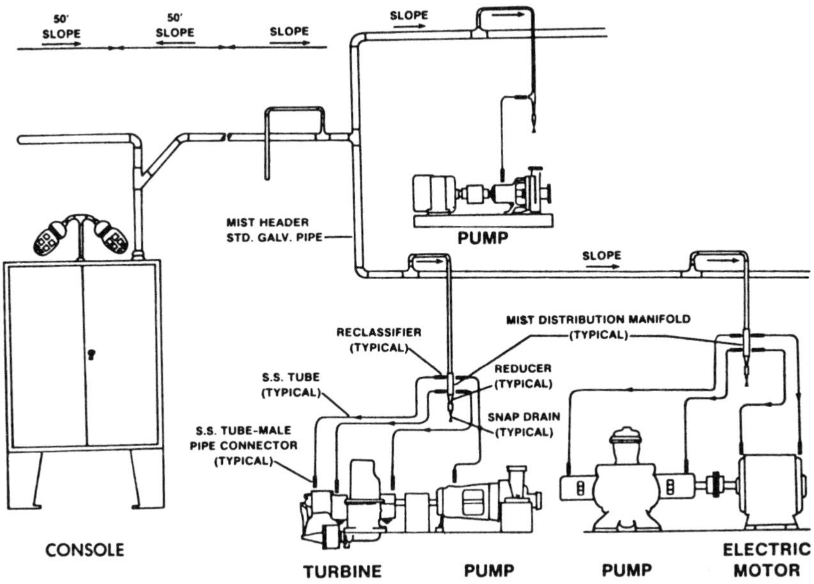

A. Mist Console: Install in an accessible location, centrally located in an area to be serviced. This will greatly be determined by pipe routing.

B. Piping: Normally piping is routed in existing pipe racks. If possible, the pipe header should be 2" Standard Galvanized Threaded and Coupled Pipe. Drops should be a minimum of ½" Standard Galvanized Threaded and Coupled Pipe. CAUTION: DO NO USE TEFLON TAPE AT SCREWED JOINTS. The use of clean pipe and clean installation procedures is highly recommended.

C. Sloping: All header piping should be sloped toward the mist console for a minimum of 50' after leaving the console. All other piping and tubing should be sloped to prevent traps in accordance with recommended sloping procedures

Figure A-2. Mist distribution system.

2.7 For record purposes, the oil-mist system vendor shall provide tabulations per Figure A-3, A-4, or A-5 with equipment and designated as to pure or purge mist, and listing reclassifier, air consumption, and related data as shown.

3.0 References

*3.1 Piping, installation and system design shall meet the requirements of the following standards (purchaser to provide tabulation, if special requirements should be called for at his location).

4.0 Design

Mist Generators

4.2 Mist generators shall be sized to furnish at least 125% of the calculated cfm requirements. They shall also be capable of proper misting operating down to 50% of rated flow.

OIL MIST LUBRICATION SUMMARY

OIL MIST CONSOLE DESIGNATION: MIS/GEN 5 LOCATION: BLOCK 7, AREA D

NOTE 1: Use Directed Oil if Bearing Pitch Line Velocity Exceeds 2000 FPM

NOTE 2: Dry Sump Oil Mist Lubrication is not Suitable for Sliding (Sleeve) Bearings. Use Oil Mist for Purge

					ROLLING ELEMENT BRG CHARACTERISTICS									APPLICATION FITTINGS: INSTALLATION LOCATION AND DIAMETER, INCHES			
					INBOARD				OUTBOARD								
ITEM	MFR. AND MODEL	RPM	HP	V>2000FPM (See Note 1)	ID	OD	ROWS	BRG INCHES	ID	OD	ROWS	BRG INCHES	DIA, if Sliding Bearing (Note 2)	IN-BOARD	OUT-BOARD	COMMON TO INB & OUTB.	OIL MIST FLOW TOTAL TO ITEM (SCFM)
P-6A	Goulds (Overhung)	3560	55		1.7	3.3	1	2	1.7	3.3	2	4				.060	.30
PM-6A	A-C 365TS (Horiz)	3560	75		1.9	3.5	1	2	1.9	3.5	1	4		.032	.032		.18
P-6B	Goulds (Overhung)	3560	55		1.7	3.3	1	2	1.7	3.3	2	4				.060	.30
PM-6B	A-C 365TS (Horiz)	3560	75		1.9	3.5	1	2	1.9	3.5	1	4		.032	.032		.18
P-9A	Bingham (Overhung)	1760	15		1.8	3.3	1	2	1.8	3.3	2	4				.047	.18
PM-9A	A-C 284T (Horiz)	1760	25		1.7	3.3	1	2	1.7	3.3	1	4		.032	.032		.18
P-9B	Bingham (Overhung)	1760	15		1.8	3.3	1	2	1.8	3.3	2	4				.06	.18
PT-9B	Elliott (Horiz)	1760	25										3	.084	.060		.48
P-17A	Pacific(Betw.Brg)	3570	2600										4	.073	.060		.75
PT-17A	Westinghouse	3570	3200										4.5	.073	.073		.90

Figure A-3. Oil-mist lubrication summary.

4.1 A sufficient number of mist generator units shall be furnished to supply oil mist to all of the equipment specified.

4.3 Mist generators shall meet the requirements of Class I, Division II, Group D, as specified in the National Electrical Code.

Oil Reservoir

1. Hazardous area-rated oil heater with temperature gauge and bi-metal type thermostat and reliable, proven control.
2. Level sight glass.
3. Oil inlet connection.
4. Oil mist header pressure gauge.
5. Pressure relief valve (set to 10 psi).
6. Opening for oil level warning switches.
7. Two-inch NPT mist outlet connection.
8. Oil drain with valve piped to outside of console.
9. Cleanout opening with cover.

Generator Inlet Air System

1. Shut-off valve and appropriate line pressure gauge (purchaser to specify pressure).
2. Auto-drain air filter/moisture separator capable of filtering out any particles that might interfere with proper mist generator operation; moisture drain piped to outside of console.

OIL MIST LUBRICATION SUMMARY

OIL MIST CONSOLE DESIGNATION: LOCATION:

NOTE 1: USE DIRECTED OIL IF BEARING PITCH LINE VELOCITY EXCEEDS 2000 FPM

NOTE 2: DRY SUMP OIL MIST LUBRICATION IS NOT SUITABLE FOR SLIDING (SLEEVE) BEARINGS. USE OIL MIST FOR PURGE

				ROLLING ELEMENT BRG CHARACTERISTICS										APPLICATION FITTINGS: INSTALLATION LOCATION AND DIAMETER, INCHES			
					INBOARD				OUTBOARD								
ITEM	MFR. AND MODEL	RPM	HP	V>2000FPM (SEE NOTE 1)	ID	OD	ROWS	BRG INCHES	ID	OD	ROWS	BRG INCHES	DIA., IF SLIDING BEARING (NOTE 2)	IN-BOARD	OUT-BOARD	COMMON TO INB & OUTB	OIL MIST FLOW TOTAL TO ITEM (SCFM)

Figure A-4. Partially completed oil-mist lubrication summary.

3. Air regulator with gauge and suitable adjusting knob with internal locking devices that can be made tamper-resistant.
4. Inlet air heater with temperature gauge, bi-metal type thermostat with proven, reliable controller, and automatic cut-out on heater element for over- or under-temperature or blockage of mist flow.

Generator Inlet Oil System

1. Shut-off valve and oil-supply pressure gauge.
2. Oil filter (spin-on type) and check valve to allow servicing without interrupting the operation of the system.
3. Bulk oil fill system shall be capable of automatically maintaining a working oil level in the reservoir from a pressurized supply.

Instrumentation

1. Sensing devices and individual alarm circuits shall be furnished for high and low:
   a. Air supply temperature
   b. Reservoir oil temperature
   c. Mist header pressure

MIST BEARING LIST

Sheet No. ______
Estimate No. ______
Job No. ______

Firm ______ Date ______
Machine or Equipment ______ Reference Dwg. ______
Builder ______ Oil Type ______
Minimum Ambient Temperature — 60°F   Operating Temperature Range — 60 - 80°F   Oil Viscosity ______

REF. No.	No. LUBE POINTS	DESCRIPTION OF LUBE POINT				SCFM PER LUBE POINT	RECLASSIFIER FITTINGS PER LUBE POINT				TOTAL SCFM
		IDENTIFICATION — TYPE OF ELEMENT (NOTE 1)	SIZE (NOTE 1)	SPEED (RPM)	SERVICE (NOTE 2)		TYPE	RATING		QTY.	
								BI	SCFM		
								TOTAL SYSTEM SCFM			

NOTE 1: SIZE REQUIREMENTS–SUGGESTED ABBREVIATIONS: ANTI-FRICTION BEARINGS (AF)—List bore diameter and number of rows; PLAIN BEARINGS (PL)—List bore diameter, length and diametral clearance; SLIDES—List slide length and width; ROLLER CHAIN (ROL CH)—List chain pitch and number of rows plus small sprocket PD and RPM; SILENT CHAIN (SIL CH)—List chain width plus small sprocket PD and RPM; CONVEYOR CHAIN (CNVR CH)—List chain width and length plus drive sprocket diameter.

NOTE 2: TYPES OF SERVICE: Light (L), Moderate (M) and Heavy (H).

Figure A-5. Mist bearing list.

d. Generator oil reservoir level

*e. Mist density

Note: Alarm circuits shall be energized during normal operation, i.e. alarm relays shall deenergize to cause alarm.

2. A connection for a remote alarm to indicate loss of electrical power or actuation of any local alarm.
3. Visual warning light (red) and normal operation light (green) shall be installed so as to be visible 360° from the generator console.
4. Local panel with separate lights and first-out feature to indicate each malfunction.

General

All fittings, instruments, etc., shall be enclosed in a weather-tight steel cabinet (equal to NEMA 4 and 12). The cabinet shall be supported on steel legs at eye level, above grade. All exposed steel surfaces shall be painted with epoxy type finish. Electrical connections shall be 120 volt/1 phase/60 cycle and shall include an explosion proof on-off switch. All wiring in the cabinet shall be enclosed in conduit. (Seal-tight or approved equal). All fasteners (nuts, bolts, etc.), tube fittings and tubing shall be stainless steel.

*Alternatively, the vendor may offer aluminum enclosures for consideration by purchaser (vendor to give break-out cost).

5.0 Back-up Unit

5.1 Back-up shall have 100% capacity, shall be enclosed in a steel weatherproof cabinet and attached to the main console.

5.2 Back-up unit shall have its own air pressure regulator, mist generating assembly, oil reservoir, and relief protection.

*5.3 Back-up shall have the following instrumentation: reservoir level sight glass, mist pressure gauge. Oil and air heaters are not required unless separately specified by purchaser. If heaters are supplied, a thermometer shall be considered within the scope of supply.

6.0 Oil Make-up System

*6.1 Automatic make-up will be either from a centralized lube system or from one of the following systems, as specified:

A. An oil make-up system with a skid-mounted minimum 275 gal (approximately 1,000 l) reservoir, air-operated main and spare oil-supply pumps (purchaser to specify if spare pump is needed), level gauge and low level alarm switch. Reservoir shall be carbon steel, hot dipped galvanized after completion of fabrication, and shall have low-point drain, fill and vent connections. Fill connection shall be equipped with (purchaser to specify) filter or equal. Reservoir shall be purge-misted.

B. An oil make-up system with a standard oil drum (by purchaser) and an air-operated (specify) drum pump.

7.0 Piping Materials

7.1 Piping and tubing system components between the generator and lubrication points shall be per the following:

Component	Material	Special Requirements
Piping system	Carbon steel, galvanized Headers: 2 in. pipe Laterals & drops: 3/4 in. pipe	Schedule 40 wall Thickness, threaded end
Fittings	Malleable iron, galvanized	150 psi rating
Valves, block Valves, snap	Carbon steel Brass	Threaded end 1/4 Curtis automotive Type CA-1300 or equal
Tubing system	18 Cr Ni Type 304 or 316 (Brass fittings with 18-8 ferrules may be used.)	Tube: 1/4 O.D. with 0.035 in. wall

8.0 Oil-mist Piping Fabrication

8.1 Cut pipe or tubing ends shall be deburred and reamed where necessary so that there is no reduction of the inside diameter at the cut.

8.2 Piping shall be fabricated so that the use of piping fittings is minimized. Reducing swage nipples and reducing couplings shall be provided at oil-mist generator.

9.0 Oil-mist Piping Installation

9.1 Unless specifically detailed on drawings, all piping shall be routed, detailed, and supported in the field with all joints exposed to view. Underground piping is permitted only if sloped, executed with suitable drain provision, and approved by purchaser.

9.2 Each piece of pipe and all fittings shall be swabbed with a clean, lint-free cloth prior to joining any threaded connections. Extreme care shall be taken to keep interior of all piping, tubing, and equipment clean.

9.3 Neither PTFE tape nor any other sealant shall be used in making up pipe-threaded joints. An approved thread lubricant is acceptable.

*9.4 Oil-mist headers and branches shall be sloped, unless otherwise specified, to drain to the mist generator or to the equipment. The amount of slope (rise/run) shall be a minimum of 1 unit per 120 units for a distance of 50 ft (approximately 15 m) from the generator and 1 unit per 10 units thereafter. Greater slopes may be required where ambient temperature may be below 0°F. Low spots or pockets are not permitted unless unavoidable and accepted by purchaser's signature. Unavoidable low points shall be provided with drains. The type and location of all drains must be approved by purchaser.

9.5 Oil-mist header branch connections shall be made at the top of the header unless specifically noted otherwise on drawings and accepted by purchaser's signature.

9.6 Tubing and piping shall be properly supported. Oil-mist headers and horizontal branches shall not have low pockets. All steel materials used in fabrication or supports shall be galvanized, unless otherwise specified.

9.7 The end of each piping lateral shall be supported. The anchor shall be connected to permanent structures, piping, foundations, or other equipment that is not normally removed for maintenance.

9.8 Tubing shall be installed so that no oil will be trapped. Tube benders shall be used as required so that tubing will be installed without kinks, wrinkles, or flattening.

9.9 In making up tubing or threaded joints, thread compound or PTFE tape shall *not* be used.

9.10 Oil-mist fittings and tubing between distribution manifolds and oil-mist fittings, or between oil-mist fittings and equipment lubrication ports, shall be installed as follows:

A. Install oil-mist fittings only after cleaning and blowing has been completed. Caps, plugs, or block valves shall be installed temporarily to allow cleaning one branch at a time.

B. Remove temporary block fittings and install oil-mist fittings. Distribution manifolds shall be furnished with a snap drain or similar quick-acting valve.

C. Make connections to equipment lubrication ports after blowing is completed and approved by owner.

D. The application fittings shall be connected to the equipment by tubing arranged so that normal maintenance may be performed without requiring the application fitting and/or lateral to be removed.

9.11 Equipment lubrication ports shall remain plugged until the oil-mist lubrication system is certified clean and final connections are being made.

*9.12 Drain legs shall be fitted with a transparent catch pot, complete with overflow vent and drain valve. It shall be sized for the drainage volume anticipated by at least 60 days of continuous full-load operation, with calculation basis approved by purchaser.

10.0 Retrofit of Existing Pumps

*10.1 If bearing housing closure-type oil seals are used on existing equipment, a permanent vent hold shall be drilled in a manner to prevent the entrance of rain water. The vent hole shall be 3 mm (1/8 in.) diameter for dry-sump installations. On thrust-loaded bearings where oil mist must be made to flow through the rotating elements, the vent location should promote through-flow.

11.0 Inspection and Testing

11.1 The purchaser reserves the right to witness all tests and to inspect all work and materials for conformance to specifications.

11.2 The purchaser's representative shall be notified of any changes in work schedules in order that the necessary inspection may be made.

12.0 System Internal Cleaning and Run-in

12.1 Steam above 100 psig (approximately 685 kPa) shall be used for blowing out the piping distribution system. Blowing shall continue for one hour minimum and until a polished surface aluminum target located at the most distant point shows no pitting. The oil-mist generator must be isolated during steam blowing. Steam blowing shall be immediately followed by compressed-air blowing for the purpose of drying the piping.

*12.2 After blowing, the oil-mist system shall be run-in using the specified lubricant as follows:

A. Fill the oil-supply reservoir and oil-mist generator reservoirs with specified lubricant.

B. Connect the wiring at the terminal strip inside the mist unit.

C. Activation of air heater shall not be possible unless air is flowing through the heater.

D. Operate the mist system at 20 in. (approximately 5 kPa) water pressure and clean all plugged mist fittings. For this lubricant flushing operation to be considered complete, the mist system shall operate continuously for a minimum of 48 hours using specified lubrication without pulsating header pressure indicating low point oil accumulation in piping.

E. An updated record of all key system components and operating parameters shall be furnished to purchaser's representative. Records shall include mist head sizes, a complete tabulation of all metering orifice fittings, regulated supply air pressure and oil-mist header pressure with generator(s) set at design conditions. (See also Paragraph 2.7.)

F. Any modifications to the system must be approved in writing by purchaser's representative.

12.3 The system should be put into oil-mist service immediately after cleaning.

12.4 Bearings that have been grease lubricated and their lubrication and vent passages shall be free of grease before being connected to an oil-mist system. This clean-out task shall be the responsibility of the oil-mist system supplier.

12.5 Bearings with double shields shall have one or both shields removed before being connected to the oil-mist system. This verification task is the responsibility of the purchaser.

12.6 All bearings shall be pre-lubricated with the oil-mist lubricant before being connected to the oil-mist system.

13.0 System Adjustments

13.1 After the oil mist has been run-in with specified lubricant, it shall be connected to the equipment lubrication ports.

13.2 Reservoirs of equipment intended for wet-sump operation shall be hand-filled with lubricant to the proper level.

13.3 New cartridges shall be installed in all permanent filters that have been used during flushing.

13.4 All piping and equipment shall be checked for excessive vibration and necessary corrections shall be made.

13.5 Oil mist shall be flowing freely from the bearing oil reservoir drain or bearing housing outlets on all equipment being served by the oil mist system. All mist fittings that do not appear to be supplying an adequate flow of oil mist shall be inspected and their cleanliness and adequacy verified.

13.6 Operational adjustments, in addition to those made to the oil-mist generator, shall be made after all work is complete. This shall be done in the presence of purchaser. Adjustments shall be made to the proper levels required by the design of the system. To the extent permitted by ambient conditions, functions to be checked shall include system mist pressure, oil-mist density, oil temperature, air temperature, oil-supply pressure, air-supply pressure, and all instrumentation and alarms.

Pressure, density, temperature, and level deviations must be simulated so as to actuate alarms. After testing, all pressures, temperatures, and levels shall be reset to normal operating conditions.

14.0 Operating Manuals

14.1 The oil-mist systems vendor shall supply five operative manuals. These manuals shall be written specifically for the oil-mist equipment being furnished and shall contain a minimum of the following information:

A. List of all distribution manifolds and application fittings by size and model and description of equipment to be served by each manifold and corresponding fittings.

B. Drawings and details necessary for the installation, commissioning, operation, and shutdown of the system without the need for assistance from the vendor.

C. Settings and operating ranges of all controls and instrumentation.

D. Complete parts list, bills of material, and set of final "as-built" drawings.

E. Switching and troubleshooting instructions shall be posted inside the console cabinet doors. These instructions shall be executed in a fashion similar to Figure A-6.

NOTICE

REPORT ALL OIL MIST PROBLEMS IMMEDIATELY

IF MAIN GENERATOR MALFUNCTIONS OR FAILS TO PRODUCE OIL MIST. TURN ON BACK-UP GENERATOR AS FOLLOWS:

1. CHECK OIL LEVEL IN BACK-UP GENERATOR. FILL IF REQUIRED.
2. CLOSE AIR SUPPLY VALVE TO MAIN GENERATOR.
3. CLOSE 2" BALL VALVE ON MIST OUTLET OF MAIN GENERATOR.
4. TURN OFF ELECTRICAL POWER TO MAIN GENERATOR.
5. OPEN 2" BALL VALVE ON MIST OUTLET OF BACK-UP GENERATOR.
6. OPEN AIR SUPPLY VALVE TO BACK-UP GENERATOR.
7. TURN ON ELECTRICAL POWER TO BACK-UP GENERATOR.

SERVICE ON OIL MIST SYSTEMS FURNISHED BY:
(Name, address, telephone number of service organization)

Figure A-6. Sample switching and troubleshooting instructions.

Appendix B

Oil-mist Troubleshooting Chart

Malfunction	Possible Cause	Remedy
High manifold pressure	Air-supply pressure to mist generator set too high	Reduce regulator setting. CAUTION: Do not reduce below recommended minimum.
	Restriction in distribution system	Check pipe, tube, hose sizes. Make sure that any valves in distribution lines are full-flow types and are fully opened.
	Plugged reclassifiers	Remove and clean orifices.
	Flow restriction through bearing housing	Check venting—correct as necessary
	Oversize mist generator	Check system design and generator rating—correct as indicated.
	Undersize application fittings	Check system design and fitting sizes—correct as indicated.
	Too much bypass air	Reduce bypass air.
Low manifold pressure	Air-supply pressure to mist generator set too low	Increase regulator setting.
	Broken or disconnected line in distribution system	Connect or replace as needed.
	Undersize mist generator	Check system design and fittings sizes—correct as indicated.
No visible mist at application fittings or housing vents	With reclassifying fittings and "mist oil," might be normal operation	Check by opening generator pressure relief valve (if provided) or disconnecting a mist line. In good light, mist should be visible against a dark background. Also check rate of oil consumption from generator reservoir.
	Air-supply pressure to generator set too low	Set air supply pressure above minimum specified for generator.

(Continued)

Malfunction	Possible Cause	Remedy
	Oil with poor misting properties at operating temperature. (This is most frequent cause of no or insufficient mist.)	Make sure oil and air temperatures are suitable for viscosity of oil used. If not, change to another oil.
	Mist generator not operating	Check manufacturer's service instructions. Could be such things as improper mist density (oil flow) adjustment, clogged oil passages or intake screen.
Excessive stray mist	Low mist manifold pressure	Increase manifold pressure.
	Mist type fittings used where reclassifying fittings should be used	Use reclassifying type application fittings wherever possible—must be used on plain bearings, slides and ways, etc.
	Oil with poor reclassifying characteristics	Change oil—mist oil might be necessary.
Excessive lubricant delivery	Improper system adjustments	Check and adjust as needed: Mist pressure should be near design value. Oil flow adjustment—usually factory set at maximum. Oil and air temperatures (if heaters used).
	Oversize application fittings	Check sizes against those specified—correct as indicated. Check system design—were heavy service formulas used just to be safe?
Insufficient lubrication delivery	Improper system adjustments	Check and adjust as needed: Mist pressure should be near design value. Oil flow adjustment—has it been reduced too far? Oil or air temperatures—are they much lower than recommended?
	Undersize application fittings	Check sizes against those specified—correct as indicated. Check system design—should longer or different type application fittings be used?
	See also "No visible mist at application fittings or housing vents"	

Appendix C

Conversion Data

Table C-1
Common Usage (Informal) Conversion Data

English-Metric	Metric-English
1 micron = 1 × 10^{-6} meter = 0.001 millimeter	1 micron = 39.37 × 10^{-6} in.
1 in. = 2.54 cm	25.4 microns = 0.001 in.
1 $in.^2$ = 6.452 cm^2	1 cm = 0.3937 in.
1 $in.^3$ = 16.39 cm^3	1 cm^2 = 0.155 $in.^2$
1 ft = 12 in. = 30.481 cm = 0.30481 meter	1 cm^3 = 0.06102 $in.^3$
1 ft^3 = 28.32 liters	1 cm = 0.3937 inch = 0.0328 ft
1 fl oz (fluid ounce) = 1.8 $in.^3$ = 29.6 cm^3	1 liter = 0.3531 ft^3
1 qt = 32 fl oz = 57.75 $in.^3$ = 0.9464 liter	1 cm^3 = 0.06102 $in.^3$ = 0.0338 fl oz
1 gal (U.S.) = 231 $in.^3$ = 4 quarts = 3.785 liters	1 liter = 61.2 $in.^3$ = 33.9 fl oz = 1.057 qt
1 imperial gallon = 1.2 U.S. gallon = 4.546 liters	1 liter = 1.057 qt = 0.2642 gal (U.S.)
1 oz (ounce weight) = 28.35 grams	1 liter = 0.2642 U.S. gal = 0.317 imperial gal
1 lb = 16 ounces = 0.4536 kg	1 gram = 0.03527 oz
1 in. H_2O = 0.036 psi = 2.54 grams/cm^2 = 2.54 cm H_2O	1 kg = 35.274 oz = 2.2046 lb
1 in. Hg = .491 psi = 13.596 in. H_2O = 34.53 cm H_2O	1 cm H_2O = 1 gm/cm^2 = 0.014 psi
1 psi = 27.7 in. H_2O = 0.07 kg/cm^2	1 cm Hg = 13.596 cm H_2O = .394 in. Hg = .193 psi
	1 kg/cm^2 = 14.22 psi
$°F = \frac{9}{5}(°C + 40) - 40$	$°C = \frac{5}{9}(°F + 40) - 40$
centistokes = 0.22 × SSU (approx), above 100 SSU	SSU = 4.6 × centistokes (approx), above 20 CS
sp gravity (at 60°F) $= \frac{141.5}{131.5 + °API}$ (lighter than H_2O)	sp gravity (at 15.5°C) $= \frac{140}{130 + °Baume}$ (lighter than H_2O)
	sp gravity (at 15.5°C) $= \frac{145}{145 - °Baume}$ (heavier than H_2O)

Table C-2
Formal Conversions*

Customary Units		Preferred Units	
Length			
1 angstrom unit	=	0.100 nanometer	nm
1 inch	=	25.400 millimeters	mm
1 foot	=	0.305 meter	m
1 micron	=	1.000 micrometer	μm
1 mil	=	25.400 micrometers	μm
1 statute mile	=	1.609 kilometers	km
1 nautical mile	=	1.852 kilometers	km
1 microinch	=	25.400 nanometers	nm
Area			
1 square inch	=	6.452 square centimeters	cm^2
1 square foot	=	929.030 square centimeters	cm^2
1 square yard	=	0.836 square meters	m^2
Volume			
1 cubic inch	=	16.387 cubic centimeters	cm^3
1 cubic foot	=	0.028 cubic meter	m^3
1 fluid ounce	=	29.574 cubic centimeters	cm^3
1 pint (U.S. liquid)	=	473.177 cubic centimeters	cm^3
1 quart (U.S. liquid)	=	946.353 cubic centimeters	cm^3
1 gallon (U.S. liquid)	=	3785.412 cubic centimeters	cm^3
1 barrel (42 gal)	=	0.159 cubic meter	m^3
1 liter†	=	1000.000 cubic centimeter	cm^3
1 milliliter†	=	1.000 cubic centimeter	cm^3
Time			
minute	=	60 seconds	s
hour	=	3600 seconds	s
day	=	24 hours	h
Temperature**			
degrees Celsius	=	5/9 (°F – 32)	°C
100 degrees F	=	37.8 degrees Celsius	°C
210 degrees F	=	98.9 degrees Celsius	°C
degrees Celsius + 273.15	=	kelvin	K
Velocity			
1 inch/second	=	25.400 millimeter/second	$mm \cdot s^{-1}$
1 foot/second	=	0.305 meter/second	$m \cdot s^{-1}$
1 foot/minute	=	5.080 millimeter/second	$mm \cdot s^{-1}$
1 mile/second	=	1.609 kilometer/second	$km \cdot s^{-1}$
1 mile/minute	=	26.822 meter/second	$m \cdot s^{-1}$
1 mile/hour	=	1.609 kilometer/hour	$km \cdot h^{-1}$
1 knot	=	0.514 meter/second	$m \cdot s^{-1}$

(Continued)

Table C-2
Formal Conversions* (*Continued*)

Acceleration		
1 inch/second^2	= 25.400 millimeter/second^2	$mm \cdot s^{-2}$
1 foot/second^2	= 0.305 meter/second^2	$m \cdot s^{-2}$
Mass		
1 ounce (avoirdupois)	= 28.350 grams	g
1 pound (avoirdupois)	= 453.592 grams	g
1 ton (2000 lb)	= 907.185 kilogram	kg
Force		
1 kilogram meter/second^2	= 1.000 newton	N
1 dyne	= 10.000 micronewtons	μN
1 gram (force)	= 9.807 millinewtons	mN
1 kilogram (force)	= 9.807 newtons	N
1 ounce (force)	= 0.278 newton	N
1 pound (force)	= 4.448 newtons	N
1 ton (force)	= 8.896 kilonewtons	kN
Pressure and Stress (Force/Area)		
1 newton/meter^2	= 1.000 pascal	Pa
1 dyne/centimeter^2	= 0.100 pascal	Pa
1 atmosphere	= 101.325 kilopascals	kPa
1 millibar	= 100.000 pascals	Pa
1 bar	= 100.000 kilopascals	kPa
1 millimeter of mercury (0°C)	= 133.322 pascals	Pa
1 torr (0°C)	= 133.322 pascals	Pa
1 inch of mercury (32°C)	= 3.386 kilopascals	kPa
1 inch of water (60°F)	= 248.840 pascals	Pa
1 kilogram (force)/centimeter^2	= 90.067 kilopascals	kPa
1 kilogram (force)/millimeter^2	= 9.807 megapascals	MPa
1 pound (force)/foot^2	= 47.880 pascals	Pa
1 pound (force)/inch^2	= 6.895 kilopascals	kPa
Energy, Work, or Quantity of Heat (Force × Length)		
1 newton meter	= 1.000 joule	J
1 erg	= 0.100 microjoule	μJ
1 gram (force) centimeter	= 98.070 microjoules	μJ
1 kilogram (force) meter	= 9.807 joules	J
1 foot pound (force)	= 1.356 joules	J
1 horsepower hour	= 2.686 megajoules	MJ
1 watt hour	= 3.600 kilojoules	kJ
1 kilowatt hour	= 3.600 megajoules	MJ
1 Btu	= 1.055 kilojoules	kJ
1 gram calorie	= 4.184 joules	J
1 kilogram calorie	= 4.184 kilojoules	kJ

(*Continued*)

Table C-2
Formal Conversions* (*Concluded*)

Power and Heat Flow [(Force × Length)/Time]			
1 joule/second	=	1.000 watt	W
1 erg/second	=	0.100 microwatt	μW
1 gram (force) centimeter/second	=	98.070 microwatts	μW
1 kilogram (force) meter/minute	=	0.163 watt	W
1 foot pound (force)/second	=	1.356 watts	W
1 foot pound (force)/minute	=	22.597 milliwatts	mW
1 Btu/second	=	1.054 kilowatts	kW
1 Btu/minute	=	17.573 watts	W
1 Btu/hour	=	0.293 watt	W
1 horsepower	=	746.000 watts	W

Density			
1 gram/cubic centimeter	=	1.000 megagram/cubic meter	$Mg \cdot m^{-3}$
1 milligram/liter	=	1.000 gram/cubic meter	$g \cdot m^{-3}$
1 milligram/milliliter	=	1.000 kilogram/cubic meter	$kg \cdot m^{-3}$
1 ounce (avoirdupois)/cubic inch	=	1.730 megagram/cubic meter	$Mg \cdot m^{-3}$
1 ounce (avoirdupois)/cubic foot	=	1.001 kilogram/cubic meter	$kg \cdot m^{-3}$
1 ounce (avoirdupois)/U.S. gallon	=	7.489 kilogram/cubic meter	$kg \cdot m^{-3}$
1 pound/cubic inch	=	27.679 megagram/cubic meter	$Mg \cdot m^{-3}$
1 pound/cubic foot	=	16.018 kilogram/cubic meter	$kg \cdot m^{-3}$
1 pound/U.S. gallon	=	119.826 kilogram/cubic meter	$kg \cdot m^{-3}$

Viscosity Dynamic (Absolute) (Force × Time/Area)			
1 newton second/meter^2	=	1.000 pascal second	$Pa \cdot s$
1 poise	=	0.100 pascal second	$Pa \cdot s$
1 centipoise	=	1.000 millipascal second	$mPa \cdot s$
1 pound (force)·second/foot^2	=	47.880 pascal second	$Pa \cdot s$
1 reyn	=	6.895 kilopascal second	$kPa \cdot s$

Kinematic (Dynamic Viscosity/Density)			
1 stoke	=	1.000 square centimeter/second	$cm^2 \cdot s^{-1}$
1 centistoke	=	1.000 square millimeter/second	$mm^2 \cdot s^{-1}$
1 SUS (=0.216 cSt) (100°F)†	=	0.216 square millimeter/second	$mm^2 \cdot s^{-1}$
1 SUS (=0.214 cSt) (210°F)†	=	0.214 square millimeter/second	$mm^2 \cdot s^{-1}$
Example: A viscosity of 150 SUS at 100°F must first be converted by ASTM D 2161 to 31.90 cSt and written 31.90 $mm^2 \cdot s^{-1}$ (37.8°C).			

* Some numbers are correct to three places after the decimal.

** The SI unit for temperature is kelvin, K (do not use the ° symbol), but Celsius is allowed where necessary. Centigrade has been replaced by Celsius.

† Liter is not an SI unit, but is acceptable for volume of liquid. The symbol is L.

†† Approximate = accurate at greater than 300 SUS and within about 10% at 80% SUS.

Source: American Society of Lubrication Engineers.

(Continued)

Table C-3
AGMA & SAE Fluid Lubricant Viscosity Rating

AGMA Grade No.	Viscosity Measurement Temperature Deg F	Viscosity: Saybolt Univ Seconds (SSU)	Viscosity: Kinematic Centistokes	Viscosity: Engler Degrees	Viscosity: Redwood Std No. 1 Seconds	
1	100	180- 240	39 - 52	5.1- 6.7	157- 210	
2	100	280- 360	62 - 79	8 - 10	245- 318	
3	100	490- 700	108 - 155	14 - 20	430- 620	
4	100	700- 1,000	155 - 220	20 - 28	620- 870	
5	210	80- 105	16 - 22	2.4- 3.0	70- 92	
6	210	105- 125	22 - 27	3.0- 3.6	92- 108	
7	210	125- 150	27 - 32	3.6- 4.3	108- 132	
8	210	150- 190	33 - 41	4.3- 5.4	132- 166	
9	210	350- 550	76 - 120	10 - 16	308- 480	
10	210	900- 1,200	200 - 265	25 - 34	800- 1,060	
11	210	1,800- 2,500	400 - 550	50 - 70	1,575- 2,200	
SAE No. Engine Oil						**Approx Equivalent AGMA No.**
5W	0	max 4,000	max 900	max 115	max 3,500	
10W	0	6,000- 12,000	1,320 - 2,700	165 - 325	5,200-10,700	1
20W	0	12,000- 48,000	2,700 -10,550	330 -1,400	10,560-42,000	2

Table C-3
AGMA & SAE Fluid Lubricant Viscosity Rating (*Concluded*)

SAE No.	Viscosity Measurement	Viscosity				
Engine Oil	Temperature Deg F	Saybolt Univ Seconds (SSU)	Kinematic Centistokes	Engler Degrees	Redwood Std No. 1 Seconds	Approx Equivalent AGMA No.
20	210	45- 58	5.7- 9.6	1.5- 1.8	41- 52	2
30	210	58- 70	10 - 13	1.8- 2.1	52- 62	3
40	210	70- 85	13 - 17	2.1- 2.5	62- 75	4
50	210	85- 110	17 - 23	2.5- 3.2	75- 98	5
Gear Oil						
75	0 210	max 15,000 40 min	max 3,300 3.9 min	max 400 1.3 min	max 13,500 36 min	1
80	0 210	15,000-100,000 48 min	3,300 -22,000 6.6 min	400 -2,920 1.5 min	13,500-89,000 43 min	2,3
90	0 210	max 750,000 75- 120	max 165,000 14.2- 25.5	max 21,800 2.3- 3.4	max 667,000 66- 105	5,6
140	210	120- 200	25.5- 44	3.4- 5.6	105- 175	7,8
250	210	200 min	44 min	5.6 min	175 min	8 to 9

Table C-4
ASTM Fluid Lubricant Viscosity Rating

ASTM Grade No.	Viscosity Measurement Temperature Deg F	Viscosity			
		Saybolt Univ Seconds (SSU)	Kinematic Centistokes	Engler Degrees	Redwood Std No. 1 Seconds
32	100	29-35	0.7- 2.7	0.97- 1.2	26-33
40	100	36-44	3.0- 5.5	1.2 - 1.4	33-40
60	100	54-66	8.5- 12	1.7 - 2.0	48-58
75	100	68-82	12.5- 16	2.1 - 2.4	60-75
105	100	95-115	19 - 24	2.8 - 3.3	82-100
150	100	135-165	29 - 36	3.8 - 4.7	120-145
215	100	194-236	42 - 52	5.5 - 6.6	170-205
315	100	284-346	63 - 76	8 - 10	250-308
465	100	419-511	92 - 112	12 - 15	370-450
700	100	630-770	140 - 172	18 - 22	552-700
1,000	100	900-1,100	200 - 242	25 - 31	800-970
1,500	100	1,350-1,650	300 - 360	38 - 46	1,200-1,450
2,150	100	1,935-2,365	430 - 520	55 - 66	1,700-2,075
3,150	100	2,835-3,465	640 - 760	80 - 97	2,500-3,000
4,650	100	4,185-5,115	930 -1,175	120 -145	3,700-4,600
7,000	100	6,300-7,700	1,410 -1,700	180 -218	5,600-6,800

Table C-5
Conversion of Kinematic Viscosity to Saybolt Universal Viscosity

Kinematic Viscosity, cs	Equivalent Saybolt Universal Viscosity, sec		Kinematic Viscosity, cs	Equivalent Saybolt Universal Viscosity, sec		Kinematic Viscosity, cs	Equivalent Saybolt Universal Viscosity, sec	
	At 100°F Basic Values	At 210°F		At 100°F Basic Values	At 210°F		At 100°F Basic Values	At 210°F
2.0	32.6	32.9	21.0	102.0	102.8	41.0	190.8	192.1
2.5	34.4	34.7	22.0	106.4	107.1	42.0	195.3	196.7
3.0	36.0	36.3	23.0	110.7	111.4	43.0	199.8	201.2
3.5	37.6	37.9	24.0	115.0	115.8	44.0	204.4	205.9
4.0	39.1	39.4	25.0	119.3	120.1	45.0	209.1	210.5
4.5	40.8	41.0						
5.0	42.4	42.7						
6.0	45.6	45.9	26.0	123.7	124.5	46.0	213.7	215.2
7.0	48.8	49.1	27.0	128.1	129.0	47.0	218.3	219.8
8.0	52.1	52.5	28.0	132.5	133.4	48.0	222.9	224.5
9.0	55.5	55.9	29.0	136.9	137.9	49.0	227.5	229.1
10.0	58.9	59.3	30.0	141.3	142.3	50.0	232.1	233.8
11.0	62.4	62.9	31.0	145.7	146.8	55.0	255.2	257.0
12.0	66.0	66.5	32.0	150.2	151.2	60.0	278.3	280.2
13.0	69.8	70.3	33.0	154.7	155.8	65.0	301.4	303.5
14.0	73.6	74.1	34.0	159.2	160.3	70.0	324.4	326.7
15.0	77.4	77.9	35.0	163.7	164.9			

Table C-5 (*Continued*)
Conversion of Kinematic Viscosity to Saybolt Universal Viscosity

Kinematic Viscosity, cs	Equivalent Saybolt Universal Viscosity, sec		Kinematic Viscosity, cs	Equivalent Saybolt Universal Viscosity, sec		Kinematic Viscosity, cs	Equivalent Saybolt Universal Viscosity, sec	
	At 100°F Basic Values	At 210°F		At 100°F Basic Values	At 210°F		At 100°F Basic Values	At 210°F
16.0	81.3	81.9	36.0	168.2	169.4	Over 70.00	Saybolt seconds = centistokes × 4.635	Saybolt seconds = centistokes × 4.667
17.0	85.3	85.9	37.0	172.7	173.9			
18.0	89.4	90.1	38.0	177.3	178.5			
19.0	93.6	94.2	39.0	181.8	183.0			
20.0	97.8	98.5	40.0	186.3	187.6			

NOTE: To obtain the Saybolt Universal viscosity equivalent to a kinematic viscosity determined at t°F multiply the equivalent Saybolt Universal viscosity at 100°F by $1 + (t-100)\ 0.000064$; for example, 10 cs at 210°F are equivalent to 58.9 × 1.0070 or 59.3 sec Saybolt Universal at 210°F.

The following multipliers may be used to make *approximate* conversions from one viscosity system to another at the SAME TEMPERATURE:

Kinematic centistokes	× 0.1316	= Engler degrees
Engler degrees	× 7.599	= Kinematic centistokes
Engler degrees @ 20°C	× 35.106	= Saybolt seconds universal @ 20°C
Engler degrees @ 50°C	× 35.173	= Saybolt seconds universal @ 50°C
Engler degrees @ 100°C	× 35.353	= Saybolt seconds universal @ 100°C
Saybolt seconds universal @ 100°F	× 0.02848	= Engler degrees @ 100°F
Saybolt seconds universal @ 210°F	× 0.02829	= Engler degrees @ 210°F

Figure C-1. Comparatiave viscosity classifications.

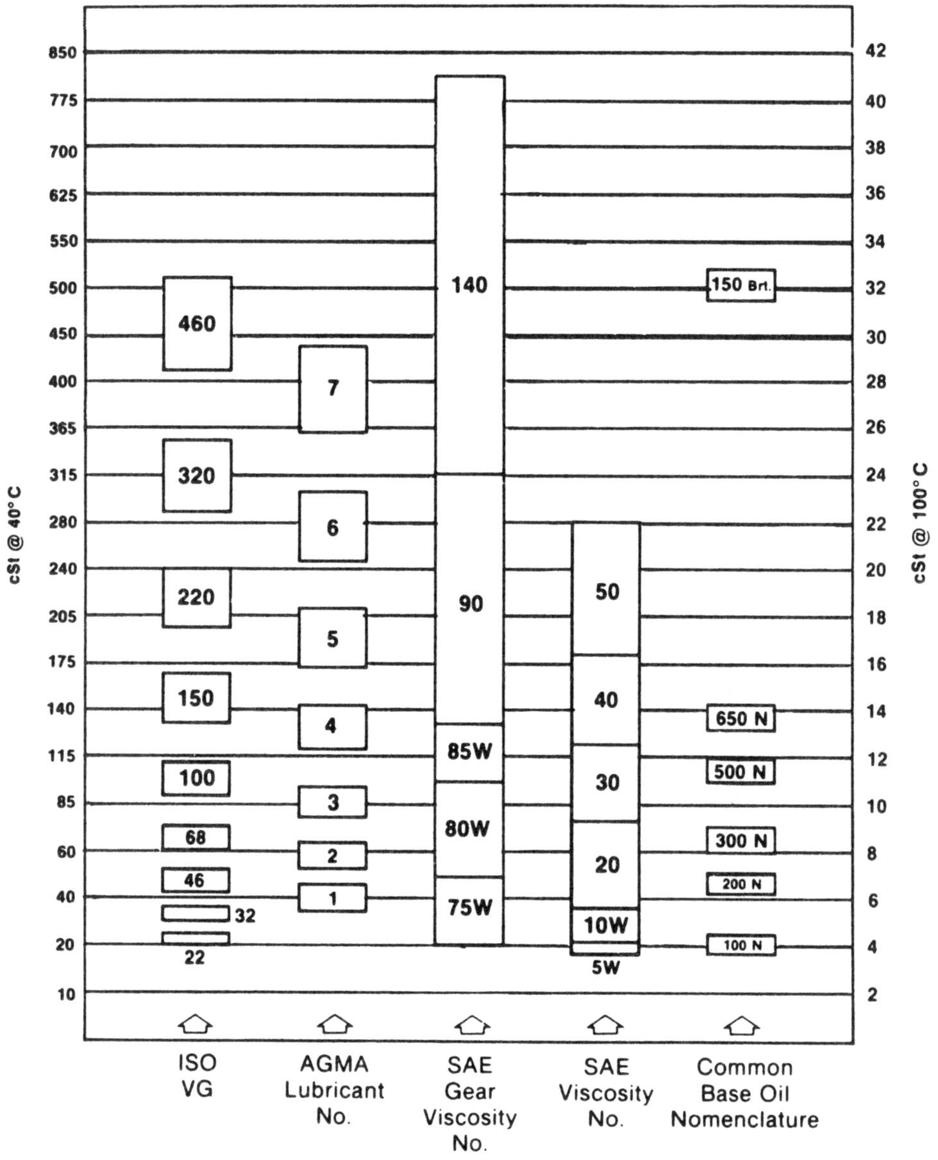

Figure C-2. Oil viscosity rating cross-reference.

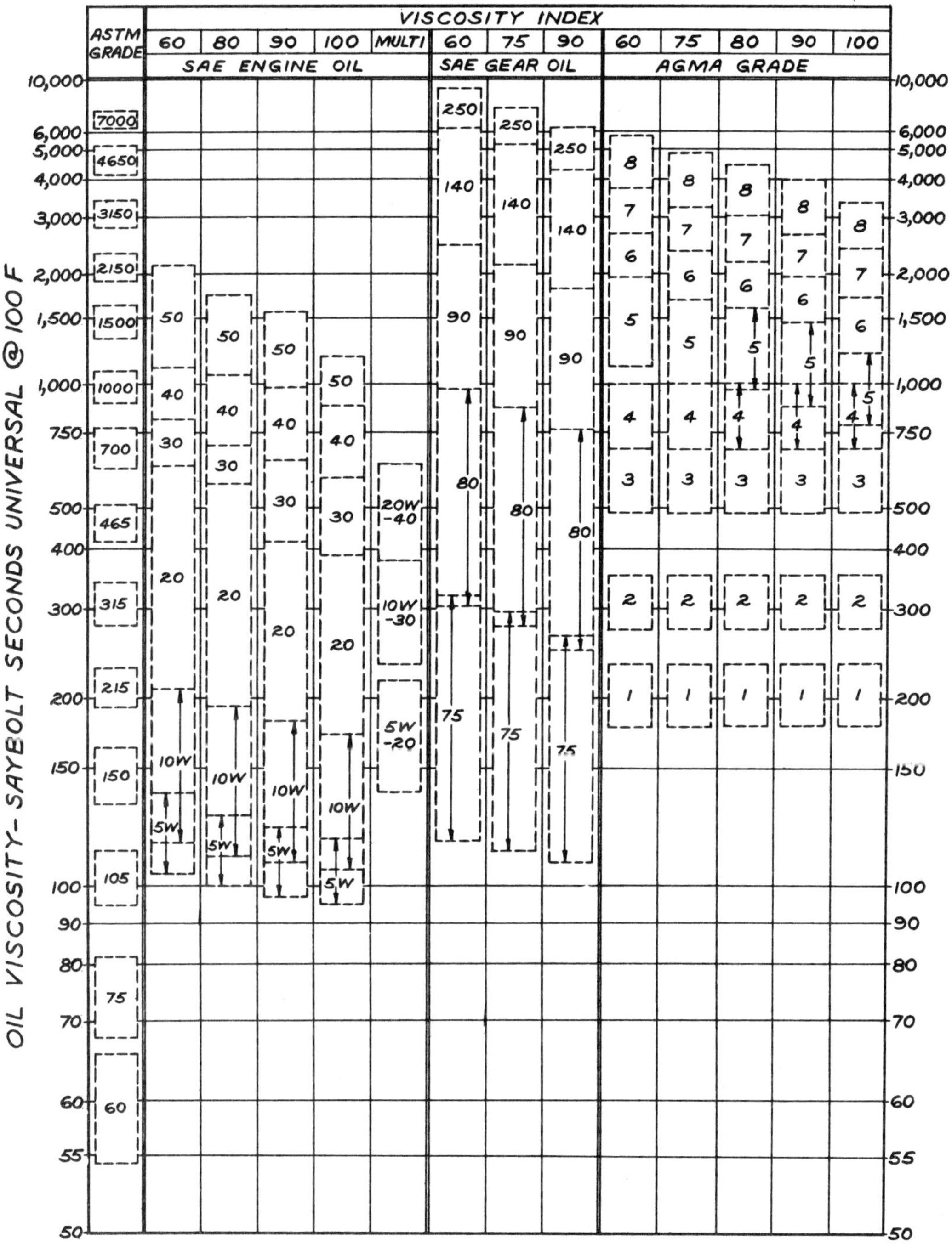

Glossary

Absolute viscosity: Kinematic viscosity corrected to overcome the variations caused by differences in specific gravity.
Aerosol: A fine suspension of liquid particles in an air stream.
Application fitting: Any fitting used to meter oil mist. *Note*: This term is often used interchangeably with the term reclassifier.
Bearing inch: An arbitrary unit of measurement used to rate the oil requirements of mechanisms lubricated by aerosols. One (1) scfm equals 20 to 40 bearing-inches, depending on what oil/air ratio is used as the basis for the bearing-inch. A value of 33 B.I. is very satisfactory for general sizing comparisons.
Branch line: Mist piping supplying one application fitting. Also, air line from plant main to the oil mist generator.
Centipoise: Unit of absolute (dynamic) viscosity—shear stress/shear rate = .01 dyne-sec/ cm^2. Water viscosity at 68°F = 1 centipoise = 1 centistoke = 30 SSU. SAE 10 oil viscosity at 60°F = 100 centipoises = 1 poise. SAE 10 oil viscosity at 60°F = 100 centipoises/0.9 gram per cm^3 = 111 centistokes.
Centistoke (CS): Absolute viscosity (centipoise) divided by density = .01 cm^2/sec. CS = .22 × SSU (approx). Applies above 100 SSU. Water viscosity at 68°F = 1 centipoise/1 gram per cm^3 = 1 centistoke = 30 SSU.
Cfm (also scfm): Cubic feet per minute of air or mist. All cfm units refer to standard conditions of 70°F and 14.7 psia unless otherwise noted. For a given size pipe, max allowable mist cfm = $7.8D^2$. (With mist velocity = 24 fps). D = internal diameter of pipe, in.
Condensate: Liquid oil in mist piping or at point of application, and water in air piping.
Condensing application fitting: An application fitting in which metering elements are baffled to separate mist into air and drops of oil.
Distribution line: A conduit used to transport oil mist to the various points of application.
Drop: Vertical line (usually 3/4 in. pipe) leading towards point to be lubricated.
Engler degrees: Time required for 200 cc of liquid to flow through Engler viscometer, divided by time for equal volume of water at 68°F.
Fluid ounce: 1 fluid ounce (fl oz) = 1.805 in^3.
Flushing: Bearings: See **Prelubrication.** Mist piping: Flush prior to installation of the application fittings. Mist piping can be flushed of scale and chips by connecting a plant air line through a filter to the mist piping, and blowing maximum air volume available through the manifold.
Fog: See **Oil fog.**
Fps: Feet per second. (a) Mist: See **Mist velocity**; (b) Speed of stock or material in process. FPS = LFM/60.
Generator: Oil-mist generator or oil-mist lubricator.
Inches, Hg: Mercury column. One (1) inch of mercury column exerts a pressure of 0.49 psi or 13.6" H_2O.

Inches, H_2O: Water column. One (1) inch of water column exerts a pressure of 0.036 psi or approximately 248 Pa.
Inches, cubic: 1 cubic inch = 0.554 fluid ounce. 1 cubic inch = 0.578 weight ounce × specific gravity.
kPa: Kilopascal, metric pressure measurement. 1 kPa = 1,000 Pa = ~ 4 in. of H_2O column = 0.145 psi.
Kinematic viscosity: The property measured when a fixed amount of oil flows through a capillary tube under the force of gravity.
Lateral: Branch line coming off the top of an oil mist header.
Lfm: Lineal feet per minute. Speed of stock of material in process.

$$LFM = 6.3 \times R \times RPM$$

R = radius of rotation, ft

To calculate LFM of shaft, bearing elements, gears, or chain operating on a different diameter than the drum, belt, or cylinder moving with the material in process:

$$LFM_{D2} = LFM_{D1} = D_2/D_1$$

D_1 = diameter of the surface at LFM D_1 velocity

D_2 = diameter of shaft, bearing elements, gears, or chain sprocket on same axis as D_1.

Lubrication unit: A unit of measurement which is numerically equal to the bearing-inch (B.I.). The lubrication unit is used to rate machine elements. All dimensions used in the lubrication unit formulas are in the metric system.
Lubricator: See **Generator.**
Machine element: Any part of a machine where surfaces in rolling or sliding contact require lubrication.
Main line: Mist pipe from oil mist generator is the primary main manifold line. Secondary mains distribute mist from the primary main to groups of branch lines. Also, refers to plant main air supply line.
Manifold: Mist distribution piping.
Manifold drop-out: Oil particles too large to be conveyed long distances, thus wetting-out and condensing.
Manifold pressure: Gauge pressure of mist in manifold.
Manometer: Gauge that indicates manifold pressure.
Micro-fog™: A Norgren trade name referring to a fine oil mist or aerosol.
Micron (μ): 1 micron = 0.001 millimeter = 0.000039 inch. 25 microns = 0.001 inch.
Mist: Oil mist is an aerosol dispersion of air and oil particles ranging in size from slightly under 1/2μ diameter to approximately 8μ diameter.
Mist fitting: An application fitting that meters oil mist with minimum conversion to oil spray or droplets.
Mist velocity: Feet per second (fps). Maximum recommended mist velocity is 24 fps; 20 fps is used in the petrochemical industry as a highly conservative value.

$$fps = 3 \times cfm/D^2$$

D = internal diameter of mist pipe, in.

Oil fog: A heterogeneous dispersion of oil mist, having particle sizes ranging from those of fine mist to small droplets.

Pa: Pascal, metric pressure measurement. 1 Pa = .000145 psi.
Peripheral speed: Lineal velocity of outer surface rotating about an axis. Usually given in lfm.
Pitch: The distance between the centers of two adjacent teeth in a toothed wheel or rack. Also, distance between centers of two adjacent chain rollers.
Pitch diameter: Diameter of an imaginary circle concentric with the axis of a toothed sprocket or gear, having a lineal speed equal to the speed of the chain or the pitch circle of mating gear.
Preload: Axial bearing load when machine is not running.
Preloaded bearing: Bearing in which the radial clearance is taken up in assembly by an axial load on the bearing. The operating temperatures of preloaded bearings must be controlled so that the thermal expansion of shaft versus housing does not increase axial load to seizure point.
Prelubrication: Since oil mist is a system that continuously supplies make-up oil, all machine elements must be preoiled before initial machine operation. Where grease was previously used, all grease must be removed before preoiling.
Pressure drop: The loss of pressure between any two points in a system or component.
Psi: Pounds per square inch gauge pressure. May also be written psig. One psi equals 6.895 kPa.
Psia: Pounds per square inch absolute pressure, psia = psi + 14.7.
PTF SAE short: Same as NPTF, except one thread shorter at the outer end.
PTF SAE special short: Same as NPTF, except one thread shorter at each end.
Radial load: Load perpendicular to axis of shaft.
Reclassifier: A special fitting or restriction used to convert "dry" mist into a wet mist. This component is used at the point of application of lubricant to a machine element. Note that the term "reclassifier" is often used interchangeably with the term "application fitting."
Regulated air pressure: Controlled gauge pressure applied to the oil-mist generating nozzle. Pressure drop across the nozzle equals regulated air pressure minus the manifold pressure in psi.
Rolling element bearing: Preferred terminology for "antifriction" bearings.
Rpm: Revolutions per minute.
Scfm: Standard cubic feet per minute. 1 scfm = .028 $m^3 min^{-1}$.
Seal, contact: Bearing seal where seal material rubs against shaft or housing. Also true for face seal and lip seal (mechanical seals). Carrier air will not vent through these seals.
Seal, labyrinth: Bearing seal where convolutions of moving part are very closely spaced from stationary part to provide barrier. Most labyrinths will vent the carrier air satisfactorily.
Slope: Refers to pitch or inclination of mist piping. Percent slope is drop of piping in units for every 100 units of length. Percent slope and angle of slope should not be confused.
Spray condensing fitting: An application fitting with the conversion efficiency of a condensing fitting and the discharge of a spray fitting.
Spray fitting: A single orifice application fitting, with sufficient bore length to provide a

turbulent region which combines the mist particles in larger sizes, so that the fitting discharge may be applied directly to moving parts.

Spray nozzle: An application fitting with two or more spray orifices. Also applies to a single orifice spray fitting, which is dimensionally interchangeable with multiple spray-orifice fittings.

SSF: Saybolt Seconds Furol; used for measuring the viscosity of heavy oils. SSF indicates the time in seconds for 60 cc of oil at a specified temperature to flow through an orifice .124 dia × .483 long. The SSF outflow time for a specific oil is about 1/10 that of SSU viscometer, as viscosities of 300-5,000 SSU.

SSU: Either SSU or SUS denotes Saybolt Universal Seconds, a method of oil viscosity measurement. SSU or SUS indicates the time in seconds for 60 cc of oil at a specified temperature to flow through an orifice .070 dia × .483 long. Oil viscosities are usually rated at 0°, 100°F, and 210°F.SSU = 4.6 × centistokes (approx). Applies above 20 centistokes. Water viscosity at 68°F = 30 SSU = 1 centistoke.

Standard air: Air at a temperature of 68°F, a pressure of 14.70 psia and a relative humidity of 36%.

Stray mist: Oil particles too small to be reclassified. Will appear as "smoke" escaping from machine elements being mist lubricated.

Sump: A pocket of oil retained in the housing of a machine element or elements to provide an oil reservoir for coating the lubricated elements at start-up. The oil may also be retained during operation of machine as additional safety factor. Note: A sump usually implies a lesser volume of oil than would be used for bath lubrication.

SUS: See **SSU**.

Thrust load: Load parallel to the axis of shaft.

Trap: Usually refers to oil pocket in mist piping that impedes or blocks the flow of mist. May also refer to oil sump in housing.

Velocity: See **Mist velocity**.

Vent: Passage provided to permit carrier air to pass to atmosphere.

Viscosity: A measure of the internal resistance of oil to flow.

Viscosity index: A number indicating the rate of change in viscosity of an oil within a given temperature range.

Working capacity of lubricator: The total usable volume of oil between the indicated maximum fill level and the minimum recommended operating oil level, after compensation for such items as switches, pumps, cups, etc., which are immersed in this volume.

References

(1) Armstrong, E.L., et al., "Evaluation of Water-Accelerated Bearing Fatigue in Oil-Lubricated Ball Bearings, "Lubrication *Engineering,* Vol. 34 (1), pp. 15—21 (1977).

(2) Grunberg, L., and Scott, D., "The Effect of Additives on the Water-induced Pitting of Bali Bearings," J. *Inst. Petrol.,* Vol. 46, pp. 259—266 (1960).

(3) Schatzberg, P., and Felsen, I.M., "Effects of Water and Oxygen During Rolling Contact Lubrication," *Wear,* Vol. 12, pp. 331—342 (19683.

(4) Ciruna, J.A., and Szieleit, H.J., "The Effect of Hydrogen on the Roiling Contact Fatigue Life of AISI 52100 and 440C Steel Balls," *Wear,* Vol. 24, pp. 107—118 (1973).

(5) Fein, R.S., "Chemistry in Concentrated-Conjunction Lubrication," *Symposium on Interdisciplinary Approach to Lubrication of Concentrated Contacts,* Vol. II, Troy, N.Y., pp. 14.1—14.69 (1969).

(6) Towne, C.A., "Practical Experience With Oil-Mist Lubrication," *Lubrication Engineering,* Vol. 39(8), pp. 496—502 (1983).

(7) Bloch, H.P., "Large-Scale Application of Pure Oil-Mist Lubrication in Petrochemical Plants," ASME Paper No. 80-C2/Lub-25, Presented at San Francisco, California, August, 1980.

(8) Chevron Marketing Services Division, "Oil-Mist Lubrication," 1970.

(9) Hartmann, L.M., "Lubricating Oil Requirements for Oil-Mist Systems," *Lubrication Engineering,* Vol. 28(1), pp. 21—25, (1972).

(10) ACGIH, "Threshold Limit Values For Chemical Substances in the Work Environment," Cincinnati, Ohio 45211, ISBN 0-936712-54-6.

(11) Costa, D.L., and Amdur, M.O., "Respiratory Response of Guinea Pigs to Oil Mists," *American Industrial Hygiene Association Journal,* Vol. 40, pp. 673—679, (August 1979).

(12) Bloch, H.P., *Practical Machinery Management for Process Plants, Vol. 1—improving Machinery Reliability, Gulf Publishing Company,* Houston, Texas, p. 248 (1982).

(13) Morrison, F.R., Zielinski, J., and James, R., "Effects of Synthetic Industrial Fluids on Ball Bearing Performance," *Trans. ASME, Journal of Energy Resource Technology,* Vol. 104, pp. 174—181 (1982).

(14) Bloch, H.P., "Dry Sump Oil Mist Lubrication for Electric Motors," *Hydrocarbon Processing,* Vol. 57(3), pp. 133—135, (March 1977).

(15) Bloch, H.P., "Storage Preservation of Machinery," Proceedings of 14th Texas A&M Turbomachinery Symposium, Houston, Texas October 1985.

(16) Bloch, H.P., and Rizo, L., "Lubrication Strategies for Electric Motor Bearings in the Petrochemical and Refining Industry," Presented at the NPRA Refinery and Petrochemical Plant Maintenance Conference, San Antonio, Texas, February 14-17, 1984.

(17) Bloch, H.P., "Criteria for Water Removal from Mechanical Drive Steam Turbine Lube Oils," ASLE Paper No. 80-A-IE-1.

(18) Bloch, H.P., "Lube Oil Reclamation Saves Money," *Chemical Processing Magazine,* February, 1982.

(19) Bloch, H.P., "Results of Plant-Wide Lube Oil Reconditioning and Analysis Program," (Presented at ASLE National Conference, Houston, Texas, April 25-28, 1983). Reprinted in Lubrication Engineering, Vol. 40 (7), pp. 402—408 (July 1984).
(20) Bloch, H.P., and Geitner, F.K., *Practical Machinery Management for Process Plants*, Vol. 2—*Machinery Failure Analysis and Troubleshooting*, Gulf Publishing Company, Houston, Texas, 1983, pp. 192—205.
(21) Stoelken, J., "Oelnebelschmierung," *Schmiertechnik Tribologie*, December 6, 1981.
(22) Bloch, H.P., "Defining Machinery Documentation Requirements for Process Plants" ASME Paper No. 81-WA/Mgt-2.
(23) Bloch, H.P., *Oil Mist Lubrication Hand Book*, First Edition, Gulf Publishing Company, Houston, (1987). [Note: Essentially Part 1 of this book]
(24) Bednarek, K., and Szydlo, Z., "Problems of Oil Mist Lubrication in Industry," *Journal of the Society of Tribologists and Lubrication Engineers*, pp. 906—910, November (1988).
(25) System Planning of LubriMist Lubricating Systems, Catalog Section MA, Manual ME 200D, Lubrication Systems Company, Houston, (1987).
(26) Micro-Fog Lubricator, C.A. Norgren Co., Littleton, Colorado.
(27) Oil Mist Application Engineering Manual, Stewart-Warner Corporation, Alemite and Instrument Division, Chicago, Illinois, (1987).
(28) Bloch, H.P., "Apply Pure Oil Mist Lubrication on a Large Scale," *Hydrocarbon Processing*, pp. 125—131, October, (1980).
(29) Towne, C.A., "Practical Experience with Oil Mist Lubrication," ASLE preprint No. 82-AM-4C-1, Presented at the 37th Annual Meeting in Cincinnati, Ohio, (1982).
(30) Gulker, E., "Oil Mist Lubrication—A Method to Design out Maintenance," *Iron and Steel Engineer, pp.* 47—5O, October, (1974).
(31) Faust, D.G., "Fog Lubrication of Machine Tools," *Journal o, the American Society of Lubrication Engineers*, Vol. 14(2), pp. 54—57 (1959).
(32) Nukiyama, S., and Tanasawa, Y., "Experiments on the Atomization of Liquids in an Air Stream," *Trans. Soc. Mech. Eng.*, Japan, Vol. 5, pp. 68—75 (1939).
(33) Rizk, N.K., and Lefabvre, A.H., "Spray Characteristics of Plainjet Air Blast Atomizers," *Trans. ASME J. Eng. Gas Turbine Power*, Vol. 106, pp. 639—644 (1984).
(34) Shamim, A., and Kettleborough, C.F., "Aerosol Aspects of Oil Mist Lubrication—Generation and Penetration in Supply Line," presented at the Energy Resources Technology Conference and Exhibition, Houston, January; Tribology Symposium ASME, PD-Vol. 72, pp. 133-140 (1995).
(35) Hinds, W.C., *Aerosol Technology*, John Wiley & Sons, Inc., New York, (1982).
(36) Anand, N.K., and McFarland, A.R., "Particle Deposition in Aerosol Sampling Lines Caused by Turbulent Diffusion and Gravitational Settling," *American Industrial Hygiene Association Journal*, Vol. 50(6), June (1989).
(37) Davies, C.N., *Aerosol Science*, Academic Press, New York, (1966).
(38) Friedlander, S.K., *Smoke, Dust and Haze*, John Wiley & Sons, Inc., New York, (1977).
(39) Liu, B.Y.H., and Agarwal, J.K., "Experimental Observation of Aerosol Deposition in Turbulent Flow," *Journal of Aerosol Science*, Vol. 5, pp. 145-155 (1974).
(40) Shamim, A., "Tribological and Aerosol Mechanics Aspects of Oil Mist Lubrica-

tion," Ph.D. Dissertation, Mechanical Engineering Department, Texas A&M University, College Station (1493).
(41) Pui, D.Y.H, Romay-Novas, F., and Liu, B.Y.H., "Experimental Study of Particle Deposition in Bends of Circular Cross Section," *Aerosol Science and Technology,* Vol. 7, pp. 301—315 (1987).
(42) Cheng, Y.S., and Wang, C.S., "Motion of Particles in Bends of Circular Pipes," *Atmospheric Environment,* Vol. 15, pp. 301—306 (1981).
(43) Fucs, N.A., *The Mechanics of Aerosol,* Macmillian Publishing Co., New York, (1964).
(44) Shamim, A., and Kettleborough, C.F., "Aerosol Aspects of Oil Mist Lubrication—Reclassification and Deposition of Oil Mist in Bearings," Presented at the Energy Week Conference and Exhibition, *Conference Proceedings,* Vol. IV, pp. 380-385 (1996).
(45) Schlichting, H., *Boundary Layer Theory,* McGraw Hill Book Company, Inc., New York (1960).
(46) Shamim, A., and Kettleborough, C.F., "New Oils for Oil Mist Lubrication to Reduce Fine Droplets in Stray Mist," Presented at the Energy Week Conference and Exhibition (1997).
(47) Wame, T.M., and Halder, C.A., "Toxicity of Lubricating Oils," *Journal of the American Society of Lubrication Engineers,* Vol. 42, No. 2, pp. 97—103 (1984).
(48) Manual Number OM-695 10M, Aercology, Inc. Custom Park, Old Saybrook, CT 06475.
(49) Ward, T.K., "Advances in Oil Mist Lubrication Technologies and Applications," Technical Publication, Lubrication Systems Company, P.O. Box 19294, Houston, Texas 7722 (19g6).
(50) Smith, J., and McCallion, J., "Oil Mist Levels Beat OSHA Standard," *Chemical Processing,* November (1995).
(51) Fowler, E.K., *Collection of Lubricating al Emissions from Gas Turbines,* Technical Publication 103/89, CECO Filters, Inc., Conshohocken, PA (1990).
(52) Shamim, A., and Kettleborough, C.F., "Tribological Performance Evaluation of Oil Mist Lubrication," *ASME Journal of Energy Resources Technology,* Vol. 116, No. 3, pp. 224—231 (1994).
(53) Shamim, A., "Design, Fabrication and Testing of a Bearing Test Rig and Preliminary Studies on Oil Mist Lubrication," M.S. Thesis, Mechanical Engineering Department, Texas A&M University, College Station (1990)
(54) FAG Rolling Bearings, FAG Bearings Corporation, Catalog WL41510/2ED, Stamford, Connecticut (1989).
(55) Rush, A.A., "Kurtosis—a Crystal Ball for Maintenance Engineers," *Iron and Steel International,* pp. 24—27 (1979).
(56) Volker, E., and Martin, H.R., "Early Detection of Damage in Rolling Bearings," *ISA Trans.,* Vol. 23(3), pp. 27—S2 (1984).
(57) McCool, J.I., "Evaluating Weibull Endurance Data by the Method of Maximum Likelihood," *ASLE Transactions,* No. 13, pp. 189—202 (1970).
(58) Johnson, L.G., "Operations Research in Lite Testing," *Industrial Mathematics,* Vol. 8 (1957).
(59) Johnson, L.G., *The Statistical Treatment of Fatigue Experiments,* Elsevier Publishing

Company, New York (1964).
(60) Erdemir, A., Ajayi, O.O., Fenske, G.R., Erck, R.A., and Hsieh, J.H., "The Synergistic Effects of Solid and Liquid Lubrication on the Tribological Behavior of Transformation-Toughened ZrO_2 Ceramics," Presented at the STLE-ASME conference on Tribology, Report No. DE91-018617, St. Louis, MO (1991).
(61) Graham, E.E., and Klaus, E.E., "Lubrication from the Vapor Phase at High Temperatures," *ASLE Transactions,* Vol. 22(2), pp. 229-234 (1986).
(62) Lauer, J.L., and Dwyer, S.R., "High Temperature Lubrication by Carbon Continuously Replenished by Surface Reaction with Carbonaceous Gases —Comparison of Metallic (Ni) and Ceramic(Si_3N_4) Surfaces," *SAE Transactions,* Vol. 99, Sec. 4, pp. 258-264 (1990).
(63) Lauer, J.L., and Dwyer, S.R., "Tribochemical Lubrication of Ceramics By Carbonaceous Vapors," *Tribology Transactions,* Vol. 33, No. 1, pp. 25—32 (1990).
(64) Bearing Installation and Maintenance Guide, SKF Industries Inc., Catalog No. 140-710, King of Prussia, Pennsylvania (1985).
(65) Pinkus, O., Decker, O., and Wilcock, D.F., "How to Save 5% of Our Energy," *Mechanical Engineering,* pp. 32—39 (1977).
(66) Wheel Flange Lubrication System, Eximport Industria E Comercio LTDA. Sao Paulo, Brasil.
(67) Bajaj, Kris K., "Oil-Mist Lubrication of Shaker Screen Bearings," *Lubrication Engineering,* pp. 852—855, October, 1988.
(68) Ward, T.K., and Paben, F., "Adaptation of Oil Mist Lubrication by a North American Pulp Mill," Technical Publication, Lubrication Systems Company, Houston, Texas. (1996).
(69) Ward, T.K., "Advances in Oil Mist Lubrication Technologies and Applications," Technical Publication, Lubrication Systems Company, Houston, Texas. (1996).
(70) Towne, C.A., and Sheppard, D.J., "Oil Mist Lubrication for Electric Motors—Where it Stands Today," IEEE Transactions on Industry Applications, Vol. IA 22, No. 6, November/December 1986.

Bibliography of Early Papers on Oil-mist (Micro-fog and/or Aerosol) Lubrication*

Adams, C.R., "Development Progress on Gas Bearings for Airborne Accessory Equipment" presented at SAE National Aeronautic Meeting, New York, N.Y., April 5-8, 1960.

Altpeter, W., "Betriebserfahrungen mit der Oelnebelschmierung in einem Huettenwerk" ("Experience with Oil-Mist Lubrication in a Steel Plant"), *Stahl& Eisen,* March, 1963.

Anderson, W.J., "Bearings," *Machine Design,* November 5, 1964, pages 165-181. Also see Nemeth, Z.N., 1965; Schuller, F.T., 1960; Coe, H.H., 1970.

Bell, D.W., and Rushforth, H., "Initial Experiences with an Oil-Mist Lubrication System and Some Thoughts on the Possibilities of Further Developments," Paper 14, Third Annual Meeting of the Lubrication and Wear Group Institute of Mechanical Engineers, 1 Birdcage Walk, Westminster, London, S. W1, England, October, 1964.

Bergner, A. See Neukirchner, J., 1970.

Besser, Helmut (De Limon Fluhme & Co., Arminstrasse 15, Dusseldorf, Germany), "Oelnebelschmierung als neuer Zweig der Schmiertechnik," *(Oil-Mist Lubrication, a New Technique), Deutsche Maschinenwelt,* Book 6, 1961.

_______, "Oelnebelschmierung: Grundlagen und Neuester Stand," Metallbearbeitung 60 (1966), Heft 2 Seite 106/111, Heft 4 Seite 218/221, also published as De Limon Fluhme Bulletin S 099. (Oil Mist Lubrication, Basic Concepts and Latest Developments).

Brehmer, John R. (Alemite Division, Stewart-Warner Corporation, Chicago, Illinois), "Oil-Mist Lubrication on Drive Gearing," presented at Gesellschaft fuer Tribologie und Schmierungstechnik Meeting, Essen, Germany, September 22-23, 1970.

Brewer, Allen F., "Iron & Steel," *Industrial Lubrication and Tribology, Vol.* 22, No. 1, pages 8-11, Jan., 1970 and Vol. 22, No. 3, page 97, March, 1970.

_______, "Machine Tool Lubrication," *Industrial Lubrication and Tribology,* Vol. 22, No. 3, pages 87-91, March, 1970.

_______, "Lubricating Printing Machinery," *Industrial Lubrication and Tribology, Vol.* 22, No. 11, pages 319-322, Nov., 1970.

_______, "Lubricating Textile Machinery," *Industrial Lubrication and Tribology,* Jan., 1971, pages 25-30.

_______, "Paper Making Machinery," *Industrial Lubrication and Tribology* March, 1971, pages 101-105.

Carr, D.W., and Knight, R.E. (C.A. Norgren Ltd., Shipston-On-Stour, Warwickshire, England), "Aerosol Lubrication and Its Application to a Wide Range of Plant and

* Courtesy of G. A Norgren Co./Littleton, Colorado.

Machinery," paper 31, Lubrication and Wear Convention [Institute of Mechanical Engineers], May, 1963.
Carr, D.W., Knight, R.E., and Gelder, R. (C.A. Norgren Ltd., Shipston-On-Stour, Warwickshire, England), "The Widening Field for Aerosol Lubrication Systems in the U.K.," International Industrial Lubrication Exhibition, London, March, 1965.
Chapman, J.T., "Some Aspects of Lubricant Development of Centralized Aerosol Lubricating Systems," *Scientific Lubrication, Vol.* 18, No. 3, pages 23-28, March, 1966, Reprinted by C. A. Norgren Ltd. as Bulletin MFT. 4-4/66.
Cichelli, A.E. (consulting engineer to Bethlehem Steel Corp., Bethlehem, Pa.), "Back-up Roll Bearings in Cold Reducing Mills," *Journal of Iron and Steel Institute, Vol.* 208, part 10, pages 894-910, Oct., 1970.
Coe, H.H., Scibbe, H.W., and Anderson, W.J., "Evaluation of Hollow (Drilled) Balls in Ball Bearings at DN Values to 2 Million," NASA Lewis Research Center, Jan., 1970, NASA TM X 52747 (N70-19336).
Ellis, E.G., "Fundamentals of Lubrication—Part 12, Industrial Applications," *Industrial Lubrication, Vol. 19,* No. 2, Feb., 1966, pages 6369.
_______, "Fundamentals of Lubrication," Chapter 11, *Industrial Lubrication, Vol. 19,* Jan., 1967, pages 37-38.
_______, "Fundamentals of Lubrication," *Scientific Publications* (G.B.) Ltd., 1968, pages 80, 85, 86.
Faust, Delbert Grant, (C.A. Norgren Co., Littleton, Colorado, U.S.A.) "Wetting Characteristics of Lubricating Aerosols," *Product Engineering,* July, 1951.
_______, "Oil-Fog Lubrication," (Alfred E. Hunt award-winning paper) *Lubrication Engineering,* August, 1952.
_______, "Fog Lubrication of Machine Tools," *Lubrication Engineering,* February, 1958.
_______, "Oil-Fog Lubrication—Past, Present and Future," *Lubrication Engineering,* August, 1961.
_______, "Current Application Design Practices for Aerosol Lubrication of Machine Tools," presented at the American Society of Lubrication Engineers Annual Meeting, St. Louis, Missouri, May, 1962.
_______, "Aerosol Lubrication," *Standard Handbook of Lubrication Engineering,* McGraw Hill Publishing Co., 1968, Library of Congress card No. 64-16489, Chapter 25, pages 45-48.
Gebauer, Georg, "*Oelnebelschmierung* in der Huettenindustrie" ("Oil-Mist Lubrication in Iron and Steel Industry"), *Schmiertechnik,* 1962, Book 1, Karl Maerklein-Verlag GmbH, Engerstrasse 21a, Dusseldorf, Germany.
Gelder, R., (C.A. Norgren Ltd., Shipston-On-Stour, Warwickshire, England), "Aerosol Lubrication in the Glass Industry," *Glass,* Glass Publications Ltd., Vol. 43, No. 6, pages 258-263, June, 1966. Also see Carr, D.W., 1965.
Goldstein, N.H., "Lubrication of Ball Bearings in High-Speed Applications," Engineering Materials and Design, July, 1965, page 470.
Green, J.I.T., "Micro-Fog Lubrication in the Steel Industry," BISRA Restricted Report PE/C/41/65, Plant and Energy Division, the British Iron and Steel Research Association, 24 Buckingham Gate, London, S.W. 1, England.
Gulker, E., "Umstellung hochbelasteter Walzlager von Fett auf Oelnebelschmierung," *SonderdruckAus VDI,* Berichte Nr. 111 1966 S 31-37. Published by De Limon Fluhme as brochure S 137. English translation available from C. A. Norgren Ltd.
_______, "Converting Heavy Duty Roller Bearings from Grease Lubrication to Aerosol

Lubrication," *Industrial Lubrication and Tribology, Vol.* 20, No. 8, Aug., 1968, pages 264-272, reprinted as C.A. Norgren Ltd. MFT 5 7/68.

Haeger, P.L., "Mounting and Lubrication of Anti-Friction Bearings for Optimum Performance, Oil Mist-Fog and Related Systems," American Society of Lubrication Engineers, April, 1961.

_______, Discussion on C.W. Southerington paper entitled, "Oil-Mist Lubrication on Anti-Friction Back-up Roll Bearings," *Iron and Steel Engineer,* December, 1961, Volume 38, page 127.

Henrikson, K.G., "Mist Lubrication," American Society of Lubrication Engineers Annual Meeting, May 6-9, 1968.

Hoffman, Nelson M., "Proposed Standard on Machine Tool Lubrication—Part A," *Plant Engineering,* October 2, 1969, "Part B," October 16, 1969.

Howell, P.G., "Aerosol Lubrication of Gearboxes," Associated Electrical Industries Ltd., Central Research Laboratory Confidential Report L5245, March, 1966.

_______, "Experiments in the Aerosol Lubrication of Power Gearing," Institution of Mechanical Engineers Lubrication and Wear Convention, May, 1967.

Johnson, R.L. and Manganiello, E.J., "Aerospace Lubrication for Advanced Vehicles," presented at German Society of Tribology Annual Meeting, Essen, Germany, Sept. 22-23, 1970 NASA TM X 52867 (N70-34302).

Knight, R.E., (C.A. Norgren Ltd., Shipston-On-Stour, Warwickshire England), "Micro-Fog Lubrication of Anti-Friction Bearings," Lecture to Hoffman Manufacturing Co. Ltd., at Chelmsford, 1962.

_______, "Aerosol Lubricators—Their Place in Industry," *Control Review* No. 1, 1966, pages 56-61. Also see Carr, D.W., 1963 and 1965, and Pass, P. J., 1960.

Kortzfleisch, B., "Oil-Mist Droplet Lubrication on a High-Speed Rod Mill," *Schmiertechnik,* Vol. 12, No. 3, May/June, 1965, pages 145147.

Lang, Klaus E., "Zentralschmiereinrichtungen fuer Fett und Oel," ("Centralized Lubrication Systems for Grease and Oil), *Betriebsbuecher* 19, Carl Hanser Verlag, Munich, 1965.

_______, "Schmiereinrichtungen und ihre Ueberwachung," ("Lubricating Systems and their Supervision") VDI, Berichte Nr. 141, 1970—in German only.

Lyth, W.W., (Eaton Fluid Power Division, Eaton, Yale & Towne, Inc., Cleveland, Ohio), "The Vortex Generator, Newest Approach to Oil Mist," 26th Annual Meeting of American Society of Lubrication Engineers, Boston, May 3-6, 1971.

Manganiello, E.J., See Johnson, R.L., 1970.

McCandless, O.G., (Butler Works, Armco Steel Corp., Butler, Pennsylvania), "Some Concepts of Mill Bearing Lubrication Employing Oil, Plastic Grease, and Oil Mist," *Iron and Steel Engineer,* September, 1961, Vol. 38, page 171.

McKee, L.W., "Ultra High-Speed Ball Bearings, Their Selection and Application," Missile Design and Development, August, 1960.

McCoy, W.E., West, C.H., and Wilks, P.E., "New Mist Lubrication Concepts for Tapered Roller Bearings Used on High Speed Rolling Mill Back-up Rolls," Proceedings of the Iron & Steel Institute "War on Wear" Conference, 1969. ISI p. 125 *Tribology in Iron & Steel Works,* Feb., 1970.

Morton, I.S., "Lubricant and Coolant Systems for Machine Tools," Industrial Lubrication and Tribology Symposium, London, Nov., 1969.

Mosier, H., "Bearings for High-Speed Rotors of Small Electric Motors," *Konstruction* 21, pages 391-399, October, 1969—In German.

Munnich, H. and Strafe, G., "Roll Neck Bearings Under Severe Operating Conditions," Proceedings of Conference on Tribology in Iron and Steel Works, Sept. 22-25, 1969 (Iron & Steel Institute and Institution of Mechanical Engineers).

Nemeth, Z.N., and Anderson, W.J., "Effect of Speed, Load and Temperature on Minimum Oil Flow Requirements of 30mm and 75mm Bore Ball Bearings," National Aeronautics and Space Administration, NASA TN D-2908, July, 1965.

Neukirchner, J . and Bergner, A., " Zur Problematik der Oelnebelschmierung," ("On the Problems of Oil Fog Lubrication"), Karl Marx-Stadt, *Schmierungstechnik* 1 (1970) 11.

Newman, L.V., "Reports on Lubrication and Wear Problems at Abbey and Margan Works—British Steel Corporation South Wales Group," Proceedings of the Iron & Steel Institute "War on Wear" Conference, 1969. ISI p. 125, *Tribology in Iron &0Steel Works,* Feb., 1970.

Nica, A., "Theory and Practices of Lubrication Systems," Scientific Publications (G.B.), Ltd., 1969.

Ortman, G., "Lubrication of Foundry Equipment," *Giesserei, Vol.* 55, May 9, 1968, pages 232-238. English translation ref. T1343, May, 1970 by British Cast Iron Research Association.

Palei, L. Ya, "Anti-friction Bearings for Surface Grinding Machine Spindles," *Machines & Tooling, Vol. 39,* No. 3, pages 15-16.

Pass, P.J. and Knight, R.E., (C.A. Norgren Ltd., Shipston-On-Stour, Warwickshire, England), "Fog Lubrication of Machines and Tools," Paper 4, Second European Fluid Power Conference, London, April 25-29, 1960.

Rugg, P.J., (C.A. Norgren Ltd., Shipston-On-Stour, Warwickshire, England), "Micro-Fog Lubrication applied to Machine Tools," *Metal Working Equipment News,* January, 1967, pages 4-6.

Rushforth, H. See Bell, D.W., 1964.

Saverskii, A.S., "Oil-Mist Lubrication," *Russian Engineering Journal,* 1960 40 (No. 1) 8 Translated from *Vestnik Mashinostroaniya,* page 11.

Schmemann, Alfred, (De Limon Fluhme & Co., Arminstrasse 15, Dusseldorf, Germany), " Latest Developments in Aerosol Lubrication Practice in German Steel Works," Paper 11, Third Annual Meeting of the Lubrication and Wear Group (Institute of Mechanical Engineers), October, 1964.

Schneider, H.G., (De Limon Fluhme & Co., Industriestrasse 1, Dusseldorf, Germany), "Oelnebelschmierung von Gelenkspindeln an Walzgeruesten" (Oil-Mist Lubrication of Drive Spindles on Rolling Mills), Baender Bleche Rohre 12 (1971) Nr. 5, pages 205/207. In German. De Limon Fluhme publication S514. English translation available from C.A. Norgren Ltd., England.

Schuller, F.T. and Anderson, W.J., (Lewis Research Center, NASA, Cleveland, Ohio), "Operating Characteristics of 75mm Bore Ball Bearings at Minimum Oil Flow Rates over a Temperature Range of 500°F." The 15th American Society of Lubrication Engineers Annual Meeting, April, 1960.

Scibbe, H.W. See Coe, H.H., 1970.

Seaton, J.J., "The Performance of Oil-Mist Lubrication in the Steel Industry." Presented at the Annual Convention of the National Lubricating Grease Institute, San Francisco, October 24-27, 1965.

Simon, John, (National Tube Division, U.S. Steel Corp., McKeesport, Pa.), "Spindle Coupling Lubrication Methods," presented at the 16th American Society of Lubrication Engineers Annual Meeting in Philadelphia, Pennsylvania, April, 1962,

Lubrication Engineering, June, 1962.

Smith, A.C., "Oil Mist (Oil-Fog) Lubrication," *The Application of Lubricants,* Shell International Petroleum Company Limited, London, Chapter 5, Publication 1723/20.65/16M, January, 1966.

Southerington, C.W., (Dallas Division, Revere Copper & Brass, Inc., Chicago, Illinois), "Oil-Mist Lubrication on Anti-Friction Backup Roll Bearings," *Iron and Steel Engineer,* December, 1961, Vol. 38, page 121. Also see Haeger, P.L., 1961, for a discussion of this paper.

Standley, Harold B., (C.A. Norgren Co., 5400 South Delaware, Littleton, Colorado 80120, U.S.A.), "The Effects of Air and Oil Heat on Mist Lubrication," November, 1971, C. A. Norgren Co. Technical Paper NTP-1.

Steiner, K., (Hoerbiger Pneumatik, Germany), "Druckluftoeler-Ergebnisse einer Versuchsreihe ueber Typenwahl und Anwendung," *Pneumatic Digest,* Heft 1, Febr., 1971, 5 Jahrgang.

Strafe, G. See Munnich, H., 1969.

Index